PREFACE

This book covers the syllabus on Mathematics for the Class 2 Marine Engineers' Certificate of Competency of the Department of Transport (DTp) and is a useful aid to students on Business and Technical Education Council (BTEC) and Scottish Vocational Education Council (SCOTVEC) courses.

Basic principles are dealt with, commencing at a fairly elementary stage. Each chapter has fully worked examples interwoven into the text, test examples are set at the end of each chapter for the student to work out, and finally there are some typical examination questions included.

The author has gone beyond the normal practice of merely supplying bare answers to the test examples and examination questions by providing fully worked step by step solutions leading to the final answers.

CONTENTS

INTRODUCTION TO SI UNITS

SI is the abbreviation for Systême International d'Unités, the version of the Metric System now in international use and adopted by British industry.

It is built upon the six basic measurements: the metre as the unit of length; kilogramme, unit of mass; second, unit of time; kelvin, unit of temperature; ampere, unit of electrical current; candela, unit of luminous intensity.

It is a coherent system, this means that the product or quotient of any two quantities produces the unit of the resultant quantity. Thus, to give a few examples, unit length (1 metre) multiplied by unit length (1 metre) gives unit area (1 square metre). Unit length or distance (1 metre) divided by unit time (1 second) produces unit velocity (1 metre per second). When unit mass (1 kilogramme) is multiplied by unit acceleration (1 metre per second^2) the result is unit force (1 newton). If unit force (1 newton) acts through unit distance (1 metre), unit work (1 joule) is done.

In order to keep the number of names of units to a minimum, multiples and sub-multiples of the fundamental units, in chosen powers of ten, represented in each case by a prefix, is combined with the symbol of the unit.

Multiplication Factor	Standard Form	Prefix	Symbol
1 000 000 000 000	10^{12}	tera	T
1 000 000 000	10^{9}	giga	G
1 000 000	10^{6}	mega	M
1 000	10^{3}	kilo	k
100	10^{2}	hecto	h
10	10^{1}	deca	da
0·1	10^{-1}	deci	d
0·01	10^{-2}	centi	c
0·001	10^{-3}	milli	m
0·000 001	10^{-6}	micro	μ
0·000 000 001	10^{-9}	nano	n
0·000 000 000 001	10^{-12}	pico	p

No more than one prefix is used with each unit. When a prefix is attached to a unit, it becomes a new unit symbol on its own account. Multiples of 10^3 are recommended but others are recognised because of convenient sizes and established usage and custom. These will be introduced and explained in the text when the occasion arises.

It will take time for Engineers brought up on the Imperial system to become sufficiently familiar with SI to be able to think entirely in this system. It may be of assistance therefore to compare the magnitudes of SI units with their corresponding Imperial units. A selection of those used in this book are listed below, showing their abbreviations in brackets. Note that the symbols representing units are in lower case letters except those named after famous persons when capital letters are used. The symbol is always in the singular and no full stops are inserted as in other abbreviated words.

LENGTH

1 inch (in)	25·4 millimetres (mm)
1 foot (ft)	304·8 mm = 0·3048 metre (m)
1 mile	1·609 kilometres (km)
1 nautical mile (international)	1·852 km

AREA

1 square inch (in^2)	645·2 square millimetres (mm^2)
1 square mile ($mile^2$)	2·59 square kilometres (km^2)

VOLUME

1 cubic inch (in^3)	16 390 cubic millimetres (mm^3)
1 gallon (gal)	4·546 litres (1)
	1 litre = 1 cubic decimetre (dm^3)

VELOCITY

1 foot per second (ft/s)	0·3048 metre per second (m/s)
1 mile per hour (mile/h)	0·447 m/s
1 international knot (= 1 naut. mile per hour)	1·852 kilometres per hour (km/h)

MASS

1 pound (lb)	0·4536 kilogramme (kg)
1 ton	1·016 tonne (t)
	(1 tonne = 1 000 kg)

FORCE

1 pound (lbf)	4·448 newtons (N)
1 ton (tonf)	9·964 kilonewtons (kN)

PRESSURE

1 pound per square inch (lbf/in^2)	6·895 kilonewtons per square metre (kN/m^2)
1 ton per square inch ($tonf/in^2$)	15·44 meganewtons per square metre (MN/m^2)

ENERGY

1 foot pound (ft/lbf)	1·356 joules (J)
1 British thermal unit (Btu)	1·055 kilojoules (kJ)

POWER

1 horse power (hp)	0·7457 kilowatt (kW)

CHAPTER I

ARITHMETIC

POWERS

An index is a short method of expressing a quantity multiplied by itself a number of times, thus,

4×4 is written 4^2, this is the 'second power' of 4 commonly called the 'square' of 4.

$x \times x \times x$ is written x^3, this is the 'third power' of x commonly called the 'cube' of x.

$10 \times 10 \times 10 \times 10 \times 10 \times 10$ is written 10^6, and so on.

TO MULTIPLY POWERS OF THE SAME QUANTITY, add their indices, thus,

$$2^3 \times 2^4$$

$$\text{written right out} = 2 \times 2 \times 2 \times 2 \times 2 \times 2 \times 2$$

$$= 2^7$$

The same result is obtained by adding the indices 3 and 4.

$$2^3 \times 2^4$$

$$= 2^{3+4}$$

$$= 2^7$$

TO DIVIDE POWERS OF THE SAME QUANTITY, subtract their indices, thus,

$$3^5 \div 3^2$$

$$= \frac{3 \times 3 \times 3 \times 3 \times 3}{3 \times 3}$$

$$= 3 \times 3 \times 3$$

$$= 3^3$$

The same result is obtained by subtracting the indices,

$$3^5 \div 3^2$$

$$= 3^{5-2}$$

$$= 3^3$$

NEGATIVE INDICES

Consider the example $5^2 \div 5^6$

written right out $= \dfrac{5 \times 5}{5 \times 5 \times 5 \times 5 \times 5 \times 5}$

$$= \frac{1}{5^4}$$

or, subtracting the indices,

$$5^{2-6}$$

$$= 5^{-4}$$

Hence we see that a quantity with a negative index, in this case 5^{-4}, is equal to the reciprocal of that quantity with a positive index, that is

$$5^{-4} = \frac{1}{5^4}$$

Similarly, $x^{-2} = \dfrac{1}{x^2}$

$$10^{-3} = \frac{1}{10^3}$$

Likewise, $\dfrac{1}{x^{-3}} = x^3$

$$\frac{1}{10^{-9}} = 10^9$$

POWER OF UNITY. Any quantity to a power of unity is the quantity itself, thus,

$$3^3 \div 3^2 = 3^{3-2} = 3^1$$

therefore $3^1 = 3$

ZERO POWER. Any quantity to a power of zero is equal to unity, thus,

$$2^3 \div 2^3 = 2^{3-3} = 2^0$$

therefore $2^0 = 1$

similarly, $x^2 \div x^2 = x^{2-2} = x^0$

$$\therefore x^0 = 1$$

TO RAISE A POWERED QUANTITY TO A FURTHER POWER, multiply their indices, thus,

$(2^2)^3$

written right out $= 2^2 \times 2^2 \times 2^2$

$= 2 \times 2 \times 2 \times 2 \times 2 \times 2$

$= 2^6$

therefore $(2^2)^3$

$= 2^{2\times 3}$

$= 2^6$

ROOTS

A root is the opposite of a power and the root symbol is $\sqrt{}$.

$\sqrt[2]{}$ represents the 'square root' of a quantity and means that we are required to find the number which, when squared, will be equal to that quantity.

Thus, $\sqrt[2]{25} = 5$ because 5^2 is 25

$\sqrt[3]{}$ represents the 'cube root' of a quantity, this means that it is required to find the number whose cube is equal to that quantity.

Thus, $\sqrt[3]{27} = 3$ because $3^3 = 27$

Similarly, $\sqrt[4]{16} = 2$

$\sqrt[5]{243} = 3$

The square root of a quantity is usually written without any figure indicating the root, thus the square root of 64 is usually written $\sqrt{64}$ instead of $\sqrt[2]{64}$. The value of any other root must of course be clearly stated.

Another method of indicating a root of a quantity is by expressing it as a power equal to the reciprocal of the root, for example,

the square root of 49 may be written,

$\sqrt{49}$ or $49^{\frac{1}{2}}$, which is 7;

the cube root of 27 may be written,

$\sqrt[3]{27}$ or $27^{\frac{1}{3}}$, which is 3;

the fourth root of x may be written,

$\sqrt[4]{x}$ or $x^{\frac{1}{4}}$

The square of the cube root of 64 may be written,

$(\sqrt[3]{64})^2$ or $64^{\frac{2}{3}}$, which is 16.

POWERS AND ROOTS OF VULGAR FRACTIONS

The power or root of a vulgar fraction is equal to the power or root of the numerator divided by the power or root of the denominator.

$$\left(\frac{6}{2}\right)^2 = 3^2 = 9$$

$$\text{or} \quad \frac{6^2}{2^2} = \frac{36}{4} = 9$$

$$\sqrt{\frac{64}{4}} = \sqrt{16} = 4$$

$$\text{or} \quad \frac{\sqrt{64}}{\sqrt{4}} = \frac{8}{2} = 4$$

The advantage of applying the above rule will be seen in the following examples,

$$\left(\frac{3}{8}\right)^2 = \frac{3^2}{8^2} = \frac{9}{64}$$

$$\left(1\tfrac{1}{2}\right)^2 = \left(\frac{3}{2}\right)^2 = \frac{3^2}{2^2} = \frac{9}{4} = 2\tfrac{1}{4}$$

$$\sqrt[3]{\frac{125}{27}} = \frac{\sqrt[3]{125}}{\sqrt[3]{27}} = \frac{5}{3} = 1\tfrac{2}{3}$$

SURDS

A root which does not work out exactly is termed a surd, such as:

$$\sqrt{2} = 1{\cdot}41421\ldots \text{etc.}$$
$$= 1{\cdot}414 \text{ to nearest four figures.}$$

Similarly, $\sqrt{3}$, $\sqrt{5}$, $\sqrt{6}$, are all surds.

It will be found very useful if, at least, the values of $\sqrt{2}$, $\sqrt{3}$ and $\sqrt{5}$ to the nearest four figures, were committed to memory. Many useful manipulations can be performed easily and quickly when solving or simplifying expressions which include these quantities as factors,

Memorise:

$$\sqrt{2} = 1{\cdot}414$$
$$\sqrt{3} = 1{\cdot}732$$
$$\sqrt{5} = 2{\cdot}236$$

Example: $\sqrt{18} = \sqrt{2} \times \sqrt{9}$
$= 1{\cdot}414 \times 3$
$= 4{\cdot}242$

Example:
$$\frac{2 \times \sqrt{2}}{\sqrt{6}}$$

$$= \frac{2 \times \sqrt{2}}{\sqrt{2} \times \sqrt{3}}$$ $\sqrt{2}$ cancels top and bottom

$$= \frac{2}{\sqrt{3}}$$ now multiply top and bottom by $\sqrt{3}$

$$= \frac{2 \times \sqrt{3}}{\sqrt{3} \times \sqrt{3}}$$ on the bottom we have $\sqrt{3} \times \sqrt{3} = 3$

$$= \frac{2 \times \sqrt{3}}{3}$$

$$= \frac{2 \times 1{\cdot}732}{3}$$

$$= \frac{3{\cdot}464}{3}$$

$= 1{\cdot}155$ (to nearest four figures).

RATIO

A ratio is a comparison of the magnitude of one quantity with another quantity of the same kind; it expresses the relationship of one to the other and therefore may be stated in fractional form. Being a means of comparison it is often convenient to express a ratio in terms of unity. The ratio sign is the colon:

Thus, if the lengths of two bars are 250 millimetres and 2 metres respectively, the ratio of one to the other may be expressed,

250 : 2000 (note both quantities must be expressed in the same units in this case both in millimetres)
or, 1 : 8

or, the former is $\frac{1}{8}$ of the length of the latter.

Example. The mass of a solid shaft 300 mm diameter is 1200 kg. Another shaft is hollow, 300 mm outside diameter and 150 mm bore and of the same length as the solid shaft. If the mass of the hollow shaft is 900 kg, express the ratio of the mass of the hollow shaft to that of the solid shaft.

Ratio of masses, hollow : solid

$$900 : 1200$$
$$3 : 4$$
$$\tfrac{3}{4} : 1 .$$

PROPORTION

Proportion is an equation of ratios, that is, it expresses that the ratio of one pair of quantities is equal to the ratio of another pair. The proportion sign is the double colon :: but the equal sign may be used. For example, the ratio of 5 to 10 is equal to the ratio of 20 to 40, and this may be expressed,

$$5 : 10 :: 20 : 40$$
$$\text{or, } 5 : 10 = 20 : 40$$
$$\text{or, } \frac{5}{10} = \frac{20}{40}$$

It can also be seen that the two inside terms multiplied together is equal to the two outside terms multiplied together, thus,

$$10 \times 20 = 5 \times 40$$

this is sometimes expressed in the manner "*the product of the means is equal to the product of the extremes*", and is a useful method to apply when one of the terms is unknown and we require to find that missing term.

Example. A pump takes 55 minutes to deliver 4400 litres of water. Under similar conditions, how long would it take to deliver 6000 litres?

Let x = time (in minutes) to deliver 6000 litres,

Ratio of times taken :: Ratio of quantities delivered

$$55 : x :: 4400 : 6000$$
$$x \times 4400 = 55 \times 6000$$

dividing both sides by 4400,

$$x = \frac{55 \times 6000}{4400}$$
$$x = 75 \text{ minutes}$$

Example. The ratio of the areas of two circles is equal to the ratio of the squares of their diameters. The diameter of one circle is 20 millimetres and is area is 314 mm^2, another circle is 30 millimetres diameter, find its area.

Let x represent the area of the second circle,

$$\text{Ratio of areas} :: \text{Ratio of diameters}^2$$

$$314 : x :: 20^2 : 30^2$$

$$314 : x :: 400 : 900$$

$$x \times 400 = 314 \times 900$$

$$x = \frac{314 \times 900}{400} = \frac{314 \times 9}{4}$$

$$x = 706{\cdot}5 \text{ mm}^2$$

INVERSE PROPORTION. The above examples are cases of *direct* proportion because, in the pump question for instance, an *increase* in time of running results in an *increase* in quantity of water delivered, thus, the quantity of water delivered *varies directly* as the time. There are however, many cases where the increase in one quantity causes a *decrease* in another, this is *inverse* proportion where one quantity is stated to *vary inversely* as the other. Suppose one pump could empty a tank in 20 minutes then two similar pumps drawing from this tank could empty it in half the time, i.e., 10 minutes, or three pumps would do the work in one-third of the time. Here we see that the *greater* the number of pumps, the *less* the time taken; the time *varies inversely* as the number of pumps. In setting down such a problem by the proportion method, one of the pairs of ratios must be reversed.

METHOD OF UNITY

Many students will find it more convenient to apply the method of unity to proportion problems, especially when dealing with compound proportion in which there are more than two pairs of quantities. The following examples demonstrate working by this method.

Example. If the cost of 84 tonne of fuel is £2352, what would be the cost of 189 tonne?

Cost of 84 tonne = 2352 pounds

$$\text{Cost of 1 tonne} = \frac{2352}{84} \text{ pounds}$$

$$\text{Cost of 189 tonne} = \frac{2352 \times 189}{84}$$

$$= £5292$$

Example. A ship travelling at 12 knots can complete a certain voyage in 16 days. How many days would the ship take to do the same voyage at a speed of 15 knots?

At a speed of 12 knots, time $= 16$ days

„ „ „ 1 knot, „ $= 16 \times 12$ days

„ „ „ 15 knots, „ $= \dfrac{16 \times 12}{15}$

$= 12{\cdot}8$ days

(note that the above is an example of inverse proportion).

Example. If 8 men can erect 2 engines in 18 days, how long would it take 12 men, working at the same rate, to erect 5 similar engines?

Time for 8 men to erect 2 engines $= 18$ days

„ „ 1 man „ „ 2 „ $= 18 \times 8$ days

„ „ 1 man „ „ 1 engine $= \dfrac{18 \times 8}{2}$ „

„ „ 12 men „ „ 1 „ $= \dfrac{18 \times 8}{2 \times 12}$ „

„ „ 12 men „ „ 5 engines $= \dfrac{18 \times 8 \times 5}{2 \times 12}$

$= 30$ days

Example. A general service pump can pump out a tank in 10 hours and the ballast pump can pump out the same tank in 5 hours. If both pumps are working together, how long will it take to empty the tank?

Quantity pumped by G.S. pump in 10 hours = whole tank

,, ,, ,, ,, ,, ,, 1 hour = $\frac{1}{10}$ of tank

,, ,, ,, ballast ,, ,, 5 hours = whole tank

,, ,, ,, ,, ,, ,, 1 hour = $\frac{1}{5}$ of tank

Quantity pumped in one hour by both pumps working together

$$= \tfrac{1}{10} + \tfrac{1}{5} \text{ of the tank}$$

$$= \tfrac{3}{10} \text{ of the tank}$$

$$\therefore \text{ Time to empty} = \tfrac{10}{3} \text{ hours}$$

$$= 3\tfrac{1}{3} \text{ hours}$$

VARIATION

Variation is a further step in ratio and proportion. As previously expalined, when an increase of one quantity depends upon the increase of another, one is said to *vary directly* as the other; when an increase in one quantity depends upon the decrease of another, one is said to *vary inversely* as the other. The variation sign is $\propto$, thus, if the cost of casting a propeller varies as its mass, this may be written:

$$\text{cost} \propto \text{mass}$$

and, if the time to travel a certain distance varies inversely as the speed, this may be written:

$$\text{time} \propto \frac{1}{\text{speed}}$$

Taking a simple numerical example on the first case: The cost of a propeller of 2 tonne mass is £800, the cost of another propeller of 3 tonne mass is £1200,

1st propeller, $\dfrac{\text{cost}}{\text{mass}} = \dfrac{800}{2} = £400$ per tonne

2nd propeller, $\dfrac{\text{cost}}{\text{mass}} = \dfrac{1200}{3} = £400$ per tonne hence, $\dfrac{\text{cost}}{\text{mass}} = \text{a constant amount}$

therefore, $\dfrac{\text{cost}_1}{\text{mass}_1} = \dfrac{\text{cost}_2}{\text{mass}_2}$

For the second case take the example: A train travels a certain distance in 5 hours when travelling at an average speed of 60 kilometres per hour; another train takes 4 hours over the same journey travelling at an average speed of 75 kilometres per hour.

1st train, speed $\times$ time $= 5 \times 60 = 300$ kilometres

2nd train, speed $\times$ time $= 4 \times 75 = 300$ kilometres

hence, speed $\times$ time $=$ a constant (distance)

therefore, $\text{speed}_1 \times \text{time}_1 = \text{speed}_2 \times \text{time}_2$

Summing up, (i) when one quantity varies directly as another, their *quotient* is constant; (ii) when one quantity varies inversely as another, their *product* is constant.

Example. Within a certain range, the power of an engine varies directly as the mass of fuel burned per hour. An engine uses 180 kilogrammes of oil per hour when the power developed is 800 kilowatts, what will be the fuel consumption when the power is 900 kilowatts?

$$\text{fuel per hour} \propto \text{power}$$

$$\therefore \frac{\text{fuel per hour}}{\text{power}} = \text{a constant}$$

$$\therefore \frac{\text{fuel}_1}{\text{power}_1} = \frac{\text{fuel}_2}{\text{power}_2}$$

$$\frac{180}{800} = \frac{\text{fuel}_2}{900}$$

$$\therefore \text{fuel}_2 = \frac{180 \times 900}{800}$$

$$= 202{\cdot}5 \text{ kilogrammes per hour.}$$

Example. If a gas is maintained at the same temperature its pressure varies inversely as its volume. Six cubic metres of gas at atmospheric pressure is compressed until its pressure is four times as much, the temperature remaining unaltered, find the volume of the compressed gas.

$$\text{pressure} \propto \frac{1}{\text{volume}}$$

$$\text{pressure} \times \text{volume} = \text{a constant}$$

$$\therefore p_1 \times V_1 = p_2 \times V_2$$

$$1 \times 6 = 4 \times V_2$$

$$V_2 = \frac{1 \times 6}{4}$$

$$V_2 = 1{\cdot}5 \text{ cubic metres}$$

Example. The resistance of a conducting wire varies directly as its length and inversely as its cross-sectional area. A wire 100 metres long and one square millimetre cross sectional area has a resistance of 2 ohms, what would be the resistance of a wire of similar material 250 metres long and 0·5 square millimetre cross-sectional area?

$$\text{resistance} \propto \text{length}$$

also,

$$\text{resistance} \propto \frac{1}{\text{area}}$$

$$\therefore \frac{\text{resistance} \times \text{area}}{\text{length}} = \text{a constant}$$

$$\frac{R_1 \times A_1}{l_1} = \frac{R_2 \times A_2}{l_2}$$

$$\frac{2 \times 1}{100} = \frac{R_2 \times 0{\cdot}5}{250}$$

$$R_2 = \frac{2 \times 1 \times 250}{0{\cdot}5 \times 100}$$

$$R_2 = 10 \text{ ohms}$$

PERCENTAGE

Percentage is another method of expressing a ratio in fractional form, using 100 as the denominator, and substituting the denominator of 100 with the percentage symbol %

Thus, the ratio of 4 to 25

can be written $\frac{4}{25}$ in fractional form

which is equal to $\frac{16}{100}$ on a denominator of 100

and written 16% in percentage form.

Since a fraction is part of unity, and a percentage is part of one hundred, a fraction is converted into a percentage by multiplying the fraction by 100. Conversely, a percentage is converted into a fraction by dividing by 100. Thus:

$$\tfrac{1}{2} = 50\%$$

$$\tfrac{1}{3} = 33\tfrac{1}{3}\%$$

$$\tfrac{3}{8} = 37\tfrac{1}{2}\%$$

To express percentage increase or percentage decrease:

$$\%\ \text{increase} = \frac{\text{total increase}}{\text{original amount}} \times 100$$

$$\%\ \text{decrease} = \frac{\text{total decrease}}{\text{original amount}} \times 100$$

Example. (i) A spring of original length 80 millimetres, is stretched to a length of 100 millimetres, what is the percentage increase in length?

(ii) Another spring of original length 100 millimetres, is compressed to a length of 80 millimetres, what is the percentage decrease in length?

(i) actual increase $= 20$ mm

$$\% \text{ increase} = \frac{20}{80} \times 100$$

$$= 25\%$$

(ii) actual decrease $= 20$ mm

$$\% \text{ decrease} = \frac{20}{100} \times 100$$

$$= 20\%$$

Example. The ratio of the masses of engine A to engine B is 5 to 3. Find (i) by what percentage is A heavier than B, (ii) by what percentage B is lighter than A.

$$\text{Difference} = 5 - 3 = 2$$

(i) $\% \text{ increase of A over B} = \frac{2}{3} \times 100 = 66\tfrac{2}{3}\%$

$\therefore$ A is $66\tfrac{2}{3}\%$ heavier than B

(ii) $\% \text{ decrease of B over A} = \frac{2}{5} \times 100 = 40\%$

$\therefore$ B is 40% lighter than A

CONSTITUENT PARTS

The parts which constitute a whole, such as the quantities of metals mixed together to form an alloy, may be expressed in the form of ratios, fractions or percentages, and it is often convenient to convert one form into another when solving problems. Consider a brass consisting of two parts copper and one part zinc, in every 3 grammes of brass there will be 2 grammes of copper and 1 gramme of zince, therefore two-thirds (i.e. $66\tfrac{2}{3}\%$) of the total mass of the brass is copper and the remaining one-third (i.e. $33\tfrac{1}{3}\%$) is zinc. Hence to convert ratio of parts into fractional amounts, add the ratios together to produce a common denominator, and then each ratio forms its own respective numerator. The fractions are multiplied by 100 to convert them into percentages.

Example. A fusible metal is composed of 5 parts bismuth, 3 parts lead and 2 parts tin. Express this ratio as fractional quantities and percentages, and find the mass of each element required to produce 1·5 kilogrammes of fusible metal.

Ratio, bismuth : lead : tin $= 5 : 3 : 2$

Sum of ratios $= 5 + 3 + 2 = 10$

$\therefore$ mass of bismuth $= \frac{5}{10} = \frac{1}{2} = 50\%$ of total mass

„ „ lead $= \frac{3}{10} = 30\%$ „ „

„ „ tin $= \frac{2}{10} = \frac{1}{5} = 20\%$ „ „

To produce 1·5 kilogrammes = 1500 grammes of metal:

mass of bismuth $= \frac{1}{2} \times 1500 = 750$ grammes

„ „ lead $= \frac{3}{10} \times 1500 = 450$ „

„ „ tin $= \frac{1}{5} \times 1500 = 300$ „

„ „ fusible metal = sum = 1500 grammes.

Example. In a certain two-cylinder engine, the power developed in No. 1 cylinder is 20% more than the power developed in No. 2 cylinder. What percentage of the total power of the engine is developed in each cylinder?

Ratio of powers, No. 1 : No. 2

$$= 120 : 100$$

$$= 1{\cdot}2 : 1$$

Sum of ratios $= 1{\cdot}2 + 1 = 2{\cdot}2$

$\therefore$ percentage of total power developed in No. 1 cylinder

$$= \frac{1{\cdot}2}{2{\cdot}2} \times 100 = 54{\cdot}55\%$$

$\therefore$ percentage of total power developed in No. 2 cylinder

$$= \frac{1}{2 \cdot 2} \times 100 = 45 \cdot 45\%$$

Example. Of the total heat in the fuel burned in the cylinders of a diesel engine, 40% is converted into indicated power, 30% is carried away by the cooling water, 2% is lost by radiation, and the remainder is carried away in the exhaust gases. (i) What percentage of the total heat of the fuel does the heat in the exhaust gases represent? (ii) If 80% of the indicated power is usefully imparted to the crank shaft as brake power and the remainder lost in friction, what percentage of the total heat of the fuel does the brake power represent? (iii) If 40% of the heat in the exhaust gases is recovered in an exhaust gas boiler and the remainder carried away up the funnel, what percentage of the total heat in the fuel is recovered in the boiler?

The full amount of heat in the fuel is represented by 100%
Accounted for:

Heat converted to ip = 40%

Heat carried away in cooling water = 30%

Radiation loss = 2%

total = 72%

Remainder = 100 − 72 = 28

$\therefore$ Exhaust gases carry away 28% of the total heat (i)

$$\text{bp} = 80\% \text{ of } 40\%$$

$$= \frac{8}{10} \text{ of } 40\%$$

$$= 32\%$$

Heat equivalent of bp = 32% (ii)
(Friction loss represents 40 − 32 = 8% of the total heat)

$$\text{Recovery in boiler} = 40\% \text{ of } 28\%$$

$$= \frac{4}{10} \text{ of } 28\%$$

$$= 11{\cdot}2\% \text{ (iii)}$$

(Heat carried away up the funnel = 28 − 11·2 = 16·8%)

A "heat-balance" diagram can now be drawn to illustrate the distribution of the heat in the fuel:

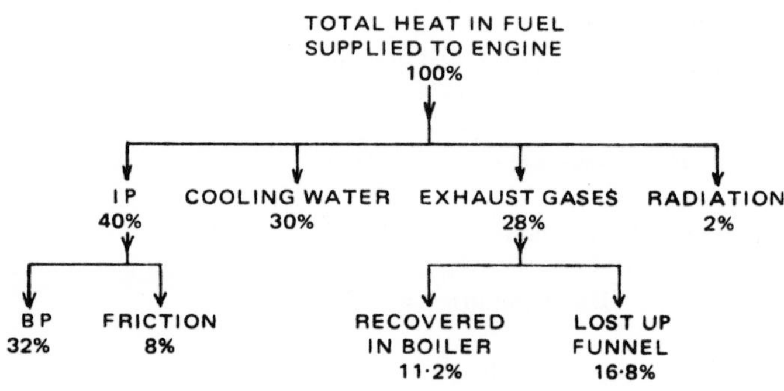

Fig. 1

AVERAGES

The average (or *mean*) value of a group of collected quantities is obtained by dividing the total of the quantities by the number of quantities in the group.

Example. The distances covered by a ship on four successive days were 320, 300, 310 and 330 nautical miles respectively. Find the average day's run.

$$\text{Average day's run} = \frac{\text{total distance}}{\text{total number of days}}$$

$$= \frac{320 + 300 + 310 + 330}{4}$$

$$= \frac{1260}{4}$$

$$= 315 \text{ nautical miles.}$$

Example. A ship steams a distance of 15 nautical miles up-river against the current, in 72 minutes, and then 15 nautical miles down-river with the current, in 50 minutes.

Find (i) the average speed up-river;
(ii) the average speed down-river;
(iii) the average speed over the double journey.

$$\text{Speed} = \frac{\text{distance}}{\text{time}}$$

$$\text{Speed up-river} = 15 \text{ naut. miles} \div \frac{72}{60} \text{ hours}$$

$$= \frac{15 \times 60}{72}$$

$$= 12\tfrac{1}{2} \text{ nautical miles per hour}$$

$$= 12\tfrac{1}{2} \text{ knots (i)}$$

$$\text{Speed down-river} = 15 \text{ naut. miles} \div \frac{50}{60} \text{ hours}$$

$$= \frac{15 \times 60}{50}$$

$$= 18 \text{ naut. miles per hour}$$

$$= 18 \text{ knots (ii)}$$

Average speed over the double journey

$$= \frac{\text{total distance}}{\text{total time}}$$

$$= 30 \text{ naut. miles} \div \frac{122}{60} \text{ hours}$$

$$= \frac{30 \times 60}{122}$$

$$= 14{\cdot}75 \text{ knots (iii)}$$

The above example illustrates speed being obtained by dividing distance by time. Units of speed are usually expressed in kilometres per hour (kilometres of distance divided by hours of time), or metres per second (metres of distance divided by seconds of time).

Speeds of sea-going craft are usually measured in knots which are nautical miles per hour, and obtained by dividing nautical miles of distance by hours of time. One international nautical mile is equal to 1·852 kilometres.

It is important to note how essential it is to divide one total by another total to obtain an average. In the last example, total distance was divided by total time to obtain the average speed over the double journey, it would be wrong to take the average of the average speeds up and down river.

TEST EXAMPLES 1

1. Find the value of $\dfrac{5^3 \times 5 \times 5^5}{5^4 \times 5^2}$

2. Find the value of $\dfrac{10^{\frac{3}{2}} \times 10^4 \times 10^{\frac{3}{4}}}{10^3 \times 10^{\frac{1}{4}} \times 10^2}$

3. Find the value of $\dfrac{4^2 \times 4^{\frac{2}{3}} \times 4^{-2}}{4^{\frac{1}{2}} \times 4^{\frac{1}{6}}}$

4. Find the value of $\dfrac{8^{-2}}{8^{-5}}$

5. Find the value of $\dfrac{(2^2)^3 \times (2^{\frac{1}{2}})^4 \times 2^2}{(2^{\frac{2}{3}})^3 \times (2^3)^2 \times 1}$

6. Find the value of $\dfrac{(3^{\frac{2}{5}})^{\frac{5}{3}} \times \sqrt{3} \times 3}{3^{-\frac{4}{3}} \times 3^3}$

7. Find the value of $(1\frac{2}{3})^2 + \sqrt{\frac{4}{81}} + \sqrt{200}$

8. Find the value of $\sqrt[\frac{3}{2}]{27} + \sqrt{6\frac{1}{4}}$

9. Find the value of $\dfrac{(2^2)^3 \times \sqrt{3} \times 27^{\frac{2}{3}} \times \sqrt{12}}{\sqrt[4]{81} \times 4^{\frac{3}{2}}}$

10. The ratio of the circumferences of two circles is equal to the ratio of their diameters. The diameter of one circle is 35 mm and its circumference is 110 mm, find the circumference of a circle whose diameter is 105 mm.

11. The ratio of the volumes of two spheres is equal to the ratio of the cubes of their diameters; the volume of one sphere is 24·25 cubic centimetres, find the volume of another sphere whose diameter is twice as much.

12. A pump can empty a tank in 12 hours, another pump can empty the same tank in 4 hours, and another can empty this tank in 9 hours. If all three pumps are set working together on this tank, how long would it take to empty it?

13. If 5 machines can be assembled by 6 craftsmen in 10 hours, estimate the time it would take 8 craftsmen to assemble 24 machines.

14. The resistance of a conducting wire varies directly as its length and inversely as the square of its diameter. A wire 1000 metres long and 1 mm diameter has a resistance of 25 ohms, calculate the length of wire of similar material 0·5 mm diameter to have a resistance of 150 ohms.

15. The mass of a round steel shaft, 100 mm diameter and 4·5 m long is 280 kilogrammes. Find the mass of a shaft of similar material, 90 mm diameter and 5 m long, given that the mass varies directly as the length and the square of the diameter.

16. The strength of a beam varies directly as its breadth, directly as the square of its depth, and inversely as its length. A beam is 5 metres long, 40 mm broad, and 100 mm deep; find the breadth of another beam of similar material, 3 metres long and 80 mm deep, to have equal strength.

17. A piece of mild steel 50 mm long between gauge points and 80 mm^2 cross-sectional area, was tested in a tensile testing machine and broke when the gauge length was 62·5 mm and cross-sectional area 48 mm^2. Find the percentage elongation and the percentage reduction in cross-sectional area.

18. The ratio of the strengths of a solid shaft and a hollow shaft in torsion is 1 to $\frac{15}{16}$ and the ratio of their masses is 1 to $\frac{3}{4}$, when the hollow shaft has an external diameter equal to the diameter of the solid shaft and an internal diameter equal to half the external diameter. Find:

(i) by what percentage the solid shaft is stronger than the hollow shaft;
(ii) by what percentage the hollow shaft is weaker than the solid shaft;
(iii) by what percentage the solid shaft is heavier than the hollow shaft;
(iv) by what percentage the hollow shaft is lighter than the solid shaft.

19. In a certain three cylinder engine, the power developed in No. 1 cylinder is 15% more than in No. 3, and 5% less power is developed in No. 2 than in No. 3. What percentage of the total engine power is developed in each cylinder? Give the answers to the nearest one-tenth per cent.

20. A White Metal for lining bearings is composed of 10 parts tin, one part copper and one part antimony, by mass. Find the mass of copper and antimony required to combine with 112 kg of tin to make White Metal. Express the given proportions as percentages.

21. A leaded yellow brass is composed of 71% copper, 1% tin, 3% lead and the remainder zinc. Find the mass required of each constituent, to make 500 kg of this alloy.

22. The speed of a ship's propeller through the water is a steady 15 knots but, due to slip, the ship covers a distance of only 324 nautical miles in 24 hours. What percentage of the propeller speed does the slip represent?

23. The heights of an indicator diagram measured at regular intervals along its length are as follows: 27, 39, 47, 51, 48, 32, 20, 11, 8, 5 mm respectively. Find the mean height of the diagram in millimetres.

24. At 8.00 a.m. the reading of the engine counter was 312 460 and at noon the reading was 333 520. What was the mean speed of the engine in revolutions per minute if the clock was put forward 6 minutes during the watch? If the speed of the engine is increased by 10% at noon and the new speed maintained during the whole of the following watch, what is the counter reading at 4 p.m.?

25. 200 tonne of oil were bought at one port at £60 per tonne and 600 tonne of oil at another port at £70 per tonne. What was the average cost of oil per tonne?

26. A ship covers the "measured mile" (one nautical mile) against the current, in 3 minutes 20 seconds, and then in the opposite direction over the same distance with the current in 3 minutes exactly. Find, in knots,

(i) the speed against the current;

(ii) the speed with the current;

(iii) the average speed, correct to two decimal places.

27. A train is scheduled to do a journey of 384 kilometres at an average speed of 64 kilometres per hour. Due to slower running through patches of fog the average speed over the first half of the journey was 48 kilometres per hour, what speed should the train average over the second half of the journey to arrive at its destination on time?

CHAPTER 2

ELEMENTARY ALGEBRA

Algebra is a convenient system of short-hand arithmetic using letters of the alphabet to represent certain articles, quantities or hidden numbers. For instance if, in a box containing 25 bolts, 34 nuts and 47 washers, we add 15 bolts and 18 washers and take away 14 nuts, there would then be in the box 40 bolts, 20 nuts and 65 washers. This could be set down in the following manner:

$$\begin{array}{rl} & 25b + 34n + 47w \\ \text{add} & 15b - 14n + 18w \\ \hline \text{result} & 40b + 20n + 65w \end{array}$$

If the sign of the first term of an expression is positive it is usually omitted as in the above example. Similarly for any single term, thus $5x$ means $+5x$.

The number which multiplies a term is called a coefficient. In the above example, the result line shows $+40$ as the coefficient of b, $+20$ as the coefficient of n, and $+65$ as the coefficient of w.

An algebraic expression is a collection of terms. One term may be a single quantity or a group of quantities multiplied (or divided) together, terms are separated from each other by plus or minus signs. When a term consists of quantities multiplied together, it is usual to omit the multiplication signs, $x \times y \times z$ is written xyz.

In the expression $2x^2y + 4xy + 6xy^2$
there are three terms,

$$2 \times x^2 \times y$$

$$4 \times x \times y$$

$$\text{and } 6 \times x \times y^2$$

An algebraic expression which consists of two terms is called a *binomial* expression, and one consisting of three terms is called a *trinomial* expression.

ADDITION

Set the expressions down with similar terms under each other, re-arranging the terms if necessary (they can be written in any order), then add the numbers of each column separately in the usual way.

Example. Add together $2a + 7b + 3c$ and $6c + 5a - 2b$.

	$2a + 7b + 3c$
add	$5a - 2b + 6c$
result	$7a + 5b + 9c$

Example. Add together $3x - 5y + 6z$ and $4x + 2y - 8z$.

	$3x - 5y + 6z$
add	$4x + 2y - 8z$
result	$7x - 3y - 2z$

SUBTRACTION

To subtract numbers or algebraic terms, change the sign of the term to be subtracted and then add.

Examples.

(i) From $8x$ take away $5x$

written $8x - 5x$

result $= 3x$

(ii) From $8x$ take away $-5x$

written $8x - (-5x)$

$= 8x + 5x$

result $= 13x$

(iii) From $-8x$ take away $5x$

written $-8x - 5x$

result $= -13x$

(iv) From $-8x$ take away $-5x$

$$\begin{aligned} \text{written} \quad & -8x - (-5x) \\ &= -8x + 5x \\ \text{result} \quad &= -3x \end{aligned}$$

Example.

Subtract $6c - 3b + 4a - 7d$ from $5b + 3d + 2a - 9c$

For expressions consisting of a number of terms, set down as shown below with like terms under each other; to subtract, change the signs of the subtracting expression mentally, add and write down the results.

From	$2a + 5b - 9c + 3d$
Subtract	$4a - 3b + 6c - 7d$
Result	$-2a + 8b - 15c + 10d$

COLLECTION OF TERMS

An expression may contain a number of similar terms, the first process in solving an algebraic problem is to simplify such an expression by collecting like terms together and writing down the expression in its simplest form.

Example.

Simplify $2x + 5y + 4x - 3z - 2y - z - x$

Re-arranging, $2x + 4x - x + 5y - 2y - 3z - z$

Collecting like terms, $5x + 3y - 4z$

POWERS AND ROOTS

The same rules apply with regard to powers and roots as for ordinary numbers, therefore when observing the following examples, refer if necessary to the explanations given in Chapter 1.

Examples.

$$x \times x = x^2$$

$$2a \times a \times 4a = 8a^3$$

$$x^2 \times x^3 = x^{2+3} = x^5$$

$$x^5 \div x^2 = x^{5-2} = x^3$$

$$x^2 \div x^5 = x^{2-5} = x^{-3} = \frac{1}{x^3}$$

$$a^3 \div a^7 = a^{3-7} = a^{-4} = \frac{1}{a^4}$$

$$(x^2)^3 = x^{2\times 3} = x^6$$

$\sqrt{x}$ may be written $x^{\frac{1}{2}}$

$(\sqrt[3]{x})^2$,, ,, ,, $x^{\frac{2}{3}}$

$$x^3 \div x^2 = x^{3-2} = x^1 = x$$

$$x^2 \div x^2 = x^{2-2} = x^0 = 1$$

MULTIPLICATION AND DIVISION

When algebraic quantities are to be multiplied together, the multiplication sign may be omitted, thus,

$x \times y$ may be written xy

$2 \times a \times b$ may be written $2ab$

$(3x + y) \times (x - 4y)$ may be written $(3x + y)(x - 4y)$

When one quantity is to be divided by another, such as x divided by y, it may be written in any of the following forms:

$$x \div y \quad \text{or} \quad \frac{x}{y} \quad \text{or} \quad x/y$$

The signs are very important, the above are ordinary positive values but care must be taken when any of the terms are negative. The rule is: *like signs multiplied or divided give a positive answer, unlike signs multiplied or divided give a negative answer.*

Examples.

$+2x$	multiplied by	$+4x$	equals	$+8x^2$
$+2x$	,, ,,	$-4x$	,,	$-8x^2$
$-2x$	,, ,,	$+4x$	,,	$-8x^2$
$-2x$	,, ,,	$-4x$	,,	$+8x^2$
$+8x^2$	divided by	$+2x$	,,	$+4x$
$-8x^2$	,, ,,	$+2x$	,,	$-4x$
$-8x^2$	,, ,,	$-2x$	,,	$+4x$
$+8x^2$	,, ,,	$-2x$	,,	$-4x$

Example.

Multiply $(2x + 3y)$ **by** $(3x - 2y)$

Procedure:

(i) set down in long multiplication style;

(ii) multiply $2x + 3y$ by one term of the multiplier, say $3x$;

(iii) Multiply $2x + 3y$ by the other term, $-2y$, putting like terms under each other;

(iv) collect terms by addition.

$$\begin{array}{lll} 2x & + 3y & \\ 3x & - 2y & \\ \hline 6x^2 & + 9xy & \\ & - 4xy & - 6y^2 \\ \hline 6x^2 & + 5xy & - 6y^2 \end{array}$$

When the two expressions to be multiplied together each have only two terms, the multiplication can quite easily be done mentally. In the above example we have,

$$(2x + 3y)(3x - 2y) = 6x^2 + 5xy - 6y^2$$

Note that the first term of the answer is the product of the first terms of the two expressions, i.e., $2x \times 3x = 6x^2$; the last term of the answer is the product of the last terms, i.e., $3y \times -2y = -6y^2$; the middle term of the answer is the sum of the product of the 'means' and the product of the 'extremes', i.e., $3y \times 3x$ plus $2x \times -2y =$ $9xy - 4xy = +5xy$.

Now check the following:

$$(x + 4)(x + 5) = x^2 + 9x + 20$$
$$(4x + 3)(2x - 1) = 8x^2 + 2x - 3$$
$$(2x - 5y)(x + 3y) = 2x^2 + xy - 15y^2$$
$$(3x - 4y)(5x - 6y) = 15x^2 - 38xy + 24y^2$$
$$(a + b)(a + b) = a^2 + 2ab + b^2$$
$$(a - b)(a - b) = a^2 - 2ab + b^2$$
$$(a + b)(a - b) = a^2 - b^2$$
$$(a^2 + b^2)(a^2 - b^2) = a^4 - b^4$$

Example. Multiply $4x + 3y - 2z$ by $x - 2y$

$$\begin{array}{lllll} 4x & + 3y & - 2z & & \\ & x & - 2y & & \\ \hline 4x^2 & + 3xy & - 2xz & & \\ & - 8xy & & - 6y^2 & + 4yz \\ \hline 4x^2 & - 5xy & - 2xz & - 6y^2 & + 4yz \end{array}$$

Example. Divide $6x^3 - x^2 - 14x + 8$ by $3x - 2$

Procedure:

(i) set down in long division style;

(ii) see how many times the first term of the divisor goes into the first term of the dividend and write this down as the first term of the answer: thus, $3x$ into $6x^3$ goes $2x^2$ times.

(iii) multiply the divisor by $2x^2$ and set down under the dividend, like terms under each other;

(iv) subtract, this gives $3x^2$;

(v) bring down the next term of the dividend (i.e., $-14x$);

(vi) repeat the operations as from (ii);

$$
\begin{array}{r}
3x - 2)\,6x^3 - x^2 - 14x + 8\,(2x^2 + x - 4 \quad \text{Answer} \\
\underline{6x^3 - 4x^2} \qquad\qquad\qquad\qquad\qquad\qquad \\
3x^2 - 14x \qquad\qquad\qquad\qquad\quad \\
\underline{3x^2 - 2x} \qquad\qquad\qquad\qquad\quad \\
-12x + 8 \qquad\qquad\qquad\qquad \\
\underline{-12x + 8} \qquad\qquad\qquad\qquad \\
\cdot \quad \cdot \qquad\qquad\qquad\qquad
\end{array}
$$

Example. Divide $2x^2 + 5x - 7$ by $x + 4$

$$
\begin{array}{r}
x + 4)\,2x^2 + 5x - 7\,(2x - 3 \\
\underline{2x^2 + 8x} \qquad\qquad\quad \\
-3x - 7 \qquad\quad \\
\underline{-3x - 12} \qquad\quad \\
+5 \qquad\quad
\end{array}
$$

(i) Proceed as in the previous example by setting down as long division (see above).

(ii) The first term of the divisor, x, into the first term of the dividend, $2x^2$, goes $2x$ times, write this as the first term of the answer.

(iii) Multiply the divisor by $2x$, which gives $2x^2 + 8x$ and set this down under the dividend.

(iv) Subtract to get $-3x$.

(v) Bring down the next term of the dividend, that is -7, to give $-3x - 7$.

(vi) Repeat procedure as from (ii), i.e., x into $-3x$ goes -3 times, put -3 into the answer. Multiply the divisor by -3 to get $-3x - 12$ and set this down. Subtract, and this leaves $+5$. There is now nothing more to bring down from the dividend and x will not go into $+5$, hence this is left over and written as a remainder:

$$(2x^2 + 5x - 7) \div (x + 4) = (2x - 3) \text{ remainder } + 5$$

or

$$2x^2 + 5x - 7 = (x + 4)(2x - 3) + 5$$

EFFECT OF ZERO IN MULTIPLICATION AND DIVISION

When any quantity is multiplied by zero, the result is zero.

Examples.

$$3 \times 0 = 0$$
$$1 \times 2 \times 3 \times 4 \times 5 \times 0 = 0$$
$$a \times b \times c \times 0 = 0$$
$$(x + y) \times 0 = 0$$
$$0 \times (a - b) = 0$$

When zero is divided by any quantity, the result is zero.

Examples.

$$0 \div 4 = 0$$
$$\frac{0}{a + b} = 0$$
$$0/(x^2 - y^2) = 0$$

When any quantity is divided by zero, the result is infinity.

Examples.

$$5 \div 0 = \infty$$
$$\frac{2x + 3y}{0} = \infty$$

When an expression works out to be zero divided by zero, the result is indefinite, that is, it can have any value and indicates that the problem requires to be solved by a different method.

Example. Solve $\dfrac{a^2 - b^2}{a - b}$ when $a = 3$, and $b = 3$

One method, $\dfrac{3^2 - 3^2}{3 - 3}$

$$= \frac{0}{0}$$

Other method, $\dfrac{a^2 - b^2}{a - b}$

$$= \frac{(a + b)(a - b)}{a - b} \qquad (a - b) \text{ cancels}$$

$$= a + b$$
$$= 3 + 3$$
$$= 6$$

REMOVAL OF BRACKETS

A quantity immediately in front (or behind) a bracket indicates that every term inside the bracket must be multiplied by that quantity on removing the brackets. Care must be taken with regard to the signs, if the multiplier is positive, the signs of the terms in the brackets will remain the same, if the multiplier is negative all the signs of the terms will be changed. If the bracket is preceded by a sign only, e.g., + or −, it indicates that the multiplier is unity.

Examples.

$$2(x^2 + 3x - 5)$$
$$= 2x^2 + 6x - 10$$

$$-3(2a^2 - 7ab + 3b^2)$$
$$= -6a^2 + 21ab - 9b^2$$

$$4(2x - y) - 2x(x + 3y) - (3x + 4y)$$
$$= 8x - 4y - 2x^2 - 6xy - 3x - 4y$$
$$= -2x^2 + 8x - 3x - 6xy - 4y - 4y$$
$$= -2x^2 + 5x - 6xy - 8y$$

BRACKETS CONTAINED WITHIN BRACKETS

The common sets of brackets are,

( ) small, used as inner brackets,

{ } intermediate brackets,

[ ] square, used as outer brackets.

When simplifying expressions composed of bracketed quantities within other brackets, it is usual to remove the inner brackets first, then the intermediate brackets, and finally the outer brackets.

Example. $y(1 + 3x) + [3x(x - y) - \{-2(x - 1) + 3x^2\} - y]$
$= y + 3xy + [3x^2 - 3xy - \{-2x + 2 + 3x^2\} - y]$
$= y + 3xy + [3x^2 - 3xy + 2x - 2 - 3x^2 - y]$
$= y + 3xy + 3x^2 - 3xy + 2x - 2 - 3x^2 - y$
$= 3x^2 - 3x^2 + 3xy - 3xy + 2x + y - y - 2$
$= 2x - 2$

Example. $(a + 2)^2 - (3a + 1)(2a - 4)$
$= (a^2 + 4a + 4) - (6a^2 - 10a - 4)$
$= a^2 + 4a + 4 - 6a^2 + 10a + 4$
$= -5a^2 + 14a + 8$

FACTORISATION

Factorising is the reverse of multiplying, that is, it is the process of finding the numbers or quantities which, when multiplied together, will constitute the expression given to be factorised. Generally, expressions to be factorised fall into one or more of four types; (i) those made up of terms with common multipliers, (ii) difference between two squares, (iii) perfect squares, and (iv) those whose factors have to be found by trial and error.

(i) Factorising expressions which contain common multipliers. If every term contains a common factor, divide throughout by this factor and express the factor as a multiplier to the resultant terms enclosed in a bracket, thus,

$2 \times 3 + 2 \times 4 - 2 \times 5$
$= 2(3 + 4 - 5)$

This is the reverse process to removal of brackets.

Example. $3x + 2xy - xz$
$= x(3 + 2y - z)$

Example. $\dfrac{2a^2}{b} - \dfrac{2a}{3b} + \dfrac{4ac^2}{b}$

$= \dfrac{2a}{b}(a - \frac{1}{3} + 2c^2)$

(ii) Factorising expressions which are the difference between two squares. Examine first the following cases and note the similarity of the factors and the results of the multiplications:

$$(x + 3)(x - 3) = x^2 - 9$$
$$(x + 1)(x - 1) = x^2 - 1$$
$$(x + y)(x - y) = x^2 - y^2$$
$$(2a + 3b)(2a - 3b) = 4a^2 - 9b^2$$

Hence we see that if the *sum* of two terms is multiplied by the *difference* between those two terms, the result is the difference between the squares of the two terms. Therefore any expression which consists of the difference between two squares can be readily factorised.

Factors of $a^2 - b^2 = (a + b)(a - b)$

" " $y^2 - 16 = (y + 4)(y - 4)$

" " $16D^2 - 25d^2 = (4D + 5d)(4D - 5d)$

(iii) Factorising expressions which are perfect squares. Firstly, check the following and note the form of the result in each case.

$$(x + 3)^2 = (x + 3)(x + 3) = x^2 + 6x + 9$$
$$(x - 5)^2 = (x - 5)(x - 5) = x^2 - 10x + 25$$
$$(x - 4y)^2 = (x - 4y)(x - 4y) = x^2 - 8xy + 16y^2$$

The above are simple cases where the coefficient of x^2 is unity, such expressions can be easily recognised as perfect squares because, in each case, the third term is equal to the square of half the coefficient of x. Note that,

$+ 9$ is the square of half of $+ 6$

$+25$ " " " " -10

$+16y^2$ " " " " $- 8y$

Examples. $\sqrt{(a^2 + 4a + 4)} = a + 2$

$\sqrt{(x^2 - 12x + 36)} = x - 6$

Now check and examine the following slightly more difficult cases,

$\sqrt{(9x^2 + 24x + 16)} = 3x + 4$

because, $(3x + 4)(3x + 4) = 9x^2 + 24x + 16$

$\sqrt{(16x^2 - 48x + 36)} = 4x - 6$

because, $(4x - 6)(4x - 6) = 16x^2 - 48x + 36$

$$\sqrt{25x^2 + 70xy + 49y^2)} = 5x + 7y$$

because, $$(5x + 7y)(5x + 7y) = 25x^2 + 70xy + 49y^2$$

Expressions which are perfect squares can therefore be recognised at once and their square root becomes obvious. Note particularly the effect of the minus sign.

(iv) Factorising by trial and error.

Check the following and note the form of the expressions:

$$(x + 2)(x + 3) = x^2 + 5x + 6$$
$$(2x + 5)(3x - 4) = 6x^2 + 7x - 20$$
$$(3x + 2y)(4x - 5y) = 12x^2 - 7xy - 10y^2$$

Note that the first term of the expression is the product of the first terms of the factors, the last term of the expression is the product of the last terms of the factors, the middle term of the expression is the sum of the product of the mean terms and the product of the extremes.

Example. Factorise $x^2 + 5x + 6$

The factors of x^2 are x and x

The factors of $+6$ can be $+6$ and $+1$
or -6 and -1
or $+2$ and $+3$
or -2 and -3

Although any of these pairs *multiplied* together will give the third term, i.e., $+6$, only one of the pairs *added* together will give the coefficient of the middle term, i.e., $+2$ added to $+3$ is $+5$, hence the factors of $x^2 + 5x + 6 = (x + 2)(x + 3)$

Example. Factorise $6x^2 + 7x - 20$

The factors of $6x^2$ can be $6x$ and x
or $2x$ and $3x$

The factors of -20 can be $+20$ and -1
or -20 and $+1$
or $+10$ and -2
or -10 and $+2$
or $+5$ and -4
or -5 and $+4$

We can now try combining the pairs of factors of $6x^2$ with the pairs of factors of -20 until it is seen that the combination which will produce the correct middle term is $(2x + 5)$ and $(3x - 4)$ hence, the factors of $6x^2 + 7x - 20 = (2x + 5)(3x - 4)$.

After a little practice, if we keep an eye on the middle term, obvious misfits can quickly be eliminated and the correct factors spotted without much trouble.

LOWEST COMMON MULTIPLE

As in arithmetic, the L.C.M. of a group of algebraic terms is the lowest quantity into which each of the terms will divide.

For example, the L.C.M. of $4x^2y$, $6xy^2$, $2xyz$ and x^2z^2 is $12x^2y^2z^2$

FRACTIONS

To add and subtract algebraic fractions, bring them all to a common denominator by finding the L.C.M. of the given denominators of the fractions, then add and subtract, like terms being combined together. The procedure is demonstrated in the following example.

Simplify $$\frac{3x + 1}{2x} - \frac{x + 2}{3x} - \frac{2x - 3}{6x}$$

The L.C.M. of the denominators is $6x$.

First fraction: $6x \div 2x = 3$, thus the denominator of $2x$ is to be multiplied by 3 to make it $6x$, therefore the numerator must also be multiplied by 3.

Second fraction: $6x \div 3x = 2$, the denominator of $3x$ is to be multiplied by 2, therefore the numerator must also be multiplied by 2.

Third fraction: the denominator is already $6x$ therefore the numerator also remains unchanged.

$$\frac{3x + 1}{2x} - \frac{x + 2}{3x} - \frac{2x - 3}{6x}$$

$$= \frac{3(3x + 1)}{6x} - \frac{2(x + 2)}{6x} - \frac{(2x - 3)}{6x}$$

usually written $$\frac{3(3x + 1) - 2(x + 2) - (2x - 3)}{6x}$$

$$= \frac{9x + 3 - 2x - 4 - 2x + 3}{6x}$$

$$= \frac{5x + 2}{6x}$$

Note that the given expression contained no brackets, the reason being that division lines of fractions act as a double purpose, to indicate that the numerator is divided by the denominator but also to indicate that everything above (or below) the line is one complete term, thus

$$\frac{3x + 1}{2x} \text{ could be written } \frac{(3x + 1)}{2x}$$

However, when proceeding with the simplification of the expression, it is necessary to insert the brackets before multiplying out to ensure that all the quantities constituting the term are so affected.

The final answer could be written $\frac{(5x + 2)}{6x}$ or $\frac{5x + 2}{6x}$

EVALUATION

Evaluation is the process of substituting the numerical values of the algebraic symbols and working out the value of the whole expression.

The substitution of the algebraic symbols by their numerical values may be done in the original expression to be evaluated, or, the expression may be simplified first and numerical values substituted later, whichever is the more convenient.

The usual rules of arithmetic apply, i.e., quantities enclosed in brackets should be solved first, and multiplication and division must be performed before addition and subtraction.

Example. Evaluate $3xy + x^2 - 4y$
when $x = 2$, and $y = 3$

$$3xy + x^2 - 4y$$
$$= 3 \times 2 \times 3 + 2^2 - 4 \times 3$$
$$= 18 + 4 - 12$$
$$= 10$$

Example. Evaluate $3a^2 - 3b^2 - 4c^2$
when $a = 4$, $b = -2$, and $c = 3$

$$3a^2 - 3b^2 - 4c^2$$
$$= 3 \times 4^2 - 3 \times (-2)^2 - 4 \times 3^2$$
$$= 3 \times 16 - 3 \times 4 - 4 \times 9$$
$$= 48 - 12 - 36$$
$$= 0$$

Example. Evaluate $(a + b)^2 - (c + d) + x^3 - y$
when $a = 3$, $b = 4$, $c = 5$, $d = 6$, $x = -2$, $y = -3$

$$(a + b)^2 - (c + d) + x^3 - y$$
$$= (3 + 4)^2 - (5 + 6) + (-2)^3 - (-3)$$
$$= (7)^2 - (11) + (-8) - (-3)$$
$$= 49 - 11 - 8 + 3$$
$$= 33$$

Example. Evaluate $\dfrac{3x + 2y}{z} + \dfrac{x - z}{y} - \dfrac{2y + z}{x}$
when $x = 4$, $y = 3$, $z = -2$

$$\frac{3x + 2y}{z} + \frac{x - z}{y} - \frac{2y + z}{x}$$
$$= \frac{3 \times 4 + 2 \times 3}{-2} + \frac{4 - (-2)}{3} - \frac{2 \times 3 + (-2)}{4}$$
$$= \frac{12 + 6}{-2} + \frac{4 + 2}{3} - \frac{6 - 2}{4}$$
$$= \frac{18}{-2} + \frac{6}{3} - \frac{4}{4}$$
$$= -9 + 2 - 1$$
$$= -8$$

REMAINDER THEOREM

The *remainder theorem* states that if a polynomial in x is divided by $x - a$, the remainder is equal to the result obtained when a is substituted for x in the polynomial.

A *polynomial* is an expression consisting of a number of terms, each term being a multiple of an integral power of a quantity such as x. The *degree* of the polynomial signifies the highest power.

For example: $3x^2 - 4x + 2$ is a polynomial of the second degree in x; $4y^3 + 2y^2 + 3y - 5$ is a polynomial of the third degree in y, and so on.

A *function* of a quantity is an expression which depends upon that quantity. For example: $2x^3 + 3x^2 - 2x - 5$ is a function of x and denoted by $f(x)$; $2a^3 + 3a^2 - 2a - 5$ is a function of a and denoted by $f(a)$; $2(2)^3 + 3(2)^2 - 2(2) - 5$ is a function of 2 and denoted by $f(2)$

Hence, using the above notations, and R for the remainder, the remainder theorem may be briefly written:

$$R = f(a) \text{ when } f(x) \div (x - a)$$

Note that if $f(a) = 0$, the remainder is zero, which means that $x - a$ is then a factor of $f(x)$. This is the *factor theorem*.

Example. Find the remainder when $2x^2 + 5x - 7$ is divided by $x + 4$.

$$\begin{aligned} \text{Let } x - a &= x + 4 \\ \text{then } a &= -4 \\ f(x) &= 2x^2 + 5x - 7 \\ R = f(a) &= 2(-4)^2 + 5(-4) - 7 \\ &= 32 - 20 - 7 \\ &= 5 \text{ Ans.} \end{aligned}$$

Example. Find the remainder when $2x^3 - 3x^2 - 3x + 5$ is divided by $x - 2$.

$$\begin{aligned} \text{Let } x - a &= x - 2 \\ \text{then } a &= +2 \\ f(x) &= 2x^3 - 3x^2 - 3x + 5 \\ R = f(a) &= 2(2)^3 - 3(2)^2 - 3(2) + 5 \\ &= 16 - 12 - 6 + 5 \\ &= 3 \text{ Ans.} \end{aligned}$$

Example. Find the value of y in the expression $x^2 - 5x + y$ if $(x - 7)$ is a factor.

Factor theorem: If $(x - a)$ is a factor of $f(x)$, the remainder is zero, and $f(a) = 0$

$$
\begin{aligned}
\text{Let } x - a &= x - 7 \\
\text{then } a &= +7 \\
x - 7 \text{ is a factor}, \therefore f(+7) &= 0 \\
f(x) &= x^2 - 5x + y \\
f(+7) &= (7)^2 - 5(7) + y \\
0 &= 49 - 35 + y \\
y &= -14 \text{ Ans.}
\end{aligned}
$$

Example. Find the values of b and c if $x^3 + bx^2 + cx - 6$ is divisible by $(x + 1)(x - 2)$ with no remainder.

Factor theorem: If $(x - a)$ is a factor of $f(x)$, the remainder is zero, and $f(a) = 0$

$$
\begin{aligned}
\text{Let } x - a &= x + 1 \\
\text{then } a &= -1 \\
x + 1 \text{ is a factor}, \therefore f(-1) &= 0 \\
f(x) &= x^3 + bx^2 + cx - 6 \\
f(-1) &= (-1)^3 + b(-1)^2 + c(-1) - 6 \\
0 &= -1 + b - c - 6 \\
c &= b - 7 \quad \ldots \quad \ldots \quad \ldots \quad \ldots \quad \text{(i)}
\end{aligned}
$$

$$
\begin{aligned}
\text{Let } x - a &= x - 2 \\
\text{then } a &= +2 \\
x - 2 \text{ is a factor}, \therefore f(+2) &= 0 \\
f(x) &= x^3 + bx^2 + cx - 6 \\
f(+2) &= (2)^3 + b(2)^2 + c(2) - 6 \\
0 &= 8 + 4b + 2c - 6 \\
2c &= 6 - 8 - 4b \\
c &= -1 - 2b \quad \ldots \quad \ldots \quad \ldots \quad \ldots \quad \text{(ii)}
\end{aligned}
$$

From (i) and (ii):

$$
\begin{aligned}
b - 7 &= -1 - 2b \\
3b &= 6 \\
b &= 2
\end{aligned}
$$

From (i):

$$c = 2 - 7 = -5$$

Values of b and c are, 2 and -5 Ans.

Example. Find the constants b and c if, when $x^3 + 2x^2 + bx + c$ is divided by $(x + 3)$ there is no remainder, and when divided by $(x + 2)$ the remainder is 11.

Factor theorem: If $(x - a)$ is a factor of $f(x)$, the remainder is zero, and $f(a) = 0$

$$\begin{aligned}
\text{Let } x - a &= x + 3 \\
\text{then } a &= -3 \\
x + 3 \text{ is a factor, hence } f(-3) &= 0 \\
f(x) &= x^3 + 2x^2 + bx + c \\
f(-3) &= (-3)^3 + 2(-3)^2 + b(-3) + c \\
0 &= -27 + 18 - 3b + c \\
c &= 3b + 9 \quad \dots \quad \dots \quad \dots \quad \dots \quad \text{(i)}
\end{aligned}$$

Remainder theorem:

Remainder $= f(a)$ when $f(x)$ is divided by $(x - a)$

$$\begin{aligned}
\text{Let } x - a &= x + 2 \\
\text{then } a &= -2 \\
f(x) &= x^3 + 2x^2 + bx + c \\
R = f(a) &= (-2)^3 + 2(-2)^2 + b(-2) + c \\
11 &= -8 + 8 - 2b + c \\
c &= 2b + 11 \quad \dots \quad \dots \quad \dots \quad \dots \quad \text{(ii)}
\end{aligned}$$

From (i) and (ii),

$$\begin{aligned}
3b + 9 &= 2b + 11 \\
b &= 2
\end{aligned}$$

Substituting value of b into (i),

$$c = 3 \times 2 + 9 = 15$$

Constants are 2 and 15 Ans.

TEST EXAMPLES 2

1. (i) Add together $3x + 4y - 5z$ and $-2x - 5y + 4z$
   (ii) ,, ,, $2a^2b - ab + 3ab^2$ and $5ab^2 - a^2b + ab$
   (iii) Subtract $2x + 5y - 3z$ from $5x - 4z + 3y$
   (iv) ,, $-4c - 5b - a$,, $3a + 6c - 2b$

2. Simplify the following expressions by collecting terms,
   (i) $5x - 3z - 4x - 2y + 4y + 2z - y$
   (ii) $2{\cdot}5a + c - 1{\cdot}2a + 2{\cdot}5b - 3c + b + 1{\cdot}7a$
   (iii) $b^2 - 3ab^2 + 2a^2b - 4a^2 - 2b^2 + 5a^2 - 2ab^2$

3. Simplify the following,
   (i) $\dfrac{x \times x^3 \times x^5}{x^2 \times x^4}$
   (ii) $x^3 \div x^5$
   (iii) $x^5 \times x^{-3} \times x^{-2}$
   (iv) $\sqrt{x} \times x^{\frac{1}{2}} \times x^{-\frac{1}{3}}$
   (v) $\sqrt[2]{(x^4y^2)}$
   (vi) $\dfrac{x^3}{x^3 - 0{\cdot}5x^3}$
   (vii) $(x^2)^3 \times \sqrt[3]{x^{\frac{1}{2}}} \times x^{-5}$

4. (i) Multiply $(x + 2y)$ by $(2x + y)$
   (ii) ,, $(2x + y)$ by $(3x - 2y)$
   (iii) ,, $(3x - 4y)$ by $(2x - 3y)$

5. Work out the following:
   (i) $(a + b)^2$
   (ii) $(a - b)^2$
   (iii) $(a + b)^3$
   (iv) $(a - b)^3$

6. (i) Divide $(8a^2 - 8ab - 6b^2)$ by $(2a - 3b)$
   (ii) ,, $(9x^3 - 9x^2y - 10xy^2 + 8y^3)$ by $(3x - 4y)$
   (iii) ,, $(x^3 - y^3)$ by $(x - y)$

7. Simplify the following by removing brackets and collecting terms:
   (i) $(a + b) + (c - d) - (a - b) - (c + d) + (a - b)$
   (ii) $2\{a - 3(a + 2) + 4(2a - 1) + 5\}$
   (iii) $2x - [2x - \{2x - (2x - 2) - 2\} - 2] - 2$

8. Factorise the following expressions:
   (i) $3b^2 - 6b + 9$
   (ii) $pv + pvx$
   (iii) $ax^3 - 2bx^2 + 3cx$
   (iv) $12a^3b^3c^3 - 8a^2b^2c^2 + 4abc$

9. Factorise the following expressions:
   (i) $D^2 - d^2$
   (ii) $1 - a^2$
   (iii) $4x^2y^2 - 9z^2$
   (iv) $T_1^4 - T_2^4$

10. Factorise the following expressions:
   (i) $a^2 + 8a + 16$
   (ii) $d^2 - 10d + 25$
   (iii) $9v^2 + 12v + 4$
   (iv) $4x^2 - 12xy + 9y^2$

11. Factorise the following expressions:
   (i) $x^2 + 3x + 2$
   (ii) $a^2 - a - 6$
   (iii) $x^2 + xy - 6y^2$
   (iv) $6x^2 - 9xy - 6y^2$

12. Simplify the following:
   (i) $\dfrac{x}{2} + \dfrac{x}{4} - \dfrac{2x}{3}$
   (ii) $\dfrac{2a - 1}{b} - \dfrac{2a + 3}{2b} - \dfrac{3a - 8}{3b}$
   (iii) $\dfrac{4}{x + 2} + \dfrac{18}{x^2 - 2x - 8} - \dfrac{3}{x - 4}$

13. Find the value of:

$$3a^3 - 4a^2 - 2a - 15$$

when $a = 5$

14. Find the value of:

$$2x^3 - x^2y + xy^2 - 3y^3$$

when $x = 2$, and $y = -2$

15. Simplify the following and find the value if

$x = -2$, and $y = -3$,

$$3[4x + 2\{x - 2y - (3x + y)\} - 3x]$$

16. Find the remainder when $2x^3 - 4x^2 + 5x - 6$ is divided by $(x - 3)$.

17. Find the value of b if, when $x^3 + 6x^2 + bx + 9$ is divided by $(x + 5)$ the remainder is 4.

18. Find the value of c if $(x + 4)$ is a factor of $x^3 + x^2 - 10x + c$.

19. Find the constants b and c in the expression $x^3 + 3x^2 - bx - c$ if, when it is divided by $(x + 1)$ there is no remainder, and when divided by $(x + 2)$ the remainder is 15.

CHAPTER 3

LOGARITHMS

Calculators have replaced logarithms (and slide rules) for numerical calculations. The reader may quickly scan read this chapter, noting the mathematical principles described on Pages 53 and 54 (which have applications in technical subjects) and attempt a selective few test examples using the calculator.

The logarithm of a number to a given base is the power to which the base must be raised to be equal to that number.

$2^3 = 8$, in this case, 3 is the logarithm of 8 to the base of 2, and would be written $\log_2 8 = 3$.

$x^n = N$, this represents n as the logarithm of N to the base x and would be written $\log_x N = n$.

LOGARITHMS TO BASE 10

For general use, logarithms to base 10 are adopted because of our metric system of numbers, and these are usually called Common Logarithms.

$10^2 = 100$, hence 2 is the logarithm of 100 to base 10. This may be written $\log_{10} 100 = 2$, but since this is the common system, the suffix denoting base 10 is usually omitted.

Examples.

$$10^0 = 1 \quad \therefore \log 1 = 0$$
$$10^1 = 10 \quad \therefore \log 10 = 1$$
$$10^2 = 100 \quad \therefore \log 100 = 2$$
$$10^3 = 1000 \quad \therefore \log 1000 = 3$$

The values of logarithms to cover all numbers have been calculated to many places of decimals and books of tables are available to four places, five places, seven or more, their use depending upon the degree of accuracy required. Logarithms to four places, termed Four-Figure logs are in general use for ordinary problems in engineering and these only will be dealt with here.

The following list gives the logarithms of the simple numbers from 1 to 10.

log of	1	to base 10 is	0·0000	because 10^{0}	= 1
,,	2	,, ,, ,,	0·3010	,, $10^{0.3010}$	= 2
,,	3	,, ,, ,,	0·4771	,, $10^{0.4771}$	= 3
,,	4	,, ,, ,,	0·6021	,, $10^{0.6021}$	= 4
,,	5	,, ,, ,,	0·6990	,, $10^{0.6990}$	= 5
,,	6	,, ,, ,,	0·7782	,, $10^{0.7782}$	= 6
,,	7	,, ,, ,,	0·8451	,, $10^{0.8451}$	= 7
,,	8	,, ,, ,,	0·9031	,, $10^{0.9031}$	= 8
,,	9	,, ,, ,,	0·9542	,, $10^{0.9542}$	= 9
,,	10	,, ,, ,,	1·0000	,, 10^{1}	= 10

We have seen in the previous chapters that when multiplying powers of the same quantity, we add their indices, thus,

$$x^2 \times x^3 = x^{2+3} = x^5$$

All logarithms being powers of the same quantity (i.e., 10) the same rule applies, for instance,

$$2 \times 4 = 8$$

may be written, $10^{0.3010} \times 10^{0.6021} = 10^{0.9031}$

0·9031 is the logarithm of 8
8 is the antilogarithm of 0·9031

The rules for using logarithms are therefore the same as those for other indices, hence,

(i) to multiply numbers, add their logarithms,
(ii) to divide numbers, subtract their logarithms,
(iii) to raise a number to a power, multiply the logarithm of the number by the power,
(iv) to extract a root of a number, divide the logarithm of the number by the root.

CHARACTERISTIC AND MANTISSA

Since the common logarithm of a number is the power to which 10 is raised to equal that number, it is obvious that the logarithms of numbers from 1 to 10 are between 0 and 1,

e.g., log 4 = 0·6021 because $10^{0.6021} = 4$

Logarithms of numbers from 10 to 100 are between 1 and 2,

$$\text{e.g., } \log 40 = 1{\cdot}6021 \text{ because } 10^{1\cdot6021} = 40$$

Logarithms of numbers from 100 to 1000 are between 2 and 3,

$$\text{e.g., } \log 400 = 2{\cdot}6021 \text{ because } 10^{2\cdot6021} = 400$$

Now consider numbers less than unity,

$$0{\cdot}4 = 4 \times \tfrac{1}{10} = 4 \times 10^{-1}$$
$$\therefore 0{\cdot}4 = 10^{0\cdot6021} \times 10^{-1}$$

Note that the index 0·6021 is positive while the index −1 is negative, adding these indices +0·6021 and −1 we get −0·3979, but is is more convenient in logarithms to write $\bar{1}{\cdot}6021$ with the minus sign above the figure 1 only, indicating only that part is negative while the ·6021 is positive.

By similar reasoning,

$$0{\cdot}04 = 4 \times \tfrac{1}{100} = 4 \times 10^{-2}$$
$$\therefore 0{\cdot}04 = 10^{0\cdot6021} \times 10^{-2}$$
$$= 10^{\bar{2}\cdot6021}$$
$$\therefore \log 0{\cdot}04 = \bar{2}{\cdot}6021$$

also,

$$\log 0{\cdot}004 = \bar{3}{\cdot}6021$$

and,

$$\log 0{\cdot}0004 = \bar{4}{\cdot}6021$$

Thus, only the whole number part of the logarithm is different in all the above cases and its value depends upon the position of the decimal point. The whole number part of the logarithm is called the *characteristic* and careful examination of the above will show that the characteristic is found purely by inspection. For a number greater than unity, the characteristic of the logarithm is positive and its value is one less than the number of figures before the decimal point. For a number less than unity, the characteristic of its logarithm is negative and its value is one more than the number of noughts between the decimal point and the first figure.

The decimal part of the logarithm is called the *mantissa*, it is always positive and its value is read off a table of logarithms; the method of reading these tables will be dealt with later, in the meantime study carefully the following examples to clarify the method of determining the value of the characteristic.

NUMBER	NUMBER WRITTEN IN STANDARD FORM	NUMBER WRITTEN AS POWER OF 10	LOGARITHM AS COMMONLY WRITTEN
4357	$4{\cdot}357 \times 10^{3}$	$10^{0{\cdot}6392} \times 10^{3}$	3·6392
435·7	$4{\cdot}357 \times 10^{2}$	$10^{0{\cdot}6392} \times 10^{2}$	2·6392
43·57	$4{\cdot}357 \times 10^{1}$	$10^{0{\cdot}6392} \times 10^{1}$	1·6392
4·357	$4{\cdot}357 \times 10^{0}$	$10^{0{\cdot}6392} \times 10^{0}$	0·6392
0·4357	$4{\cdot}357 \times 10^{-1}$	$10^{0{\cdot}6392} \times 10^{-1}$	$\bar{1}{\cdot}6392$
0·04357	$4{\cdot}357 \times 10^{-2}$	$10^{0{\cdot}6392} \times 10^{-2}$	$\bar{2}{\cdot}6392$
0·004357	$4{\cdot}357 \times 10^{-3}$	$10^{0{\cdot}6392} \times 10^{-3}$	$\bar{3}{\cdot}6392$
0·0004357	$4{\cdot}357 \times 10^{-4}$	$10^{0{\cdot}6392} \times 10^{-4}$	$\bar{4}{\cdot}6392$

READING FOUR-FIGURE LOG TABLES

As previously stated, the mantissa (decimal part) of the logarithm of a number is read from tables, the reader should now have a set of four-figure log tables at hand to consult while studying the following.

Example. To find the log of 125·7.

The characteristic is 2 (being one less than the number of figures before the decimal point).

The mantissa is read from the tables thus:

Look up the first two figures of the number (12) in the left hand column, then read horizontally along in the column headed by the third figure (5), this reads 0969. Now take the number in the right hand margin, on the same horizontal line and in the column headed by the fourth figure (7), this reads 24 and is to be added to 0969.
0969 + 24 = 0993.

The log of 125·7 is therefore 2·0993

Example. To find the log of 0·005623.

The characteristic is $\bar{3}$ (because there are two noughts between the decimal point and the first figure).

Look up the mantissa for 5623 in a similar manner as explained above, this is 7497 + 2 = 7499

The log of 0·005623 is therefore $\bar{3}$·7499.

Now find the logarithms of the following numbers from the tables and check against the answers given.

GIVEN NUMBER	LOGARITHM
4189	3·6221
0·7854	$\bar{1}$·8951
0·01048	$\bar{2}$·0204
99·99	2·0000
0·00009487	$\bar{5}$·9771

ANTILOGARITHM TABLES

These are the reverse to log tables and are used to read off the number corresponding to the given logarithm. Turn to the pages of antilogarithms and study the procedure in the following example.

To find the antilog of $\bar{2}$·5737 (that is, to find the number whose logarithm is $\bar{2}$·5737).

Look up the first figures of the mantissa (57) in the left hand column, read horizontally along to the number under the column headed by the third figure (3), this reads 3741. Along the same horizontal line read the number in the margin under the fourth figure (7), this reads 6. Add this 6 to 3741 to get 3747. The characteristic ($\bar{2}$) indicates the position of the decimal point, in this case there is one nought between the decimal point and the first figure of the number, therefore the answer is 0·03747.

Now take the following examples of logarithms, find their corresponding numbers from the antilog tables and check the answers with those given.

GIVEN LOGARITHM	NUMBER READ FROM ANTILOG TABLES
2·3675	233·1
$\bar{1}$·4971	0·3142
3·4057	2545
0·4323	2·706
$\bar{6}$·8325	0·0000068

MULTIPLICATION BY LOGS

To multiply numbers, add their logarithms, the resultant sum is the logarithm of the answer.

Example. Find the value of 0·04218 × 4750

log of 0·04218 = $\bar{2}$·6251
„ 4750 = 3·6767 (add)

Sum = 2·3018

Antilog of 2·3018 = 200·4 Ans.

Note that anything "carried over" from the addition of the mantissa is a positive value. In the above +1 is carried which is included in the addition of the characteristics, thus, $\bar{2}$, 3, and 1 added together gives 2.

Example. Find the value of 24·18 × 0·3217 × 2·935

log of 24·18 = 1·3834
„ 0·3217 = $\bar{1}$·5074
„ 2·935 = 0·4676

Sum = 1·3584

Antilog of 1·3584 = 22·82 Ans.

DIVISION BY LOGS

To divide numbers, subtract the logarithm of the divisor from the logarithm of the dividend, the resultant difference is the logarithm of the answer.

Example. Divide 240 by 4345

$$\begin{aligned} \text{log of } 240 &= 2{\cdot}3802 \\ \text{,, } 4345 &= 3{\cdot}6380 \\ \text{difference} &= \bar{2}{\cdot}7422 \\ \text{Antilog of } \bar{2}{\cdot}7422 &= 0{\cdot}05524 \text{ Ans.} \end{aligned}$$

Example. Divide 1970 by 0·06401.

$$\begin{aligned} \text{log of } 1970 &= 3{\cdot}2945 \\ \text{,,}\, 0{\cdot}06401 &= \bar{2}{\cdot}8063 \\ \text{difference} &= 4{\cdot}4882 \\ \text{Antilog of } 4{\cdot}4882 &= 30770 \text{ Ans.} \end{aligned}$$

Note, after subtracting the bottom mantissa from the top, we are left with +1 to carry to the bottom characteristic of $\bar{2}$. +1 and $\bar{2}$ gives $\bar{1}$. Subtracting $\bar{1}$ from +3 (change the sign and add) gives +4.

COMBINED MULTIPLICATION AND DIVISION

Example. Evaluate the value of E in the following,

$$E = \frac{0{\cdot}5 \times I \times (N_1^2 - N_2^2)}{0{\cdot}85 \times l}$$

$$\begin{aligned} \text{where } I &= 1{\cdot}612 \\ N_1 &= 1026 \\ N_2 &= 994 \\ l &= 91{\cdot}07 \end{aligned}$$

Simplify quantity in brackets first, note that this is the difference between two squares and the easiest way to deal with it is to factorise.

$$\begin{aligned} N_1^2 - N_2^2 &= (N_1 + N_2)(N_1 - N_2) \\ &= (1026 + 994)(1026 - 994) \\ &= 2020 \times 32 \end{aligned}$$

Substituting this and remainder of values,

$$E = \frac{0{\cdot}5 \times 1{\cdot}612 \times 2020 \times 32}{0{\cdot}85 \times 91{\cdot}07}$$

Adding logs of numerators:		Adding logs of denominators:	
log 0·5 =	$\bar{1}$·6990	log 0·85 =	$\bar{1}$·9294
log 1·612 =	0·2074	log 91·07 =	1·9593
log 2020 =	3·3054		1·8887
log 32 =	1·5051		
	4·7169		

Subtracting log of denominator from log of numerator:

4·7169
1·8887

2·8282

Antilog of 2·8282 = 673·3 Ans.

POWERS

To raise a number to a power, multiply the logarithm of the number by the power and the resultant product is the logarithm of the answer.

Example. Find the value of $(4{\cdot}189)^2$

log of 4·189 = 0·6221
multiply by the power 2

1·2442

Antilog of 1·2442 = 17·55 Ans.

Example. Find the value of $(0{\cdot}818)^3$

log of 0·818 = $\bar{1}$·9128
3

$\bar{1}$·7384

Antilog of $\bar{1}$·7384 = 0·5475 Ans.

Note, after multiplying the mantissa by 3, there is +2 to carry; multiplying the characteristic $\bar{1}$ by 3 gives $\bar{3}$, adding +2 to $\bar{3}$ gives $\bar{1}$.

Example. Find the value of $(13{\cdot}41)^{1{\cdot}37}$

$$
\begin{array}{rcl}
\text{log of } 13{\cdot}41 & = & 1{\cdot}1274 \\
 & & \phantom{1{\cdot}1}1{\cdot}37 \\
\hline
 & & 78918 \\
 & & 33822 \\
 & & 11274 \\
\hline
 & & 1{\cdot}544538
\end{array}
$$

To nearest four decimal places $= 1{\cdot}5445$

Antilog of $1{\cdot}5445 = 35{\cdot}03$ Ans.

Example. Find the value of $(0{\cdot}04601)^{1{\cdot}28}$

log of $0{\cdot}04601 = \bar{2}{\cdot}6629$ to be multiplied by $1{\cdot}28$

When the logarithm contains a negative characteristic and the power is not a simple single number, it is more convenient to express the logarithm as an all-minus quantity instead of dealing with it in its present form of part positive and part negative, there is then no complication in "carrying over". The following working demonstrates this method. First change the log into an all-minus quantity, multiply by the given power, then bring the result back into the usual lcg form (positive mantissa) to enable its antilog to be read from the tables.

$\bar{2}{\cdot}6629$ is really -2 and $+0{\cdot}6629$
the collective value is $-1{\cdot}3371$ (all minus),

multiplying this by the power $1{\cdot}28$, by long multiplication,

$$
\begin{array}{r}
-1{\cdot}3371 \\
1{\cdot}28 \\
\hline
106968 \\
26742 \\
13371 \\
\hline
-1{\cdot}711488 \text{ (all minus)}
\end{array}
$$

To nearest four places of decimals $= -1{\cdot}7115$

This has now to be expressed as a number with a positive decimal part therefore the characteristic must be the next highest whole number to $1{\cdot}7115$ and negative in value, that is $\bar{2}$ The mantissa is then obtained by subtracting $1{\cdot}7115$ from 2 to give $0{\cdot}2885$.

Hence, $-1{\cdot}7115$ is equal to -2 and $+0{\cdot}2885$
and written $\bar{2}{\cdot}2885$

Antilog of $\bar{2}{\cdot}2885 = 0{\cdot}01943$ Ans.

ROOTS

To find the root of a number, divide the logarithm of the number by the root and the result is the logarithm of the answer.

Example. Find the square root of 7365.

log of 7365 $= 3{\cdot}8672$
$3{\cdot}8672$ divided by 2 $= 1{\cdot}9336$
Antilog of $1{\cdot}9336 = 85{\cdot}82$ Ans.

Example. Find the value of $\sqrt[3]{0{\cdot}0003657}$

log of $0{\cdot}0003657 = \bar{4}{\cdot}5631$ to be divided by 3

In this example the characteristic is negative and if 3 were divided into $\bar{4}$ there would be a remainder of minus 1 which cannot be carried over to the positive mantissa. Therefore make the characteristic such that the divisor 3 will divide into it exactly with no remainder. To do this we add minus 2 to the characteristic to make it minus 6 (then 3 divides into $\bar{6}$ to give $\bar{2}$), but if we add minus 2 we must also add plus 2 so that the total value of the logarithm is not altered, the plus 2 is added to the mantissa, thus:

$\bar{4}{\cdot}5631$

which is -4 and $+0{\cdot}5631$
add -2 and $+2$ to get,

-6 and $+2{\cdot}5631$

Now divide by 3 (to get cube root), this gives,

-2 and $+0{\cdot}8544$

which is written $\bar{2}{\cdot}8544$

Antilog of $\bar{2}{\cdot}8544 = 0{\cdot}07152$ Ans.

Example. Find the value of $\sqrt[1{\cdot}35]{0{\cdot}002586}$

log of $0{\cdot}002586 = \bar{3}{\cdot}4126$ to be divided by $1{\cdot}35$.

Here we have the more difficult case involving a logarithm having a negative characteristic and a divisor being more than a single fingure. A similar case was explained in the last example on powers and this is dealt with in the same manner.

Express the log as an all-minus quantity,
-3 and $+0{\cdot}4126$ is equal to $-2{\cdot}5874$

Divide this by 1·35. If we do this by long division, it is more convenient to make the divisor a whole number and shifting the decimal point in the dividend through an equal number of places, i.e., $-258{\cdot}74 \div 135$.

This gives $-1{\cdot}9166$ to four decimal places.
(This step could be done by logs instead of long division.)

The quantity $-1{\cdot}9166$ is now brought back into log form consisting of a negative characteristic and a positive mantissa:

$$-1{\cdot}9166 = -2 \text{ and } +0{\cdot}0834$$

written $\overline{2}{\cdot}0834$

$$\text{Antilog} = 0{\cdot}01212 \text{ Ans.}$$

More intricate problems involving logarithms will be explained after the student has learnt how to manipulate equations and transpose formulae.

RELATION BETWEEN LOGARITHMS TO DIFFERENT BASES

As previously explained, the logarithm of a number to a given base is the power to which the base must be raised to be equal to that number.

Thus, if x is the log of N to the base a
then, $a^x = N$
and, if y is the log of N to the base b
then, $b^y = N$

$$\text{Hence,} \quad a^x = b^y$$

$$b = a^{x/y}$$

$$\frac{x}{y} = \log_a b$$

$$y = \frac{x}{\log_a b} \text{ or } \log_b N = \frac{\log_a N}{\log_a b}$$

Example. To find the log of 8 to base 2 using common log tables.

$$\log_2 8 = \frac{\log_{10} 8}{\log_{10} 2} = \frac{0{\cdot}9031}{0{\cdot}3010} = 3$$

which we know is correct because $2^3 = 8$

NAPIERIAN LOGARITHMS

In more advanced work, logarithms to the base 2·718 28 are of special importance since they appear naturally in the mathematics. This is the system discovered and used by the Scottish mathematician John Napier, and called Natural or Napierian logarithms. The number 2·718 28 is represented by the Greek letter ϵ (epsilon). To distinguish these logs from common logs to base 10, the Napierian log of N may be written,

ln N (abbreviation for natural log of N) i.e.
(log of N to base ε).

In the previous paragraph we saw the relation between logs to different bases:

$$\log_b N = \frac{\log_a N}{\log_a b}$$

Applying this to the relation between Napierian logs (to base 2·718 28) and common logs (to base 10):

$$\ln N = \frac{\log_{10} N}{\log_{10} 2{\cdot}71828}$$

$$= \frac{\log_{10} N}{0{\cdot}4343} \text{ or } 2{\cdot}3026 \times \log_{10} N$$

Hence, the Napierian log of a number may be obtained by multiplying the common log of that number by 2·3026. However, most sets of mathematical tables include a table of Napierian logarithms, these may be headed Natural Logarithms, Hyperbolic or Napierian Logarithms, and the student should now turn to these tables for reference.

The table gives logs for numbers 1 to 10 and include the characteristic. Note that the characteristic is shown only in the first column (under the zero heading) and care must be taken to include it when reading the logs from other columns.

Now, from the tables, find the Napierian logs of the following given numbers and check with those shown below.

GIVEN NUMBER	TO FIND NAPIERIAN LOG
N	$\ln N$
2·2	0·7885
3·53	1·2613
7·754	2·0482
9·999	2·3025

For numbers greater than 10 and less than 1 we can proceed as shown in the following examples.

(i) To find the Napierian log of 14:

$$\begin{aligned} 14 &= 1{\cdot}4 \times 10 \\ \ln 14 &= \ln 1{\cdot}4 + \ln 10 \\ &= 0{\cdot}3365 + 2{\cdot}3026 \\ &= 2{\cdot}6391 \end{aligned}$$

(ii) To find the Napierian log of 4189:

$$\begin{aligned} 4189 &= 4{\cdot}189 \times 10^3 \\ \ln 4189 &= \ln 4{\cdot}189 + \ln 10^3 \\ &= 1{\cdot}4325 + 3 \times 2{\cdot}3026 \\ &= 1{\cdot}4325 + 6{\cdot}9078 \\ &= 8{\cdot}3403 \end{aligned}$$

(iii) To find the Napierian log of 0·2418:

$$\begin{aligned} 0{\cdot}2418 &= 2{\cdot}418 \times 10^{-1} = 2{\cdot}418 \div 10 \\ \ln 0{\cdot}2418 &= \ln 2{\cdot}418 - \ln 10 \\ &= 0{\cdot}8829 - 2{\cdot}3026 \\ &= 0{\cdot}8829 + \bar{3}{\cdot}6974 \\ &= \bar{2}{\cdot}5803 \end{aligned}$$

(iv) To find the Napierian log of 0·04357

$$\begin{aligned} 0{\cdot}04357 &= 4{\cdot}357 \times 10^{-2} = 4{\cdot}357 \div 10^2 \\ \ln 0{\cdot}04357 &= \ln 4{\cdot}357 - \ln 10^2 \\ &= 1{\cdot}4718 - 2 \times 2{\cdot}3026 \\ &= 1{\cdot}4718 - 4{\cdot}6052 \\ &= 1{\cdot}4718 + \bar{5}{\cdot}3948 \\ &= \bar{4}{\cdot}8666 \end{aligned}$$

SUPPLEMENTARY TABLES. Most sets of Napierian tables include a supplementary table at the foot of the first page of the main table, giving the values of $\ln 10^n$ thus:

HYPERBOLIC OR NAPIERIAN LOGARITHMS OF 10^{+n}

n	1	2	3	4	5
$\ln 10^n$	2·3026	4·6052	6·9078	9·2103	11·5129

6	7	8	9
13·8155	16·1181	18·4207	20·7233

Hence, in the above example (ii), the value of $\ln 10^3$ would be directly read off this supplementary table as 6·9078 instead of having to multiply 2·3026 by 3. In example (iv), $\ln 10^2$ could be read directly as 4·6052, and so on.

In some sets of tables there is also another supplementary table at the foot of the second page of the main table giving values of $\ln 10^{-n}$ as follows:

HYPERBOLIC OR NAPIERIAN LOGARITHMS OF 10^{-n}

n	1	2	3	4	5
$\ln 10^{-n}$	$\bar{3}$·6974	$\bar{5}$·3948	$\bar{7}$·0922	$\overline{10}$·7897	$\overline{12}$·4871

6	7	8	9
$\overline{14}$·1845	$\overline{17}$·8819	$\overline{19}$·5793	$\overline{21}$·2767

From this table the value of $\ln 10^{-1}$ in example (iii) would be read directly as $\bar{3}$·6974, and $\ln 10^{-2}$ in example (iv) would be read directly as $\bar{5}$·3948.

ANTILOGS. Since the main table of Napierian logs contain numbers 1 to 10 and their corresponding logs from 0·0000 to 2·3026, figures within this range are simply read from the main table. Now check the following list by finding the numbers whose Napierian logs are given:

GIVEN NAPIERIAN LOG	TO FIND NUMBER
ln N	N
0·3365	1·4
0·5008	1·65
1·0794	2·943
2·0912	8·095

For logs greater than 2·3026 we can refer to the supplementary table of positive characteristics, and for logs less than 0·0000 we can refer to the supplementary table of negative characteristics, as well as the main table.

To find the antilog of 8·4304

This is a log greater than those contained in the main table, therefore refer to the first supplementary table and look for the next figure below 8·4304. This is 6·9078, which is 3 × 2·3026 and corresponds to 10^3.

Subtract 6·9078 from the given log 8·4304,

$$\begin{array}{rr} & 8{\cdot}4304 \\ \text{subtract} & \underline{6{\cdot}9078} \\ & 1{\cdot}5226 \end{array}$$

and look up the antilog of this difference in the main table, thus, antilog of 1·5226 reads 4·584.

The antilog of 8·4304 is therefore:

$$4{\cdot}584 \times 10^3 = 4584$$

To find the antilog of $\bar{4}{\cdot}7177$

This is less than the logs in the main table, therefore refer to the second supplementary table and look for the next figure lower than $\bar{4}{\cdot}7177$. This is $\bar{5}{\cdot}3948$ (note that minus 5 is lower in value than minus 4), and corresponds to 10^{-2} the log of which is $-2 \times 2{\cdot}3026 = -4{\cdot}6052 = \bar{5}{\cdot}3948$.

Subtract $\bar{5}{\cdot}3948$ from the given log $\bar{4}{\cdot}7177$

$$\begin{array}{rr} & \bar{4}{\cdot}7177 \\ \text{subtract} & \underline{\bar{5}{\cdot}3948} \\ & 1{\cdot}3229 \end{array}$$

and look up the antilog of this difference in the main table, thus, antilog of 1·3229 reads 3·754.

The antilog of $\bar{4}$·7177 is therefore:

$$3{\cdot}754 \times 10^{-2} = 0{\cdot}03754$$

Now check the following list by finding the numbers whose Napierian logs are given below:

GIVEN NAPIERIAN LOG $\ln N$	TO FIND NUMBER N
2·4485	11·57
5·8966	363·8
$\bar{2}$·5713	0·2396
$\bar{7}$·1784	0·00109

Example. Evaluate the following using Napierian logs:

$$\frac{1{\cdot}212 \times 69 \times 0{\cdot}005}{0{\cdot}1909 \times 0{\cdot}04^{1{\cdot}4}}$$

Finding the logs:

$$\begin{aligned}
\ln 1{\cdot}212 &= 0{\cdot}1922 \\
\ln 69 &= \ln 6{\cdot}9 + \ln 10 \\
&= 1{\cdot}9315 + 2{\cdot}3026 = 4{\cdot}2341 \\
\ln 0{\cdot}005 &= \ln 5 + \ln 10^{-3} \\
&= 1{\cdot}6094 + \bar{7}{\cdot}0922 = \bar{6}{\cdot}7016 \\
\ln 0{\cdot}1909 &= \ln 1{\cdot}909 + \ln 10^{-1} \\
&= 0{\cdot}6465 + \bar{3}{\cdot}6974 = \bar{2}{\cdot}3439 \\
\ln 0{\cdot}04^{1{\cdot}4} &= 1{\cdot}4\,(\ln 4 + \ln 10^{-2}) \\
&= 1{\cdot}4(1{\cdot}3863 + \bar{5}{\cdot}3948) \\
&= 1{\cdot}4 \times \bar{4}{\cdot}7811 \\
&= 1{\cdot}4 \times (-3{\cdot}2189) \\
&= -4{\cdot}5065 = \bar{5}{\cdot}4935
\end{aligned}$$

Adding logs of numerator:

$$0{\cdot}1922 + 4{\cdot}2341 + \bar{6}{\cdot}7016 = \bar{1}{\cdot}1279$$

Adding logs of denominator:

$$\bar{2}{\cdot}3439 + \bar{5}{\cdot}4935 = \bar{7}{\cdot}8374$$

Subtracting log of denominator from log of numerator,

$$\bar{1}{\cdot}1279 - \bar{7}{\cdot}8374 = 5{\cdot}2905$$

Finding antilog of 5·2905,

Subtract 4·6052 corresponding to 10^2,

$$5{\cdot}2905 - 4{\cdot}6052 = 0{\cdot}6853$$
$$\text{Antilog of } 0{\cdot}6853 = 1{\cdot}984$$
$$\text{Antilog of } 5{\cdot}2905 = 1{\cdot}984 \times 10^2$$
$$= 198{\cdot}4 \text{ Ans.}$$

TEST EXAMPLES 3

All the following examples are to be worked out by the use of logarithms.

1. Multiply 562·2 by 0·4323
2. Divide 1·957 by 0·486
3. Evaluate $\dfrac{11{\cdot}24 \times 0{\cdot}06836}{2254 \times 0{\cdot}00753}$
4. Find the cube of 2·54
5. Find the square of 0·3937
6. Find the value of $(0{\cdot}0807)^{2{\cdot}8}$
7. Find the cube root of 238·3
8. Find the value of $0{\cdot}0188^{\frac{1}{2}}$
9. Find the square root of 0·009025
10. Find the value of $\sqrt[1{\cdot}2]{2{\cdot}514}$
11. Find the value of $(0{\cdot}06626)^{1/1{\cdot}4}$
12. Find the value of $712{\cdot}5^{\frac{2}{3}}$
13. Find the value of V in the following,

$$V = \frac{MR(\theta + 273)}{p}$$

when $M = 0{\cdot}05$, $R = 0{\cdot}287$, $\theta = 17$, $p = 0{\cdot}97 \times 10^2$

14. Evaluate $15{\cdot}52^2 \times 1{\cdot}75(0{\cdot}766 + 0{\cdot}0434)$
15. Calculate the value of $E_1 - E_2$ in the following,

$$E_1 - E_2 = 0{\cdot}5\,mk^2(\omega_1{}^2 - \omega_2{}^2)$$

when $m = 1220$, $k = 0{\cdot}58$, $\omega_1 = 20{\cdot}94$, $\omega_2 = 18{\cdot}85$

16. Evaluate $\sqrt[3]{\dfrac{225 \times 27 \times 72^2}{1110(2 + 2{\cdot}65^2)}}$

17. Find thc value of $12{\cdot}5 \times 0{\cdot}45\,(1 + \ln 2{\cdot}55)$

18. Evaluate, $\ln \dfrac{850}{492} + \dfrac{840}{850} + 0{\cdot}6 \ln \dfrac{1160}{850}$

19. Find a multiplier to convert logs to base 10 into logs to base 2, and use this to find the log to base 2 of 1·5, 2·5 and 3·5.

20. Find the Napierian logs of the following numbers,

$$16, \ 987{\cdot}6, \ 0{\cdot}5 \ \text{and} \ 0{\cdot}008667$$

21. Find the antilogs of the following Napierian logs,

$$0{\cdot}9925, \ 2{\cdot}1453, \ 6{\cdot}8739 \ \text{and} \ \overline{9}{\cdot}9836$$

22. Evaluate $0{\cdot}06326^{-0{\cdot}25}$
   (a) using common logs,
   (b) using Napierian logs.

23. Evaluate the following by the use of Napierian logs only,

$$y = 29{\cdot}5^2\,(\ln 2{\cdot}411 + \ln 6{\cdot}234)$$

24. Find the value of k in the following expression, using Napierian logs only,

$$k = 2{\cdot}55 \times \sqrt{\frac{0{\cdot}08234}{0{\cdot}36 \times 10{\cdot}2}}$$

25. Evaluate the following,
   (a) by common logs,
   (b) by Napierian logs,

$$\frac{233{\cdot}6^{-\frac{1}{2}}}{0{\cdot}07541^{-0{\cdot}4}}$$

26. Find the value of x in the following expression,
   (a) by common logs,
   (b) by Napierian logs,

$$x = 0{\cdot}003436^{-2{\cdot}2} \times 0{\cdot}02938^{1{\cdot}4}$$

27. Find the value of V in the following,
   (a) by common logs,
   (b) by Napierian logs,

$$V = \frac{0{\cdot}02377^{-1{\cdot}35}}{0{\cdot}2455^{2{\cdot}35}}$$

CHAPTER 4

SIMPLE EQUATIONS

An equation is an expression consisting of two "sides", one side being equal in value to the other. A simple equation contains one hidden number of the first order (e.g., x, and not x^2 or x^3, etc.) which is usually referred to as the *unknown* and to *solve* the equation means to find the value of the unknown.

Since one side of the equation is equal to the other side, it follows that if, in the process of solving, we find it convenient to make some change to one side, exactly the same change must be made to the other side to ensure that the equation is still true. Hence, equality will be preserved if:

(i) the same quantity is added to both sides;

(ii) the same quantity is subtracted from both sides;

(iii) every term on both sides is multiplied by the same quantity;

(iv) every term on both sides is divided by the same quantity;

(v) both sides are raised to the same power;

(vi) the same root is taken of both sides;

and any one or more of the above steps may be necessary in solving an equation.

There can be any number of terms in an equation, some may be fractions, others may be bracketed quantities: the unknown quantity may be included in more than one term and perhaps on both sides of the equation. In such cases it will be necessary to simplify the equation, preferably one step at a time, following the usual rules of arithmetic and algebra thus:

(i) eliminate fractions; this is done by multiplying every term on both sides by the L.C.M. of all the denominators;

(ii) remove brackets; in doing so observe strictly the rules on removal of brackets explained in chapter on algebra;

(iii) place all terms containing the unknown on one side, and pure numbers on the other side; terms taken over from one side to the other must have their signs changed;

(iv) collect and summarise terms on each side;

(v) find the value of the unknown; for the simplest of equations this is done by dividing both sides by the coefficient of the unknown, other equations may require further or different treatment such as taking the same root of both sides or raising both sides to the same power.

The above is a general purpose guide to the usual steps in solving simple equations. However, other processes can be introduced where most convenient, for instance, in some equations which include algebraic fractions, it may be possible to cancel common factors in the numerator and denominator of one or more of the fractions, if this is done in the initial stages labour will be saved later.

The following examples will help to clarify the above steps.

Example. $$4x + 10 = 18$$

Take 10 to the other side and change its sign, this is the same as subtracting 10 from each side,

$$4x = 18 - 10$$
$$4x = 8$$

Divide both sides by 4 (the coefficient of x)

$$x = 2 \text{ Ans.}$$

Example. $$7(2x - 4) = 2(x + 4) - 5(3 - x)$$

Remove brackets, multiplying the terms inside each bracket by the multiplier in front.

$$14x - 28 = 2x + 8 - 15 + 5x$$

Place all terms containing x on one side and pure members on the other, not forgetting to change signs when terms are taken over,

$$14x - 2x - 5x = 8 - 15 + 28$$

Summarise each side,

$$7x = 21$$

Divide both sides by 7,

$$x = 3 \text{ Ans.}$$

Example. $$\frac{2x}{3} + 1{\cdot}5x = 16 - \frac{x}{2}$$

L.C.M. of denominators is 6, multiply every term by 6,

$$\frac{6 \times 2x}{3} + 6 \times 1{\cdot}5x = 6 \times 16 - \frac{6 \times x}{2}$$

Cancel out denominators into the 6 on the top, and simplify each term,

$$\begin{aligned} 4x + 9x &= 96 - 3x \\ 4x + 9x + 3x &= 96 \\ 16x &= 96 \\ x &= 6 \text{ Ans.} \end{aligned}$$

Example. $\dfrac{3x+2}{2} - \dfrac{5(x-1)}{6} = \dfrac{5x}{4} - \dfrac{5-x}{2}$

L.C.M. of denominators is 12, multiplying each term by 12 and cancelling denominators in one step,

$$\begin{aligned} 6(3x + 2) - 10(x - 1) &= 15x - 6(5 - x) \\ 18x + 12 - 10x + 10 &= 15x - 30 + 6x \\ 18x - 10x - 15x - 6x &= -30 - 12 - 10 \\ -13x &= -52 \\ x &= 4 \text{ Ans.} \end{aligned}$$

(Note that -52 divided by -13 is $+4$)

Example. $8 - \dfrac{9x}{2} - \dfrac{6}{x} = 2(4 - 3x)$

L.C.M. of denominators is $2x$, multiply every term by $2x$ and at the same time cancel out denominators,

$$\begin{aligned} 16x - 9x^2 - 12 &= 4x(4 - 3x) \\ 16x - 9x^2 - 12 &= 16x - 12x^2 \end{aligned}$$

Collect functions of x^2 and x on one side, numbers on the other side,

$$-9x^2 + 12x^2 + 16x - 16x = 12$$

Simplify, note that $+16x$ and $-16x$ cancel out,

$$3x^2 = 12$$

Divide both sides by 3,

$$x^2 = 4$$

Take square root of both sides,

$$x = \pm 2 \text{ Ans.}$$

Note. The square root of 4 can be +2 or −2, because −2 multiplied by −2 gives +4. Therefore when taking the square root of a number in the solution of equations we must write ±, indicating "plus or minus". If either +2 or −2 be substituted for x in the original equation, it will be seen that either value of x satisfies the equation.

Example. $$\frac{3x^2 + 10x + 3}{x + 3} = \frac{x^2 - 4x + 3}{x - 1}$$

On inspection of the numerators in this equation, it appears a possibility of their respective denominators dividing into them, this is always worth trying because it will greatly simplify the working. Instead of actual division we can of course try factorising the numerators, one of the factors being the same as the denominator. This example is such that it will factorise, cancel, and form an easy solution as shown below. If however this was not possible, the alternative procedure would be to multiply throughout by the L.C.M. of the denominators as previously explained.

$$\frac{3x^2 + 10x + 3}{x + 3} = \frac{x^2 - 4x + 3}{x - 1}$$

$$\frac{(x + 3)(3x + 1)}{x + 3} = \frac{(x - 1)(x - 3)}{x - 1}$$

$$3x + 1 = x - 3$$

$$3x - x = -3 - 1$$

$$2x = -4$$

$$x = -2 \text{ Ans.}$$

PROBLEMS INVOLVING SIMPLE EQUATIONS

A problem is a descriptive question requiring a solution. From the facts given in the problem and what is required to be found, the first step is to make up a straightforward equation; this is the sorting-out part of the solution where common sense naturally plays an important part. The following hints will assist the student to attack the problem in an orderly manner.

If the problem can be pictured in the mind, put this picture down on paper, that is, make a sketch of the problem if it is at all possible and insert on this sketch all the given data and indicate the unknown quantity which is to be found. Sketches will come naturally and in fact are practically indispensible in problems involving trigonometry and the like which will follow later, but with some, and especially at this stage, sketches may be quite scanty or even non-existent. If a sketch is not practicable it may help the beginner to write down the given quantities.

Note the quantity required to be found and give this unknown a symbol, for example, "let r = internal radius in millimetres", or "let m = mass in kilogrammes", and so on.

Now look for the equation, that is, see what single term or group of terms is equal to another group, and denote this by an equals sign.

The following worked examples will help the student to understand the above steps.

Example. A port tank and a starboard tank, each of 200 tonne capacity, are each half full of oil. Find what mass of oil must be pumped out of the port tank and into the starboard tank so that there will be four times the mass of oil in one tank as in the other.

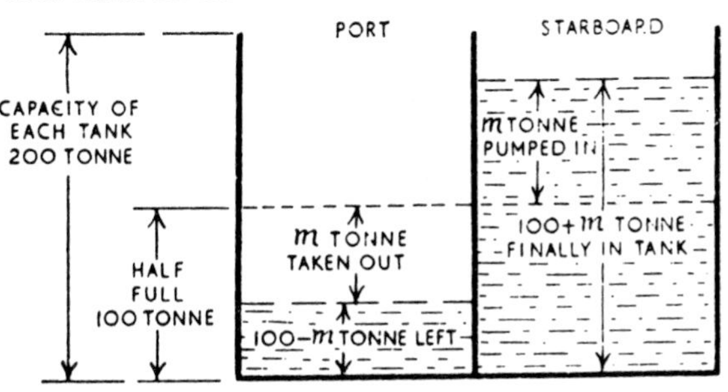

Fig. 2

Initial mass of oil in each tank = 100 tonne

Let m = mass of oil pumped over, then,

final mass of oil in port tank = $(100 - m)$ tonne

,, ,, ,, starbd. ,, = $(100 + m)$,,

Mass in the starboard tank is to be four times the mass in the port tank, therefore,

$$
\begin{aligned}
\text{Mass in starbd. tank} &= 4 \times \text{mass in port tank} \\
(100 + m) &= 4 \times (100 - m) \\
100 + m &= 400 - 4m \\
4m + m &= 400 - 100 \\
5m &= 300 \\
m &= 60
\end{aligned}
$$

∴ mass of oil to be transferred = 60 tonne. Ans.
(Then there will be 160 tonne of oil in the starboard tank and 40 tonne in the port tank.)

Example. An electric train leaves station A bound for station B at the same instant that a diesel train leaves B for A on parallel lines along the same route. The speed of the electric train is 80 kilometres per hour, that of the diesel train 64 kilometres per hour, and the distance between stations is 72 kilometres. Find the distance from A where the trains will pass each other.

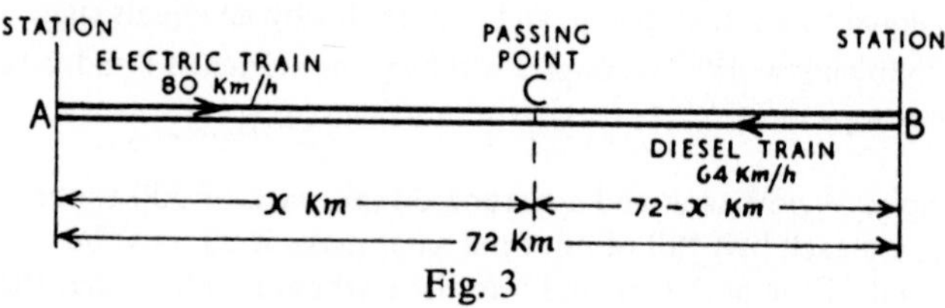

Fig. 3

Let x = distance from A to passing point C
then $(72 - x)$ = ,, ,, B ,, ,, ,, C

As the trains leave at the same time and pass each other at the same instant, then the time taken from station to passing point is the same for each train. This gives the equation'

$$\left.\begin{matrix}\text{Time for electric train}\\ \text{to travel } x \text{ kilometres}\end{matrix}\right\} = \left\{\begin{matrix}\text{Time for diesel train}\\ \text{to travel } (72 - x) \text{ kilometres}\end{matrix}\right.$$

$$\text{Substituting, Time} = \frac{\text{distance}}{\text{speed}} \text{ for each train,}$$

$$\frac{\text{distance travelled by electric train}}{\text{speed of electric train}} = \frac{\text{distance travelled by diesel train}}{\text{speed of diesel train}}$$

$$\frac{x}{80} = \frac{72 - x}{64}$$

L.C.M. = 320, multiplying every term by 320,

$$\begin{aligned} 4x &= 5(72 - x) \\ 4x &= 360 - 5x \\ 4x + 5x &= 360 \\ 9x &= 360 \\ x &= 40 \end{aligned}$$

∴ Trains pass each other at a point 40 kilometres from A. Ans.

Example. A ship travels up-river against the current for a distance of 25 nautical miles and then down-river with the current over the same distance, taking 3 hours 30 minutes over the double journey. If the normal speed of the ship in still water is 15 knots, find the speed of the current.

An equation can be found from the total time being equal to the time for the ship to go up-river plus the time for it to return down-river, and time in hours can be expressed as distance (nautical miles) divided by speed (knots) to suit the quantities given in the problem.

Let x = speed of current, in knots,

then, speed of ship against current = $(15 - x)$ knots,
" " running with " = $(15 + x)$ "
Time to go up-river + Time to go down-river = Total time

$$\frac{\text{distance up-river}}{\text{speed up-river}} + \frac{\text{distance down-river}}{\text{speed down-river}} = 3\tfrac{1}{2} \text{ hours}$$

$$\frac{25}{15 - x} + \frac{25}{15 + x} = 3{\cdot}5$$

Multiply every term by $(15 - x)(15 + x)$, the L.C.M. of the denominators,

$$\begin{aligned}
25(15 + x) + 25(15 - x) &= 3{\cdot}5(15 - x)(15 + x) \\
375 + 25x + 375 - 25x &= 3{\cdot}5(15^2 - x^2) \\
750 &= 787{\cdot}5 - 3{\cdot}5x^2 \\
3{\cdot}5x^2 &= 787{\cdot}5 - 750 \\
3{\cdot}5x^2 &= 37{\cdot}5 \\
x^2 &= 10{\cdot}71 \\
x &= \pm\sqrt{10{\cdot}71} \\
x &= \pm 3{\cdot}273
\end{aligned}$$

$\therefore$ Speed of current = $3{\cdot}273$ knots. Ans.

TRANSPOSITION AND EVALUATION OF FORMULAE

A formula is a group of algebraic symbols which expresses a rule for determining the value of one of the symbols when values are substituted for the others. To transpose means to change the order of the symbols to produce a re-arranged equation giving the value of one of the other symbols in terms of the remainder. Being an equation, the procedure is the same as for equations in general.

Example. Transpose the following formula to give p_2 in terms of the other quantities,

$$\frac{p_1 V_1}{T_1} = \frac{p_2 V_2}{T_2}$$

Note the usual practice of omitting multiplication signs between the quantities to be multiplied, for the first few examples the multiplication signs will be put in while working through the various steps.

$$\frac{p_1 \times V_1}{T_1} = \frac{p_2 \times V_2}{T_2}$$

Multiply both sides by the L.C.M. of the denominators which is $T_1 \times T_2$,

$$p_1 \times V_1 \times T_2 = p_2 \times V_2 \times T_1$$

Divide both sides by $V_2 \times T_1$, so that p_2 will be left on its own on one side,

$$\frac{p_1 \times V_1 \times T_2}{V_2 \times T_1} = p_2$$

It is usual to write expressions and formulae with the required unknown on the left side of the equation and to omit multiplication signs, thus,

$$p_2 = \frac{p_1 V_1 T_2}{V_2 T_1}$$

Example. Express S in terms of the other quantities in the following expression,

$$d = \sqrt[3]{\frac{S \times D}{C}}$$

Cube both sides,

$$d^3 = \frac{S \times D}{C}$$

Multiply both sides by C,

$$C \times d^3 = S \times D$$

Divide both sides by D,

$$\frac{C \times d^3}{D} = S$$

or,

$$S = \frac{Cd^3}{D} \text{ Ans.}$$

Example. Transpose the following expression to give the value of r in terms of E and n. Find the value of r when $E = 60$, $n = 1{\cdot}4$.

$$E = 100\left[1 - \left\{\frac{1}{r}\right\}^{n-1}\right]$$

Note, $1^{n-1} = 1$

$$\therefore E = 100\left[1 - \frac{1}{r^{n-1}}\right]$$

Divide both sides by 100,

$$\frac{E}{100} = 1 - \frac{1}{r^{n-1}}$$

Take 1 to other side leaving unknown by itself,

$$\frac{E}{100} - 1 = -\frac{1}{r^{n-1}}$$

Multiply throughout by -1 to make unknown positive,

$$1 - \frac{E}{100} = \frac{1}{r^{n-1}}$$

Cross-multiply to bring unknown on top,

$$r^{n-1} = \frac{1}{1 - \dfrac{E}{100}}$$

Take $n-1$ root of both sides,

$$r = \left\{\frac{1}{1 - \dfrac{E}{100}}\right\}^{\frac{1}{n-1}} \quad \text{Ans. (i)}$$

Insert numerical values,

$$r = \left\{\frac{1}{1 - \dfrac{60}{100}}\right\}^{\frac{1}{1\cdot4 - 1}}$$

Simplify

$$\frac{1}{1{\cdot}4 - 1} = \frac{1}{0{\cdot}4} = 2{\cdot}5$$

$$1 - \frac{60}{100} = 1 - 0{\cdot}6 = 0{\cdot}4$$

$$\frac{1}{1 - \frac{60}{100}} = \frac{1}{0{\cdot}4} = 2{\cdot}5$$

Hence,

$$\begin{aligned} r &= 2{\cdot}5^{2{\cdot}5} \\ &= 9{\cdot}88 \quad \text{Ans. (ii)} \end{aligned}$$

Example. Transpose the following expression to give V_1 in terms of the other quantities,

$$p_1 \times V_1^n = p_2 \times V_2^n$$

and find the value of V_1 when $p_1 = 1{\cdot}035$, $p_2 = 4{\cdot}14$, $V_2 = 0{\cdot}5$ and $n = 1{\cdot}3$.

$$p_1 \times V_1^n = p_2 \times V_2^n$$

Divide both sides by p_1

$$V_1^n = \frac{p_2 \times V_2^n}{p_1}$$

Take n'th root of both sides,

$$V_1 = \sqrt[n]{\frac{p_2 \times V_2^n}{p_1}}$$

$$V_1 = V_2 \times \sqrt[n]{\frac{p_2}{p_1}}$$

$$\text{or, } V_1 = V_2 \times \left(\frac{p_2}{p_1}\right)^{\frac{1}{n}} \quad \text{Ans. (i)}$$

Substituting for values,

$$\begin{aligned} V_1 &= 0{\cdot}5 \times \sqrt[1{\cdot}3]{\frac{4{\cdot}14}{1{\cdot}035}} \\ &= 0{\cdot}5 \times \sqrt[1{\cdot}3]{4} \\ &= 1{\cdot}453 \quad \text{Ans. (ii)} \end{aligned}$$

LOGARITHMIC EQUATIONS

Consider the expression $4{\cdot}576^3$. To solve this we would multiply the log of 4·576 by the power 3 and then obtain the antilog of the result, thus,

$$(\text{log of } 4{\cdot}576) \times 3 = \text{log of answer}$$
$$0{\cdot}6605 \times 3 = 1{\cdot}9815$$

and from the antilog tables we find the answer to be 95·83.

Now suppose we were given that when 4·576 is raised to a certain power, the result is 95·83 and we are required to find that power, a method similar to the above can be used,

$$\text{Let } n = \text{the power}$$
$$4{\cdot}576^n = 95{\cdot}83$$
$$(\text{log of } 4{\cdot}576) \times n = \text{log of } 95{\cdot}83$$

This form is usually referred to as a logarithmic equation and means simply that as each side of an equation is equal, then the resultant logarithm of each side must be equal.

$$(\text{log of } 4{\cdot}576) \times n = \text{log of } 95{\cdot}83$$
$$0{\cdot}6605 \times n = 1{\cdot}9815$$
$$n = \frac{1{\cdot}9815}{0{\cdot}6605}$$
$$n = 3 \text{ Ans.}$$

Note, the dividing of 1·9815 by 0·6605 may be done by long division or more conveniently by the usual method of subtracting their logs and looking up the antilog of the result.

Example. $\quad$ Given $0{\cdot}4^n = 0{\cdot}5$, find n

$$(\text{log of } 0{\cdot}4) \times n = \text{log of } 0{\cdot}5$$
$$\bar{1}{\cdot}6021 \times n = \bar{1}{\cdot}6990$$
$$n = \frac{\bar{1}{\cdot}6990}{\bar{1}{\cdot}6021}$$

Here we have numbers to be divided which are mixtures of negative and positive quantities, it is necessary to express these quantities in their true values (all minus) as previously explained.

$$\bar{1}{\cdot}6990 = -0{\cdot}3010$$
$$\bar{1}{\cdot}6021 = -0{\cdot}3979$$
$$\therefore n = \frac{-0{\cdot}3010}{-0{\cdot}3979}$$

These quantities are divided in the usual way, bearing in mind that like signs multiplied or divided results in a positive answer.

Alternatively we could, before dividing, multiply top and bottom by minus one which changes the signs.

$$n = \frac{0{\cdot}3010}{0{\cdot}3979}$$

$$n = 0{\cdot}7565 \text{ Ans.}$$

Example. Given $ab^x = cd^x$, find x when

$a = 13{\cdot}5,\ b = 1{\cdot}8,\ c = 68,\ d = 0{\cdot}55$

$$a \times b^x = c \times d^x$$

$$13{\cdot}5 \times 1{\cdot}8^x = 68 \times 0{\cdot}55^x$$

Bring all terms to the power of x to one side, and pure numbers to the other side,

$$\frac{1{\cdot}8^x}{0{\cdot}55^x} = \frac{68}{13{\cdot}5}$$

$$\left(\frac{1{\cdot}8}{0{\cdot}55}\right)^x = \frac{68}{13{\cdot}5}$$

$$3{\cdot}273^x = 5{\cdot}037$$

$$(\text{log of } 3{\cdot}273) \times x = \text{log of } 5{\cdot}037$$

$$0{\cdot}5149 \times x = 0{\cdot}7022$$

$$x = \frac{0{\cdot}7022}{0{\cdot}5149}$$

$$x = 1{\cdot}363 \text{ Ans.}$$

An alternate method could be to set the equation down as a logarithmic one before transposing,

$$13{\cdot}5 \times 1{\cdot}8^x = 68 \times 0{\cdot}55^x$$

$$\log 13{\cdot}5 + x \times \log 1{\cdot}8 = \log 68 + x \times \log 0{\cdot}55$$

$$1{\cdot}1303 + x \times 0{\cdot}2553 = 1{\cdot}8325 + x \times \bar{1}{\cdot}7404$$

$$x \times 0{\cdot}2553 - x \times \bar{1}{\cdot}7404 = 1{\cdot}8325 - 1{\cdot}1303$$

$$x(0{\cdot}2553 - \bar{1}{\cdot}7404) = 0{\cdot}7022$$

$$x \times 0{\cdot}5149 = 0{\cdot}7022$$

$$x = \frac{0{\cdot}7022}{0{\cdot}5149}$$

$$x = 1{\cdot}363 \text{ (as before)}$$

Example. Given $\frac{T_1}{T_2} = \left(\frac{p_1}{p_2}\right)^{\frac{n-1}{n}}$ find n when,

$$T_1 = 797,\ T_2 = 301,\ p_1 = 34{\cdot}4,\ p_2 = 1{\cdot}05$$

$$\frac{797}{301} = \left(\frac{34{\cdot}4}{1{\cdot}05}\right)^{\frac{n-1}{n}}$$

$$2{\cdot}648 = (32{\cdot}76)^{\frac{n-1}{n}}$$

$$\log \text{ of } 2{\cdot}648 = (\log \text{ of } 32{\cdot}76) \times \frac{n-1}{n}$$

$$0{\cdot}4229 = (1{\cdot}5154) \times \frac{n-1}{n}$$

$$n \times 0{\cdot}4229 = 1{\cdot}5154 \times (n - 1)$$

$$0{\cdot}4229n = 1{\cdot}5154n - 1{\cdot}5154$$

$$1{\cdot}5154 = 1{\cdot}5154n - 0{\cdot}4229n$$

$$1{\cdot}5154 = 1{\cdot}0925n$$

$$n = \frac{1{\cdot}5154}{1{\cdot}0925}$$

$$n = 1{\cdot}387 \text{ Ans.}$$

TEST EXAMPLES 4

1. Find the value of x which satisifies the following equation:

$$8 + 5x - 7 = 3x + 9$$

2. Find the value of a in the equation:

$$2(a + 3) + 3(2a - 4) = 4(11 - 3a)$$

3. Find the value of b in the equation:

$$(a + 5b) - 2(a - 4) = 3(2a + b) - 7(a - 3)$$

4. Solve the equation:

$$3[3 - \{x + 2(1 - x)\} - 4x] = 2[x - 3(2 + x) - 4]$$

5. Find x from the following equation:

$$\frac{2x}{3} + \frac{x}{4} - \frac{5}{6} = \frac{4x}{5} + \frac{1}{3}$$

6. Find the value of x in the following equation:

$$\frac{3}{x} + \frac{1}{6} = \frac{6}{x} - \frac{4}{3x}$$

7. Find a in the following:

$$\frac{(1 - 2a^2)}{6a} - \frac{1}{4} + \frac{a}{3} = \frac{1}{6} - \frac{5}{2a}$$

8. If $\dfrac{5}{x + 6} = \dfrac{3}{x - 4}$ find x.

9. If $\dfrac{2}{x - 3} + \dfrac{3}{x + 2} = \dfrac{4}{x^2 - x - 6}$ find x.

10. A ship's hold, A, contains 250 tonne of cargo, another hold B, contains 620 tonne. How much cargo must be taken from B and put into A so that A will contain five times as much as B?

11. A ship travelling at 17·5 knots leaves one port bound for another $4\frac{1}{2}$ hours after another ship whose speed is 16 knots leaves the same port set on the same course. After how many hours and at what distance from port will the fast ship overtake the slower one?

12. A motor boat travels up-river against the current from one point to another at a speed of 6 knots, and then down river with the current back to the original point at a speed of 9 knots, taking a total time of $2\frac{1}{2}$ hours. Assuming the speed of the current remains unchanged, find the distance between points.

13. Two aeroplanes leave one airport at the same time, bound for another, on the same route, their speeds being 672 and 960 kilometres per hour respectively. If one plane arrives at their destination 30 minutes before the other, find the distance between airports.

14. A rectangular plate is to be cut so that the length is four times the breadth and having an area of one square metre. Find the length and breadth.

15. The maximum speed of a ship in still water is 18 knots. It travels all out for a distance of 10 nautical miles directly against the current, and then all out over the return journey of the same course with the current, taking 1 hour 10 minutes to cover the double journey, find the speed of the current in knots.

16. Transpose the following equation to give θ_2 in terms of the other quantities.

$$\frac{R_2}{R_1} = \frac{R_0(1 + \alpha\theta_2)}{R_0(1 + \alpha\theta_1)}$$

and find the value of θ_2 when:

$$R_1 = 200$$
$$R_2 = 240$$
$$\theta_1 = 15$$
$$\alpha = 0{\cdot}0042$$

17. Given that $p_1 V_1^n = p_2 V_2^n$, express p_2 in terms of the other quantities, and find the value of p_2 when $p_1 = 1{\cdot}015$, $V_1 = 2{\cdot}5$, $V_2 = 0{\cdot}2$, and $n = 1{\cdot}4$.

18. Transpose the expression $p_1 V_1^n = p_2 V_2^n$ to give V_2 in terms of the other quantities, and find the value of V_2 when $p_1 = 0{\cdot}966$, $V_1 = 6{\cdot}5$, $p_2 = 34{\cdot}5$ and $n = 1{\cdot}38$.

19. The diameter in mm of coupling bolts should not be less than that given by the formula:

$$d = \sqrt{\frac{D^3}{3{\cdot}5 \times n \times R}}$$

where D = diameter of shafting in mm
n = number of bolts per coupling
R = pitch circle radius in mm

Express R in terms of the other quantities and find the pitch circle radius when $d = 82{\cdot}5$ mm, $D = 381$ mm, and $n = 8$ bolts per coupling.

20. Find the value of n in the equation:

$$3{\cdot}4^n = 5{\cdot}547$$

21. Calculate the value of x in the equation:

$$4^{x+3} = 4096$$

22. Given $p_1 V_1^n = p_2 V_2^n$ find n when $p_1 = 1{\cdot}07$, $V_1 = 2{\cdot}15$, $p_2 = 12{\cdot}43$ and $V_2 = 0{\cdot}35$.

23. Given $\dfrac{T_1}{T_2} = \left(\dfrac{p_1}{p_2}\right)^{\frac{n-1}{n}}$

find the value of n when $T_1 = 670$, $T_2 = 324$, $p_1 = 21$ and $p_2 = 1{\cdot}25$.

24. If $0{\cdot}75^n = 0{\cdot}4873$ find the value of n.

25. Given $1 - (0{\cdot}125)^{\gamma-1} = 0{\cdot}5647$ find the value of γ.

26. Find the values of x in each of the following equations:

    (a) $4^{0{\cdot}59x} = 56{\cdot}36$

    (b) $x^{1{\cdot}95} = 12{\cdot}4x^{0{\cdot}53}$

    (c) $\epsilon^{5x} = 46{\cdot}38^{2{\cdot}6}$

    (d) $\sqrt[3]{x} = \epsilon^{2{\cdot}5}$

CHAPTER 5

SIMULTANEOUS EQUATIONS

In the preceding chapter, equations containing only one unknown quantity were explained; now we come to cases where two (or more) unknown quantities are required to be found.

To find the values of two unknowns, two equations are required, each containing the same two unknown quantities, this pair of equations is called "simultaneous".

Various different methods may be used in the solution of these; three methods commonly employed will be explained here in the order, (i) by elimination, (ii) by substitution, (iii) by equating expressions of like unknowns. There is another method and that is by drawing a graph of each equation on a common base; the point of intersection of the graphs produces the solution of the unknowns; this method is included in Chapter 7 where graphs and their uses are explained.

Method (*i*). This process is to couple the two equations together in such a form that by either (*a*) adding them together, (*b*) subtracting one from the other, (*c*) multiplying them, or (*d*) dividing one by the other, one of the unknown quantities cancel out, leaving a single equation containing only one unknown which is then simply solved. When one of the quantities has thus been found its value is substituted into one of the original equations which will then become another single equation containing only the other one remaining unknown to be solved.

The following examples will serve to demonstrate some of the steps normally involved in this method of solution.

Example. When twice one number is added to five times another number, the result is 34, and when three times the latter number is subtracted from four times the former, the result is 16. Find the numbers.

Let x = one number,

and y = other number,

From the first statement, $2x + 5y = 34$ ... ... (i)

,, ,, second ,, $4x - 3y = 16$ ... ... (ii)

Multiply equation (i) by 2 so that it will contain the same multiple of x as in equation (ii), and set the pair down again,

$4x + 10y = 68$ ... ... (iii)

$4x - 3y = 16$ ... ... (iv)

By subtracting (iv) from (iii), $4x$ will cancel, leaving one equation containing only y as the unknown. Remember that when subtracting a minus quantity, we change its sign and add, thus subtracting $-3y$ from $+10y$ gives $+13y$.

$$4x + 10y = 68 \quad \ldots \quad \ldots \quad \ldots \quad \ldots \quad \text{(iii)}$$
$$\text{subtract } 4x - 3y = 16 \quad \ldots \quad \ldots \quad \ldots \quad \ldots \quad \text{(iv)}$$

$$13y = 52$$
$$y = 4$$

Substitute the value of y which we have found to be 4 into any one of the original equations which contain x and y; this will produce another equation containing only x as the unknown.

Substituting $y = 4$ into equation (i)

$$2x + 5 \times 4 = 34$$
$$2x = 34 - 20$$
$$2x = 14$$
$$x = 7$$

Therefore the numbers are 7 and 4. Ans.

Example. Given $2p + 5q = 26{\cdot}5$
and $3p - 2q = 6{\cdot}5$

find the values of p and q.

$$2p + 5q = 26{\cdot}5 \quad \ldots \quad \ldots \quad \ldots \quad \ldots \quad \text{(i)}$$
$$3p - 2q = 6{\cdot}5 \quad \ldots \quad \ldots \quad \ldots \quad \ldots \quad \text{(ii)}$$

Multiply (i) by 3, and multiply (ii) by 2,

$$6p + 15q = 79{\cdot}5 \quad \ldots \quad \ldots \quad \ldots \quad \ldots \quad \text{(iii)}$$
$$6p - 4q = 13 \quad \ldots \quad \ldots \quad \ldots \quad \ldots \quad \text{(iv)}$$

Subtract (iv) from (iii)

$$19q = 66{\cdot}5$$
$$q = 3{\cdot}5$$

Substitute $q = 3{\cdot}5$ into equation (ii)

$$3p - 2 \times 3{\cdot}5 = 6{\cdot}5$$
$$3p = 6{\cdot}5 + 7$$
$$3p = 13{\cdot}5$$
$$p = 4{\cdot}5$$

The values of p and q are 4·5 and 3·5 respectively. Ans.

Example. Find the values of x and y in the simultaneous equations,

$$\frac{x}{8} + \frac{y}{5} = \frac{7}{8}$$

$$\frac{2x}{3} - \frac{y}{2} = \frac{3}{4}$$

Simplify these equations by multiplying the first by the L.C.M. of its denominators which is 40, and the second by its L.C.M. which is 12,

$$5x + 8y = 35 \quad \dots \quad \dots \quad \dots \quad \dots \quad \text{(i)}$$
$$8x - 6y = 9 \quad \dots \quad \dots \quad \dots \quad \dots \quad \text{(ii)}$$

Multiply (i) by 3, and (ii) by 4,

$$15x + 24y = 105 \quad \dots \quad \dots \quad \dots \quad \dots \quad \text{(iii)}$$
$$32x - 24y = 36 \quad \dots \quad \dots \quad \dots \quad \dots \quad \text{(iv)}$$

Add (iii) and (iv)

$$47x = 141$$
$$x = 3$$

Substitute $x = 3$ into equation (i),

$$5 \times 3 + 8y = 35$$
$$8y = 35 - 15$$
$$8y = 20$$
$$y = 2\tfrac{1}{2}$$

The values of x and y are 3 and $2\frac{1}{2}$ respectively. Ans.

Example. Given the simultaneous equations,

$$1{\cdot}5^x \times 2^y = 18$$
$$4^x \times 1{\cdot}5^y = 54$$

find the values of x and y.

Writing the two equations in log form,

$$(\log 1{\cdot}5) \times x + (\log 2) \times y = \log 18$$
$$(\log 4) \times x + (\log 1{\cdot}5) \times y = \log 54$$

Inserting log values,

$$0{\cdot}1761x + 0{\cdot}301y = 1{\cdot}2553 \quad \dots \quad \dots \quad \dots \quad \text{(i)}$$
$$0{\cdot}6021x + 0{\cdot}1761y = 1{\cdot}7324 \quad \dots \quad \dots \quad \dots \quad \text{(ii)}$$

Dividing (i) by 0·1761, and dividing (ii) by 0·6021,

$$x + 1{\cdot}709y = 7{\cdot}127 \quad \ldots \quad \ldots \quad \ldots \quad \text{(iii)}$$
$$x + 0{\cdot}2925y = 2{\cdot}877 \quad \ldots \quad \ldots \quad \ldots \quad \text{(iv)}$$

Subtracting (iv) from (iii),

$$1{\cdot}4165y = 4{\cdot}25$$
$$y = 3$$

Substituting $y = 3$ into (iii),

$$x + 1{\cdot}709 \times 3 = 7{\cdot}127$$
$$x = 7{\cdot}127 - 5{\cdot}127$$
$$x = 2$$

The values of x and y are 2 and 3 respectively. Ans.

Method (*ii*) of solving simultaneous equations is by transposing one of the equations to express one of the unknowns in terms of the other quantities, then substituting this expression for that unknown into the other equation, thus producing a single equation containing only the other unknown which can then be solved.

Repeating the first example to be solved by this method,

$$2x + 5y = 34$$
$$4x - 3y = 16$$

Transposing the first equation to express x in terms of the other quantities.

$$2x + 5y = 34$$
$$2x = 34 - 5y$$
$$x = 17 - 2{\cdot}5y$$

Substituting this value of x into the second equation, simplifying and solving.

$$4x - 3y = 16$$
$$4(17 - 2{\cdot}5y) - 3y = 16$$
$$68 - 10y - 3y = 16$$
$$-10y - 3y = 16 - 68$$
$$-13y = -52$$
$$y = 4$$

Substituting $y = 4$ into the first equation and solving,

$$\begin{aligned} 2x + 5y &= 34 \\ 2x + 20 &= 34 \\ 2x &= 34 - 20 \\ 2x &= 14 \\ x &= 7 \end{aligned}$$

The values of x and y are 7 and 4 respectively. Ans.

Method (iii) is to transpose both equations to express the same unknown in terms of the other quantities, the two expressions are then equated together, making one equation with one unknown.

Taking the same example again,

$$2x + 5y = 34 \quad \dots \quad \dots \quad \dots \quad \dots \quad \text{(i)}$$
$$4x - 3y = 16 \quad \dots \quad \dots \quad \dots \quad \dots \quad \text{(ii)}$$

Transposing both equations, in each case expressing x in terms of the other quantities,

From (i)

$$\begin{aligned} 2x &= 34 - 5y \\ x &= 17 - 2{\cdot}5y \quad \dots \quad \dots \quad \text{(iii)} \end{aligned}$$

From (ii)

$$\begin{aligned} 4x &= 16 + 3y \\ x &= 4 + 0{\cdot}75y \quad \dots \quad \dots \quad \text{(iv)} \end{aligned}$$

Since x in one equation has the same value as in the other, then, (iii) = (iv)

$$\begin{aligned} 4 + 0{\cdot}75y &= 17 - 2{\cdot}5y \\ 2{\cdot}5y + 0{\cdot}75y &= 17 - 4 \\ 3{\cdot}25y &= 13 \\ y &= 4 \end{aligned}$$

Substituting $y = 4$ into equation (i)

$$\begin{aligned} 2x + 20 &= 34 \\ 2x &= 14 \\ x &= 7 \end{aligned}$$

$x = 7$ and $y = 4$ as before.

The method to employ in solving a given simultaneous equation will be that which is easiest to apply depending upon the form in which the equation is given.

THREE UNKNOWNS

Just as two unknowns can be found from two equations, three unknowns can be solved if three equations be given

Example. To find the values of x, y and z which satisfy the equations,

$$3x + 2y - z = 4 \quad \ldots \quad \ldots \quad \ldots \quad \ldots \quad \text{(i)}$$
$$2x + y + z = 7 \quad \ldots \quad \ldots \quad \ldots \quad \ldots \quad \text{(ii)}$$
$$x - y + z = 2 \quad \ldots \quad \ldots \quad \ldots \quad \ldots \quad \text{(iii)}$$

Eliminate one unkown at a time, take z first as it appears the easiest to get rid of.

Add equations (i) and (ii),

$$\begin{aligned} 3x + 2y - z &= 4 \\ 2x + y + z &= 7 \\ \hline 5x + 3y &= 11 \quad \ldots \quad \ldots \quad \ldots \quad \ldots \quad \text{(iv)} \end{aligned}$$

Subtract (iii) from (ii),

$$\begin{aligned} 2x + y + z &= 7 \\ x - y + z &= 2 \\ \hline x + 2y &= 5 \quad \ldots \quad \ldots \quad \ldots \quad \ldots \quad \text{(v)} \end{aligned}$$

Multiply (v) by 5 and subtract (iv) from the result,

$$\begin{aligned} 5x + 10y &= 25 \\ 5x + 3y &= 11 \\ \hline 7y &= 14 \\ y &= 2 \end{aligned}$$

Substitute $y = 2$ into (v),

$$\begin{aligned} x + 2 \times 2 &= 5 \\ x &= 5 - 4 \\ x &= 1 \end{aligned}$$

Substitute $x = 1$ and $y = 2$ into (iii),

$$\begin{aligned} 1 - 2 + z &= 2 \\ z &= 2 + 2 - 1 \\ z &= 3 \end{aligned}$$

$\therefore x = 1$, $y = 2$ and $z = 3$. Ans.

TEST EXAMPLES 5

1. The sum and difference of double one number and treble another is 16 and 4 respectively. Find the two numbers.

2. When $2\frac{1}{2}$ times one number is added to $3\frac{1}{2}$ times another the result is 19; and when $3\frac{1}{2}$ times the first number is subtracted from $2\frac{1}{2}$ times the second, the result is 3. Find the numbers.

3. Find the values of x and y in the simultaneous equations:

$$\frac{2x}{3} - \frac{3y}{5} = \frac{3}{4} \text{ and } \frac{x}{2} - \frac{y}{4} = \frac{13}{16}$$

4. Find the values of a and b which satisfy the equations:

$$a(1 + 2b) = 3 \text{ and } a(1 - 3b) = 0{\cdot}5$$

5. A man and his wife are 72 and 68 years old respectively, and have one grandson and one grand-daughter. The man's age is equal to the sum of four times the grandson's age and three times the grand-daughter's. The woman's age is equal to the sum of three times the grandson's age and four times the grand-daughter's. Find the ages of the two grandchildren.

6. The difference between two numbers is 2 and the difference between their squares is 6. Find the numbers.

7. The linear law of a simple lifting machine is given by $F = a + bm$ where m = mass lifted, F = effort applied, a and b being constants. In a certain lifting machine it was found that when m = 30 kilogrammes, F = 35 newtons, and when m = 70 kilogrammes, F = 55 newtons. Find the constants a and b, express the law of this machine, and find the effort required to lift a mass of 60 kilogrammes.

8. Two ships are 99 nautical miles apart steaming at different speeds. If they were travelling directly towards each other they would meet in 3 hours. If they were steaming in the same direction on the same course they would meet in $49\frac{1}{2}$ hours. What are the speeds of the two ships?

9. Two ships, A and B, leave one port bound for another on the same course. B leaves one hour later than A and overtakes her in 8 hours. If the speeds of each ship had been 4 knots slower, B would have overtaken A two hours earlier. Find the original speeds of the ships.

10. Willan's law expresses the relationship between the steam consumption and the power developed by a steam engine under certain conditions, and may be expressed thus:

$$m = a + bP$$

where m = mass of steam used per hour,
P = power developed,
a and b = constants.

In a certain engine it was found that $m = 2025$ kilogrammes per hour when $P = 250$ kilowatts, and $m = 1515$ kilogrammes per hour when $P = 175$ kilowatts. Find the values of the constants a and b, express Willans' law for this engine, and find m when $P = 200$.

11. Given the simultaneous equations,

$$\frac{x}{2-y} + \frac{6}{x} = 4 \text{ and } \frac{2x}{2-y} - \frac{9}{x} = 1$$

find the values of $\frac{x}{2-y}$, $\frac{1}{x}$, x and y.

12. Find the values of x and y in the simultaneous equations,

$$2^x = 4^y \text{ and } 4^{x-1} = 2^{y+1}$$

13. Find the values of x and y in the simultaneous equations,

$$1{\cdot}259^{x+1} \times 1{\cdot}175^{y-1} = 2{\cdot}323$$
$$3{\cdot}162^x \times 1{\cdot}778^y = 25{\cdot}12$$

14. Given the following relationship,

$$\frac{d^2}{x^2} = \frac{x^2}{y^2} = \frac{y^2}{D^2}$$

express x in terms of d and D, and find the values of x and y when $D = 75$ and $d = 25$.

15. Find the values of a, b and c in the simultaneous equations,

$$3a + 6b - 2c = 7{\cdot}25$$
$$2a + 3b + 4c = 26$$
$$4a - 2b + c = 10{\cdot}25$$

CHAPTER 6

QUADRATIC AND CUBIC EQUATIONS

A quadratic equation is one which contains the square of the unknown quantity, thus, $x^2 = 36$ is a quadratic equation in its very elementary form and solved simply by taking the square root of both sides, $x = \pm 6$.

The form usually associated with the title quadratic equation contains the first power of the unknown as well as its square, the general form being,

$$ax^2 + bx + c = 0$$

where a (the coefficient of x^2), b (which is the coefficient of x), and c are given quantities, and x is the unknown. These may be solved by various methods, those which will be explained here are (i) by factorisation, (ii) by completing the square, (iii) by formula. Another method is by graphical means and this is explained in Chapter 7 which deals with the plotting of graphs.

Method (i) by factorisation. This method is applied only when the expression can be readily factorised.

Example. $$x^2 - 5x + 6 = 0 \quad \ldots \quad \ldots \quad \ldots \quad \ldots \quad \text{(i)}$$

Factorise, as explained in Chapter 2,

$$(x - 2)(x - 3) = 0 \quad \ldots \quad \ldots \quad \ldots \quad \ldots \quad \text{(ii)}$$

We have previously seen that when any quantity is multiplied by zero, the result is zero. It therefore follows that if the result of the product of two quantities is zero, one of those quantities must be zero. In equation (ii) the product of $(x - 2)$ and $(x - 3)$ is zero, hence either $(x - 2)$ or $(x - 3)$ must be a zero quantity.

$$\begin{aligned} \text{If} \quad x - 2 &= 0 \\ \text{then} \quad x &= 2 \\ \text{or, if } x - 3 &= 0 \\ \text{then} \quad x &= 3 \end{aligned}$$

The value of x in the equation can therefore be 2 or 3. If we try each of these values in the original equation (i) it will be seen that either will satisfy the equation.

$$\therefore x = 2 \text{ or } 3 \text{ Ans.}$$

Example. $2x^2 - 5x - 12 = 0$

Factorise: $(2x + 3)(x - 4) = 0$

either $2x + 3 = 0$ | or $x - 4 = 0$

then $2x = -3$ | then $x = 4$

$x = -1\frac{1}{2}$

$$\text{Therefore, } x = 4 \text{ or } -1\tfrac{1}{2} \text{ Ans.}$$

Method (ii) by completing the square. Any quadratic expression, whether it can be factorised or not, can be solved by this method. It is based upon the process of changing the expression into one which is a "perfect square", that is, one whose square root can be readily obtained, then, by taking the square root, the expression so found contains only the first power of the unknown, thus producing a simple equation which can be easily solved.

Example. $4x^2 + 12x - 7 = 0$

Divide throughout by 4 to reduce coefficient of x^2 to unity,

$$x^2 + 3x - \frac{7}{4} = 0$$

We cannot take the square root of this expression, therefore move $-\frac{7}{4}$ over to the other side,

$$x^2 + 3x = \frac{7}{4}$$

and add such a quantity that will make the left hand side into a perfect square, not forgetting that whatever is added to one side the same quantity must also be added to the other side to preserve equality. In Chapter 2, section on factorisation, paragraph (iii), it was pointed out that an expression is a perfect square when the third term is equal to the square of half the coefficient of x. The coefficient of x here is 3, half of 3 is $\frac{3}{2}$, and this squared is $(\frac{3}{2})^2$. Therefore add $(\frac{3}{2})^2$ to both sides,

$$x^2 + 3x + \left(\frac{3}{2}\right)^2 = \frac{7}{4} + \left(\frac{3}{2}\right)^2$$

Now take the square root of both sides, and solve

$$x + \frac{3}{2} = \pm\sqrt{\frac{7}{4} + \frac{9}{4}}$$

$$x + \frac{3}{2} = \pm\sqrt{\frac{16}{4}}$$

$$x + \tfrac{3}{2} = \pm 2$$

$$x = \pm 2 - 1\tfrac{1}{2}$$

$$x = \tfrac{1}{2} \text{ or } -3\tfrac{1}{2} \text{ Ans.}$$

Note the ± sign. As we have previously seen, the square root of a number can be plus or minus, and both values must be taken into account.

Example. $12x^2 + x - \frac{1}{2} = 0$

Dividing throughout by 12,

$$x^2 + \frac{x}{12} - \frac{1}{24} = 0$$

Shifting third term to other side and completing the square,

$$x^2 + \frac{x}{12} + \left(\frac{1}{24}\right)^2 = \frac{1}{24} + \left(\frac{1}{24}\right)^2$$

Taking square root of both sides and simplifying,

$$x + \frac{1}{24} = \pm\sqrt{\frac{1}{24} + \frac{1}{24^2}}$$

$$x + \frac{1}{24} = \pm\sqrt{\frac{24 + 1}{24^2}}$$

$$x + \frac{1}{24} = \pm\sqrt{\frac{25}{24^2}}$$

$$x + \frac{1}{24} = \pm\frac{5}{24}$$

$$x = \pm\frac{5}{24} - \frac{1}{24}$$

$$x = \frac{4}{24} \text{ or } -\frac{6}{24}$$

$$x = \frac{1}{6} \text{ or } -\frac{1}{4} \text{ Ans.}$$

Method (iii) by formula. If we let 'a' be the coefficient of x^2, 'b' the coefficient of x, and 'c' the third term, we have the general form of a quadratic equation:

$$ax^2 + bx + c = 0$$

and by solving for x as far as we can by the method of completing the square, a ready-made formula is produced for solving any quadratic equation.

$$ax^2 + bx + c = 0$$

Dividing throughout by a,

$$x^2 + \frac{bx}{a} + \frac{c}{a} = 0$$

Taking the third term to the other side and adding to both sides the square of half the coefficient of x, to complete the square,

$$x^2 + \frac{bx}{a} + \left(\frac{b}{2a}\right)^2 = \left(\frac{b}{2a}\right)^2 - \frac{c}{a}$$

Taking square root of both sides and simplifying,

$$x + \frac{b}{2a} = \pm\sqrt{\frac{b^2}{4a^2} - \frac{c}{a}}$$

$$x + \frac{b}{2a} = \pm\sqrt{\frac{b^2 - 4ac}{4a^2}}$$

$$x + \frac{b}{2a} = \frac{\pm\sqrt{b^2 - 4ac}}{2a}$$

$$x = -\frac{b}{2a} \pm \frac{\sqrt{b^2 - 4ac}}{2a}$$

$$x = \frac{-b \pm \sqrt{b^2 - 4ac}}{2a}$$

This is a very useful formula which provides a straightforward solution to any quadratic equation and should be committed to memory.

Example. $1{\cdot}5x^2 + 2x - 10 = 0$

Applying the formula, $x = \dfrac{-b \pm \sqrt{b^2 - 4ac}}{2a}$

Substituting, $a = 1{\cdot}5$, $b = 2$, $c = -10$,

$$x = \frac{-2 \pm \sqrt{2^2 - 4 \times 1{\cdot}5 \times (-10)}}{2 \times 1{\cdot}5}$$

$$x = \frac{-2 \pm \sqrt{4 + 60}}{3}$$

$$x = \frac{-2 \pm \sqrt{64}}{3}$$

$$x = \frac{-2 \pm 8}{3}$$

$$x = \frac{6}{3} \text{ or } \frac{-10}{3}$$

$$x = 2 \text{ or } -3\tfrac{1}{3} \text{ Ans.}$$

Great care must be taken with the signs. Note that the value of c in the above example is -10, and the term $-4ac$ becomes $(-4) \times 1{\cdot}5 \times (-10)$ which is $+60$.

Another point to note is when the value of b is a minus quantity, in the next example the value of b is -9, the term $-b$ in the formula then becomes $-(-9)$ which is $+9$.

Example. $2x^2 - 9x - 35 = 0$

$$x = \frac{-b \pm \sqrt{b^2 - 4ac}}{2a}$$

where $a = 2$, $b = -9$, $c = -35$

$$x = \frac{+9 \pm \sqrt{9^2 - 4 \times 2 \times (-35)}}{2 \times 2}$$

$$x = \frac{9 \pm \sqrt{81 + 280}}{4}$$

$$x = \frac{9 \pm \sqrt{361}}{4}$$

$$x = \frac{9 \pm 19}{4}$$

$$x = \frac{28}{4} \text{ or } \frac{-10}{4}$$

$$x = 7 \text{ or } -2\tfrac{1}{2} \text{ Ans.}$$

EQUATIONS CONTAINING POWERS OTHER THAN THE SQUARE OF THE UNKNOWN WHICH CAN BE SOLVED AS QUADRATIC EQUATIONS

Equations of the form,

$$ax^{2n} + bx^{n} + c = 0$$

where n is a power of x, and $2n$ is twice that power, can be solved as a normal quadratic by substituting another symbol of the first order for x^n such as y, consequently x^{2n} would then be represented by y^2, thus,

$$ay^2 + by + c = 0$$

y is solved by one of the usual methods of solving quadratic equations then the value of x is obtained from the substitution $y = x^n$.

Example. $x^4 - 8\frac{1}{2}x^2 + 14\frac{1}{16} = 0$

Let y represent x^2, then $y^2 = x^4$,

$$y^2 - 8\tfrac{1}{2}y + 14\tfrac{1}{16} = 0$$

Solving by formula,

$$y = \frac{-b \pm \sqrt{b^2 - 4ac}}{2a}$$

where $a = 1$, $b = -8\frac{1}{2}$, $c = 14\frac{1}{16}$

$$y = \frac{8{\cdot}5 \pm \sqrt{72{\cdot}25 - 56{\cdot}25}}{2}$$

$$y = \frac{8{\cdot}5 \pm \sqrt{16}}{2}$$

$$y = \frac{8{\cdot}5 \pm 4}{2}$$

$$y = \frac{12{\cdot}5}{2} \text{ or } \frac{4{\cdot}5}{2}$$

$$y = 6{\cdot}25 \text{ or } 2{\cdot}25$$

$$v = x^2 \therefore x = \sqrt{y}$$

$$x = \pm\sqrt{6{\cdot}25} \text{ or } \pm\sqrt{2{\cdot}25}$$

$$x = \pm 2{\cdot}5 \text{ or } \pm 1{\cdot}5 \text{ Ans.}$$

Example.

$$0{\cdot}3V^{2\cdot8} - 1{\cdot}56V^{1\cdot4} + 1{\cdot}98 = 0$$

We note first that every term is divisible by 3, therefore take advantage of this by dividing every term by 0·3 to produce an equation with a simpler set of figures,

$$V^{2\cdot8} - 5{\cdot}2V^{1\cdot4} + 6{\cdot}6 = 0$$

Let $x = V^{1\cdot4}$ then $x^2 = V^{2\cdot8}$

$$x^2 - 5{\cdot}2x + 6{\cdot}6 = 0$$

By formula, $$x = \frac{-b \pm \sqrt{b^2 - 4ac}}{2a}$$

where $a = 1$, $b = -5{\cdot}2$, $c = 6{\cdot}6$

$$x = \frac{5{\cdot}2 \pm \sqrt{27{\cdot}04 - 26{\cdot}4}}{2}$$

$$x = \frac{5{\cdot}2 \pm \sqrt{0{\cdot}64}}{2}$$

$$x = \frac{5{\cdot}2 \pm 0{\cdot}8}{2}$$

$$x = \frac{6}{2} \text{ or } \frac{4{\cdot}4}{2}$$

$$x = 3 \text{ or } 2{\cdot}2$$

$$x = V^{1\cdot4} \therefore V = \sqrt[1\cdot4]{x}$$

$$V = \sqrt[1\cdot4]{3} \text{ or } \sqrt[1\cdot4]{2{\cdot}2}$$

$$= 2{\cdot}192 \text{ or } 1{\cdot}756 \text{ Ans.}$$

SIMULTANEOUS–QUADRATIC EQUATIONS

In the previous chapter methods of solving simultaneous equations were explained, but the examples were such that only simple equations were involved. Further examples can now be dealt with where quadratic equations are either given in the first place or are produced in the process of solving.

Example. Find the values of x and y which satisfy the equations,

$$2x^2 - 3y = 3$$
$$x + 5y = 4$$

Multiplying the first equation by 5 and the second equation by 3, then adding the results together to eliminate y,

$$\begin{aligned} 10x^2 - 15y &= 15 \\ 3x + 15y &= 12 \\ \hline 10x^2 + 3x &= 27 \end{aligned}$$

Re-arranging,

$$10x^2 + 3x - 27 = 0$$

Solving this quadratic by formula,

$$x = \frac{-b \pm \sqrt{b^2 - 4ac}}{2a}$$

where $a = 10$, $b = 3$, $c = -27$

$$x = \frac{-3 \pm \sqrt{9 + 1080}}{20}$$

$$x = \frac{-3 \pm \sqrt{1089}}{20}$$

$$x = \frac{-3 \pm 33}{20}$$

$$x = \frac{30}{20} \text{ or } \frac{-36}{20}$$

$$x = 1{\cdot}5 \text{ or } -1{\cdot}8$$

Substituting $x = 1{\cdot}5$ into the second equation,

$$\begin{aligned} 1{\cdot}5 + 5y &= 4 \\ 5y &= 2{\cdot}5 \\ y &= 0{\cdot}5 \end{aligned}$$

Substituting $x = -1{\cdot}8$ into the second equation,

$$\begin{aligned} -1{\cdot}8 + 5y &= 4 \\ 5y &= 5{\cdot}8 \\ y &= 1{\cdot}16 \end{aligned}$$

$$\left.\begin{aligned} x &= 1{\cdot}5 \text{ and } y = 0{\cdot}5 \\ \text{or } x &= -1{\cdot}8 \text{ and } y = 1{\cdot}16 \end{aligned}\right\} \text{Ans.}$$

An alternative method to the above is to find the value of one unknown in terms of the other from the simplest equation, and substitute this into the other equation. Thus, in the previous example,

$$2x^2 - 3y = 3 \quad \ldots \quad \ldots \quad \ldots \quad \ldots \quad \text{(i)}$$
$$x + 5y = 4 \quad \ldots \quad \ldots \quad \ldots \quad \ldots \quad \text{(ii)}$$

From (ii) $$y = \tfrac{1}{5}(4 - x)$$

Substitute this value of y into equation (i) and simplify,

$$2x^2 - 3 \times \tfrac{1}{5}(4 - x) = 3$$

Multiply throughout by 5,

$$10x^2 - 12 + 3x = 15$$
$$10x^2 + 3x - 27 = 0$$

the same quadratic to be solved as before.

Example. Find the values of x and y which satisfy the simultaneous equations:

$$2x^2 + xy - y^2 = 8 \quad \ldots \quad \ldots \quad \ldots \quad \ldots \quad \text{(i)}$$
$$3x + 2y = 5 \quad \ldots \quad \ldots \quad \ldots \quad \ldots \quad \text{(ii)}$$

Find the value of y in terms of x from (ii),

$$3x + 2y = 5$$
$$y = 2{\cdot}5 - 1{\cdot}5x \quad \ldots \quad \ldots \quad \text{(iii)}$$

Substitute for y into equation (i) and simplify,

$$2x^2 + xy - y^2 = 8$$
$$2x^2 + x(2{\cdot}5 - 1{\cdot}5x) - (2{\cdot}5 - 1{\cdot}5x)^2 = 8$$
$$2x^2 + 2{\cdot}5x - 1{\cdot}5x^2 - (6{\cdot}25 - 7{\cdot}5x + 2{\cdot}25x^2) = 8$$
$$2x^2 + 2{\cdot}5x - 1{\cdot}5x^2 - 6{\cdot}25 + 7{\cdot}5x - 2{\cdot}25x^2 = 8$$
$$2x^2 - 1{\cdot}5x^2 - 2{\cdot}25x^2 + 2{\cdot}5x + 7{\cdot}5x - 6{\cdot}25 - 8 = 0$$
$$-1{\cdot}75x^2 + 10x - 14{\cdot}25 = 0$$
$$1{\cdot}75x^2 - 10x + 14{\cdot}25 = 0$$

Solving by quadratic formula,

$$x = \frac{-b \pm \sqrt{b^2 - 4ac}}{2a}$$

$$= \frac{10 \pm \sqrt{10^2 - 4 \times 1{\cdot}75 \times 14{\cdot}25}}{2 \times 1{\cdot}75}$$

$$= \frac{10 \pm \sqrt{100 - 99{\cdot}75}}{3{\cdot}5} = \frac{10 \pm 0{\cdot}5}{3{\cdot}5}$$

$$= 3 \text{ or } 2{\cdot}714$$

From (iii) $\quad y = 2{\cdot}5 - 1{\cdot}5x$

If $x = 3$, $\quad y = 2{\cdot}5 - 1{\cdot}5 \times 3 = -2$

If $x = 2{\cdot}714$, $\quad y = 2{\cdot}5 - 1{\cdot}5 \times 2{\cdot}714 = -1{\cdot}571$

Hence, values of x and y are, respectively,

3 and -2
or, $2{\cdot}714$ and $-1{\cdot}571$ } Ans.

PROBLEMS INVOLVING QUADRATIC EQUATIONS

As explained in Chapter 4 on the solution of problems, the first procedure is to look for and compose an equation from the facts given in the problem, letting some symbol such as x representing the required unknown, the equation is then simplified and worked out. If there are two unknowns express one in terms of the other at the first opportunity.

When simplifying equations which contain terms of the unknown to other powers than unity, it is usual to arrange all terms on the left hand side in order of descending powers of the unknown, leaving zero on the other side. If this produces a quadratic equation it is solved by the most convenient of the methods just explained. If a more complicated equation is produced, other means can be employed such as solving by a graph which will be explained later.

Example. Two ships sail from one port to another, a distance of 825 nautical miles, on the same course, the speed of one ship being 4 knots faster than the other. The fast ship leaves port 2 hours after the slow ship and arrives at their destination 14 hours sooner. Find the speeds of the two ships.

Let x knots = speed of slow ship
then $(x + 4)$ knots = speed of fast ship

Difference in times for the two ships to cover the journey,

$$= 2 + 14 = 16 \text{ hours.}$$

An equation can be formed from the times taken on the journey:
Time taken by slow ship − Time taken by fast ship = 16 hours

$$\text{time} = \frac{\text{distance}}{\text{speed}}, \text{ therefore,}$$

$$\frac{\text{distance travelled by slow ship}}{\text{speed of slow ship}} - \frac{\text{distance by fast ship}}{\text{speed of fast ship}} = 16$$

$$\frac{825}{x} - \frac{825}{(x + 4)} = 16$$

Multiplying every term by $x(x + 4)$ and simplifying,

$$825(x + 4) - 825x = 16x(x + 4)$$
$$825x + 3300 - 825x = 16x^2 + 64x$$
$$3300 = 16x^2 + 64x$$
$$16x^2 + 64x - 3300 = 0$$

Dividing throughout by 16,

$$x^2 + 4x - 206\tfrac{1}{4} = 0$$

Solving this quadratic by method of completing the square,

$$x^2 + 4x = 206\tfrac{1}{4}$$
$$x^2 + 4x + (2)^2 = 206\tfrac{1}{4} + (2)^2$$
$$x + 2 = \pm\sqrt{206\tfrac{1}{4} + 4}$$
$$x + 2 = \pm\sqrt{210\tfrac{1}{4}}$$
$$x + 2 = \pm 14\tfrac{1}{2}$$
$$x = \pm 14\tfrac{1}{2} - 2$$
$$x = 12\tfrac{1}{2} \text{ or } -16\tfrac{1}{2}$$

The minus quantity is highly improbable as it indicates the ship travelling backwards, the practical value of x is therefore $12\frac{1}{2}$.

$$\left.\begin{array}{r} \text{Speed of slow ship} = 12\tfrac{1}{2} \text{ knots} \\ \text{Speed of fast ship} = 12\tfrac{1}{2} + 4 = 16\tfrac{1}{2} \text{ knots} \end{array}\right\} \text{Ans.}$$

CUBIC EQUATIONS

A cubic equation is one which contains the cube of the unknown, thus, $x^3 = 8$ is a cubic equation of the most elementary form and solved simply by taking the cube root of both sides, $x = \sqrt[3]{8} = 2$.

However, the form usually associated with the title cubic equation contains either or both of the first and second powers of the unknown as well as its cube, such as $x^3 + 2x^2 - x = 2$.

One method of solution is to bring all terms of the equation to the left hand side and leaving zero on the other side, resolve the expression into its three factors, each of these factors are then, in turn, equated to zero which produces a value of the unknown that will satisfy the given equation.

Example. $x^3 + 2x^2 - x - 2 = 0$

The three factors of this expression are,

$$(x - 1)(x + 1)(x + 2)$$

By equating each factor to zero, the three roots are obtained.

$$x - 1 = 0 \quad \therefore x = 1$$
$$x + 1 = 0 \quad \therefore x = -1$$
$$x + 2 = 0 \quad \therefore x = -2$$

The roots of this equation are $+1$, -1, and -2, and any one of these values of x will satisfy the given equation.

If the equation is not a very easy one like the above, the factors will not be readily seen. In such cases, attempt to get one root by trial, from this we get the first factor. Dividing the cubic equation by this factor will produce a quadratic equation which can be solved by factors or by quadratic formula. The following examples show how this is done.

Example. Find the roots of the equation,

$$x^3 - x^2 = 5{\cdot}75x - 7{\cdot}5$$

Re-arrange with all terms on the left hand side, in descending powers of x,

$$x^3 - x^2 - 5{\cdot}75x + 7{\cdot}5 = 0$$

Find the first root by trial, try $x = 1$,

$$1^3 - 1^2 - 5{\cdot}75 \times 1 + 7{\cdot}5$$
$$= 1 - 1 - 5{\cdot}75 + 7{\cdot}5$$
$$= 1{\cdot}75$$

Try $x = 2$,

$$\begin{aligned} & 2^3 - 2^2 - 5{\cdot}75 \times 2 + 7{\cdot}5 \\ = \; & 8 - 4 - 11{\cdot}5 + 7{\cdot}5 \\ = \; & 0 \text{ this is it.} \end{aligned}$$

$x = 2$ satisfies the equation, therefore this is a root, and $x - 2$ must be a factor. Divide this factor into the cubic expression, this will produce a quadratic whose roots can be found.

$$\begin{array}{l} x - 2)x^3 - x^2 - 5{\cdot}75x + 7{\cdot}5(x^2 + x - 3{\cdot}75 \\ \qquad x^3 - 2x^2 \\ \hline \qquad \cdot \quad x^2 - 5{\cdot}75x + 7{\cdot}5 \\ \qquad\quad\; x^2 - 2x \\ \hline \qquad\quad \cdot \; - 3{\cdot}75x + 7{\cdot}5 \\ \qquad\quad\;\; - 3{\cdot}75x + 7{\cdot}5 \\ \hline \qquad\qquad \cdot \qquad \cdot \end{array}$$

Equate the resulting quadratic to zero and, if the factors cannot be seen, solve by formula,

$$x^2 + x - 3{\cdot}75 = 0$$

$$\begin{aligned} x &= \frac{-b \pm \sqrt{b^2 - 4ac}}{2a} \\ &= \frac{-1 \pm \sqrt{1^2 - 4 \times 1 \times (-3{\cdot}75)}}{2 \times 1} \\ &= \frac{-1 \pm \sqrt{16}}{2} = \frac{-1 \pm 4}{2} \\ &= 1{\cdot}5 \text{ or } -2{\cdot}5 \end{aligned}$$

The roots of the given equation are,

2, 1·5, and −2·5 Ans.

The student could now check this result by substituting for x, in the given cubic equation, each of these roots.

Example. Solve the following equation,

$$2x^2 = 17 + \frac{3}{x}$$

Multiply throughout by the least common denominator, which is x, to eliminate the fraction,

$$2x^3 = 17x + 3$$

Arrange all terms on left hand side and equate to zero,

$$2x^3 - 17x - 3 = 0$$

Find the first root by trial, try $x = 1$,

$$2 \times 1^3 - 17 \times 1 - 3$$
$$= 2 - 17 - 3$$
$$= -18$$

Try $x = 2$,

$$2 \times 2^3 - 17 \times 2 - 3$$
$$= 16 - 34 - 3$$
$$= -21$$

Try $x = 3$,

$$2 \times 3^3 - 17 \times 3 - 3$$
$$= 54 - 51 - 3$$
$$= 0$$

Hence $x = 3$ is a root and $x - 3$ must be a factor. Divide the cubic equation by this factor to obtain a quadratic which can be solved,

$$\begin{array}{l} x - 3)2x^3 - 17x - 3(2x^2 + 6x + 1 \\ \quad\quad\;\; \underline{2x^3 - 6x^2} \\ \quad\quad\quad\quad\quad 6x^2 - 17x - 3 \\ \quad\quad\quad\quad\quad \underline{6x^2 - 18x} \\ \quad\quad\quad\quad\quad\quad\quad\quad x - 3 \\ \quad\quad\quad\quad\quad\quad\quad\quad \underline{x - 3} \\ \quad\quad\quad\quad\quad\quad\quad\quad \cdot \quad \cdot \end{array}$$

Equate the resulting quadratic to zero and solve by formula,

$$2x^2 + 6x + 1 = 0$$

$$x = \frac{-b \pm \sqrt{b^2 - 4ac}}{2a}$$

$$= \frac{-6 \pm \sqrt{6^2 - 4 \times 2 \times 1}}{2 \times 2}$$

$$= \frac{-6 \pm \sqrt{28}}{4} = \frac{-6 \pm 5{\cdot}292}{4}$$

$$= -0{\cdot}177 \text{ or } -2{\cdot}823$$

The roots of the given cubic equation are,

$$3, \ -0{\cdot}177 \ \text{and} \ -2{\cdot}823 \quad \text{Ans.}$$

Note the last two examples. In the former the last term of the cubic equation is 7·5 and easy factors of this number which come readily to mind are 1, 7·5; 2, 3·75; 3, 2·5; 5, 1·5, all plus or minus, and the first root of the equation, by trial, was one of these. In the latter example, the last term of the cubic equation is 3 and its ready factors are 1, 3; 2, 1·5, all plus or minus, and the first root of this equation, by trial, was one of these. Trial figures for the first root should therefore be chosen with this in mind.

If the first root is not a number which can readily be found by trial as in the foregoing examples, the above process could be quite laborious and the equation would probably be more easily solved by a graphical solution.

TEST EXAMPLES 6

1. Find the value of x in each of the following equations,

   (i) $(2x + 8)(3x - 5) = 0$
   (ii) $(0{\cdot}5x - 10)(0{\cdot}25x + 5) = 0$
   (iii) $(5x + 0{\cdot}5)(4x + 0{\cdot}8) = 0$

2. Solve the following equations by the method of factorising,

   (i) $x^2 + 5x + 6 = 0$
   (ii) $x^2 - 10x + 24 = 0$
   (iii) $x^2 + 2x - 15 = 0$

3. Solve the following equations by the method of factorising,

   (i) $3x^2 + 2x - 33 = 0$
   (ii) $4x^2 - 17x + 4 = 0$
   (iii) $12x^2 + 10x - 12 = 0$

4. Solve the following equations by the method of completing the square,

   (i) $x^2 - x - 3\tfrac{3}{4} = 0$
   (ii) $3x^2 + 2x - 1 = 0$
   (iii) $4x^2 - 9x + 2 = 0$

5. Solve the following equations by the quadratic formula,

   (i) $3x^2 - 2x + 0{\cdot}25 = 0$
   (ii) $5x^2 + 4x - 5{\cdot}52 = 0$
   (iii) $10x^2 - x - 0{\cdot}2 = 0$

6. Use any method of solving the quadratic equations in this and the following problems:
   Find the value of x when:
   $$\log 0{\cdot}5x = 2 \times \log(x - 6)$$

7. Find the value of x if:
   $$\log(x^2 + 2) - \log(x - 1) = 2$$

8. Find the value of b in the equation,
   $$6b^4 - 2{\cdot}46b^2 + 0{\cdot}24 = 0$$

9. Find the value of V in the equation,
   $$V^{2\cdot8} - 5{\cdot}1V^{1\cdot4} + 5{\cdot}6 = 0$$

10. Find the value of g in the equation,
    $$2g - 5\sqrt{g} - 3{\cdot}96 = 0$$

11. Find the values of x and y in the following simultaneous equations,
    $$x^2 - xy + 2y^2 = 16$$
    $$x + 2y = 8$$

12. If an increase of speed of $1\frac{1}{2}$ knots would cut 3 hours off the time over a voyage of 665 nautical miles, find the actual speed of the ship and the voyage time.

13. In 5 hours less time than it takes a certain ship to travel 330 nautical miles, another ship which is $3\frac{1}{2}$ knots faster can travel 4 nautical miles further. What are the speeds of the two ships?

14. The area of a rectangle is 76 square centimetres and the perimeter is 350 millimetres. Find the length and breadth.

15. The area of a rectangle is 27 square centimetres and the diagonal is 75 millimetres. Find the length and breadth.

16. Find the roots of the following cubic equation,
    $$2x^3 + 3x^2 - 17x = 30$$

17. Find the values of x to satisfy the following equation,
    $$2x - \frac{16}{x}\left\{2 - \frac{3}{x}\right\} = 3$$

CHAPTER 7

GRAPHS

A graph is a diagram which shows the relation between two quantities. Graphs are usually plotted on squared paper and the student should now have a supply of this paper by him. He should not be content with merely looking at examples demonstrated here but should actually work them out and plot the graphs for himself as he reads through this chapter.

Taking the related quantities to be x and y, Fig. 4 shows the elements of plotting graphs.

Both quantities vary in value throughout the graph and the value of one depends upon the value of the other. In practice we often have cases where the values of one of the quantities is entirely dependent upon how the other is varied. Take for example the stretching of a spring, if a load is hung on the spring hook a certain amount of stretch takes place, if a heavier load is hung a greater stretch occurs, thus the amount of stretch depends on the magnitude of the load. In this example the load would be referred to as the *independent variable* and the stretch as the *dependent variable.* In the drawing of a graph of a function of x such as $x^2 - 2x + 5$, we let the value of this be denoted by y and write $y = x^2 - 2x + 5$, then for a series of chosen values of x the values of y are calculated to obtain a number of plotting points to enable the graph to be drawn. Note that we choose the values of x and the values of y depend upon these, hence x is the independent variable and y the dependent variable.

The larger the scale to which the graph is drawn the more accurate the results obtained when reading points from the graph, therefore the scale chosen should be as large as possible depending upon the highest and lowest values to be represented.

The various values of x and y (or other similar pairs of variable quantities) are plotted thus, referring to Fig. 4:

The point of intersection of the horizontal and vertical base lines, marked 0, is the common zero point for both quantities and therefore represents zero value for both x and y. Positive values of x are measured horizontally to the right from 0 and negative values are measured horizontally to the left. The horizontal base line is referred to as the xx axis. Horizontal measurements are termed abscissae (singular:

abscissa). Positive values of y are measured vertically above 0 and negative values are measured vertically downwards from 0. The vertical base line is the yy axis. Vertical measurements are termed ordinates.

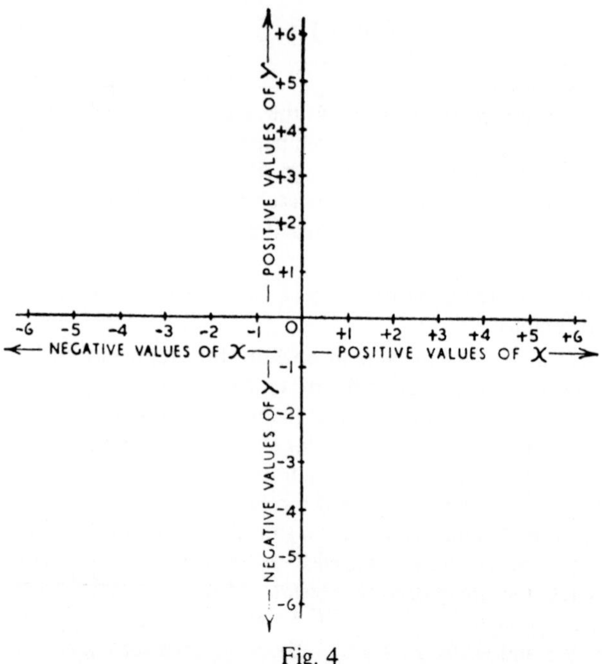

Fig. 4

PLOTTING OF STRAIGHT-LINE GRAPHS

Consider an equation of the form $y = a + bx$ where a and b are any constant quantities and may be positive or negative and as an example take such an equation where the value of the constant a is 2 and that of b is 1·5, thus

$$y = 2 + 1{\cdot}5x$$

Working out the values of y for various values of x between the extreme limits of, say, $x = -2$ and $x = +4$,

When $x = -2,$	$y = 2 - 3$	$= -1$
,, $x = -1,$	$y = 2 - 1{\cdot}5$	$= +0{\cdot}5$
,, $x = 0,$	$y = 2 + 0$	$= +2$
,, $x = +1,$	$y = 2 + 1{\cdot}5$	$= +3{\cdot}5$
,, $x = +2,$	$y = 2 + 3$	$= +5$
,, $x = +3,$	$y = 2 + 4{\cdot}5$	$= +6{\cdot}5$
,, $x = +4,$	$y = 2 + 6$	$= +8$

Plotting these points and drawing a graph through the plotted points as in Fig. 5:

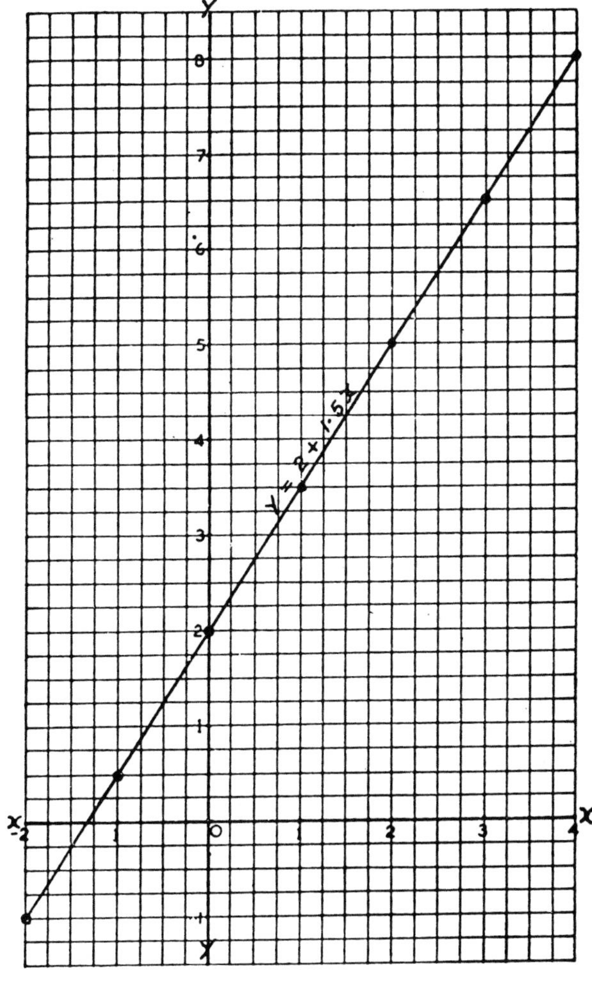

Fig. 5

Note (i) that the graph is a *straight line*, and with this knowledge the graph of any equation of the form $y = a + bx$ may be plotted with two pairs of values only to give two points through which to draw the straight line.

(ii) From left to right the graph *slopes upwards* and the value of b determines the gradient of the slope. In this case, for every unit increase of x the increase of y is 1·5; if b had been greater than 1·5 the slope would have been greater and vice-versa. The graph slopes upwards because b is *positive*, if b had been negative the graph would have sloped downwards.

(iii) The value of y when x is zero is 2, and this is the value of a.

Now consider the equations,

$$y = 4 + x$$
$$y = 3 + 0{\cdot}5x$$
$$y = 2 - 0{\cdot}25x$$
$$y = -0{\cdot}5x$$
$$y = -2 - 0{\cdot}5x$$

All of these are of the general form $y = a + bx$. Graphs of these equations are plotted in Fig. 6 between the limits $x = 0$ and $x = 12$. Examine these graphs carefully.

DETERMINING THE EQUATION TO A STRAIGHT LINE GRAPH

Instead of drawing a graph to a given equation we will now proceed to find the equation to a given straight line graph.

Example. Find the equation to a straight line graph which passes through the points (2, 4), (10, 7).

Plot the two points, $x = 2$, $y = 4$, and $x = 10$, $y = 7$, and draw the straight line through them, extending the line to cut the axis oy as shown in Fig. 7. The equation is of the form $y = a + bx$. The value of y when $x = 0$ is the constant a, from the graph we read $a = 3{\cdot}25$. The increase of y per unit increase of x is the value b, to obtain an accurate value we take two points on the graph as far apart as convenient, say at the points indicated by P_1 and P_2. From Q to P_2 is the increase of y when the increase of x is P_1 to Q. Thus y increases by 3·75 when x is increased by 10, dividing 3·75 by 10 we get 0·375, hence y will increase by 0·375 when x is increased by 1 and therefore the value of b is 0·375.

The equation to the graph is $y = 3{\cdot}25 + 0{\cdot}375x$

There are many practical applications of determining the equation to a given graph such as when performing a series of experiments on a machine, an engine, or a piece of material under a strength test, to

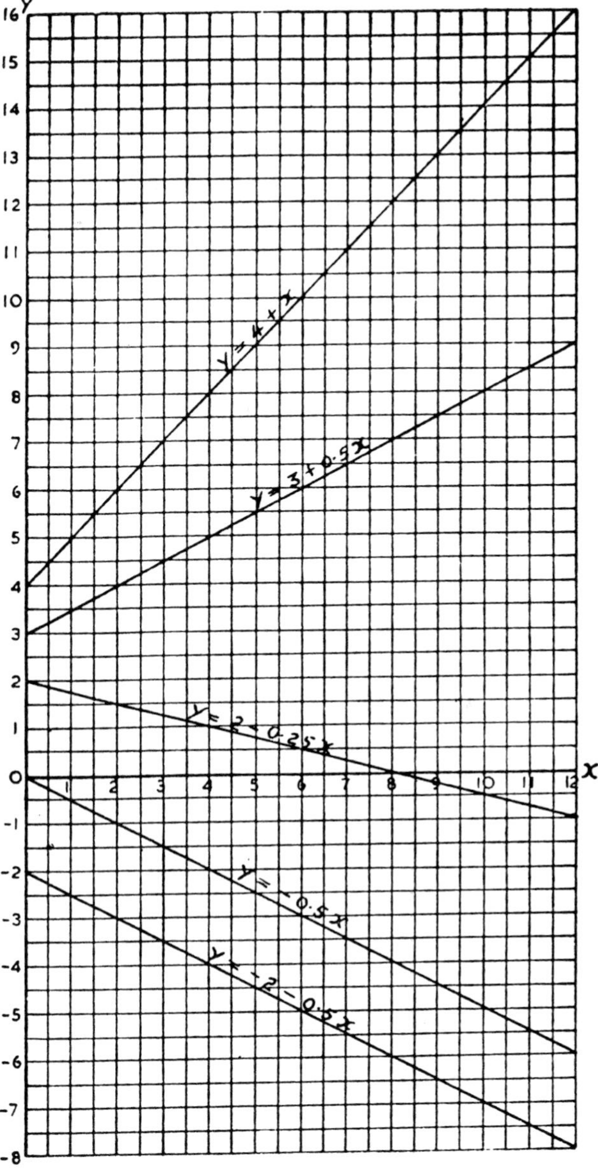
Y
16
15
14
13
12
11
10
9
8
7
6
5
4
3
2
1
0
-1
-2
-3
-4
-5
-6
-7
-8
1
2
3
4
5
6
7
8
9
10
11
12
X
Y = 4 + X
Y = 3 + 0.5X
Y = 2 - 0.25X
Y = -0.5X
Y = -2 - 0.5X

Fig. 6

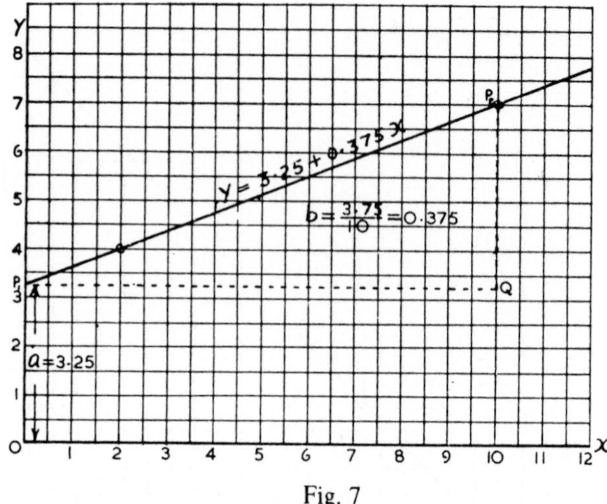

Fig. 7

ascertain how one quantity varies with another. For instance, various loads can be hung from the lifting hook of a lifting machine and the effort required to lift each load found experimentally and tabulated, a graph is then drawn showing how the effort varies with the load, the equation to the graph is determined and this is the law of that particular machine.

Example. In an experiment on a certain lifting machine the following data were observed,

Load	20	40	60	80	100	120
Effort	4·5	6·1	7·8	8·9	10·4	12·1

If the law of this machine expressing the relationship between effort applied (F) and load lifted (m) be expressed by the linear equation $F = a + bm$, plot a graph representing the above experimental values on a base of load and from it determine the linear law of this machine.

After plotting the experimental values a straight line is drawn as near as possible through these points as shown in Fig. 8. Those points not exactly on this line are probably due to irregularities of the machine or errors of observation during the experiment.

The value of a reads 3.

The value of b is the increase of effort per unit increase of load. Choosing two points on the line, P_1 and P_2, Q to P_2 measures 7·5, P_1 to Q measures 100, therefore $b = 7{\cdot}5 \div 100 = 0{\cdot}075$.

Hence the linear law of this machine is

$$F = 3 + 0{\cdot}075\, m$$

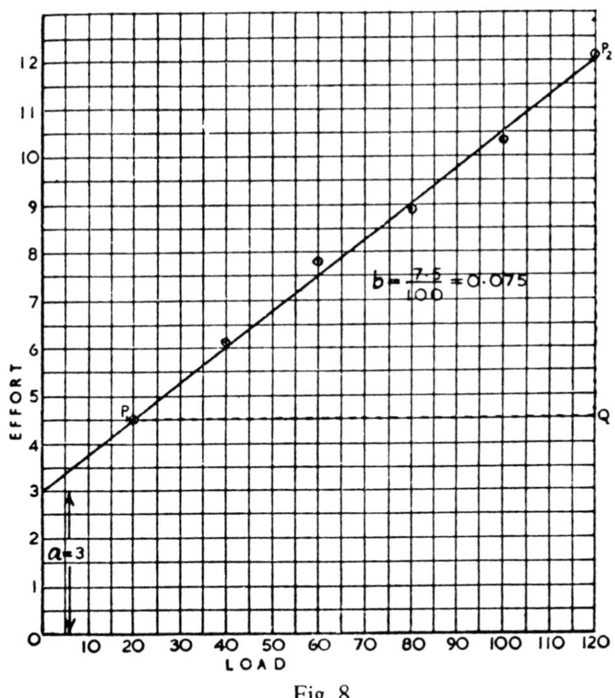

Fig. 8

GRAPHICAL SOLUTION OF SIMULTANEOUS LINEAR EQUATIONS

The method of solving simultaneous equations by graphical means can best be demonstrated by an example:

To find the values of x and y which satisfy the equations,

$$2x + 5y = 34$$
$$\text{and } 4x - 3y = 16$$

Find the value of y in the first equation,

$$2x + 5y = 34$$
$$5y = 34 - 2x$$
$$y = 6{\cdot}8 - 0{\cdot}4x \quad \ldots \quad \ldots \quad \text{(i)}$$

Find the value of y in the second equation,

$$4x - 3y = 16$$
$$-3y = 16 - 4x$$
$$3y = 4x - 16$$
$$y = 1\tfrac{1}{3}x - 5\tfrac{1}{3} \ldots \quad \ldots \quad \ldots \quad \text{(ii)}$$

Finding two plotting points for each equation,

$$y = 6{\cdot}8 - 0{\cdot}4x$$

when $x = 0$, $\quad y = 6{\cdot}8 - 0 = 6{\cdot}8$

when $x = 10$, $\quad y = 6{\cdot}8 - 4 = 2{\cdot}8$

$$y = 1\tfrac{1}{3}x - 5\tfrac{1}{3}$$

when $x = 0$, $\quad y = 0 - 5\tfrac{1}{3} = -5\tfrac{1}{3}$

when $x = 9$, $\quad y = 12 - 5\tfrac{1}{3} = 6\tfrac{2}{3}$

The points $x = 0$ and $x = 10$ for the first equation, and $x = 0$ and $x = 9$ for the second equation were chosen simply because they appeared to be easy figures for substituting and to produce reasonably sized graphs. $x = 0$ is obviously the first choice for a plotting point of any graph, the other point could be any value of x but it should be chosen with a view to produce simple figures.

The graphs are now plotted as shown in Fig. 9. At the point of intersection of the graphs we read the values $x = 7$ and $y = 4$ and these are the only values which are true for both equations.

Hence, $x = 7$ and $y = 4$ Ans.

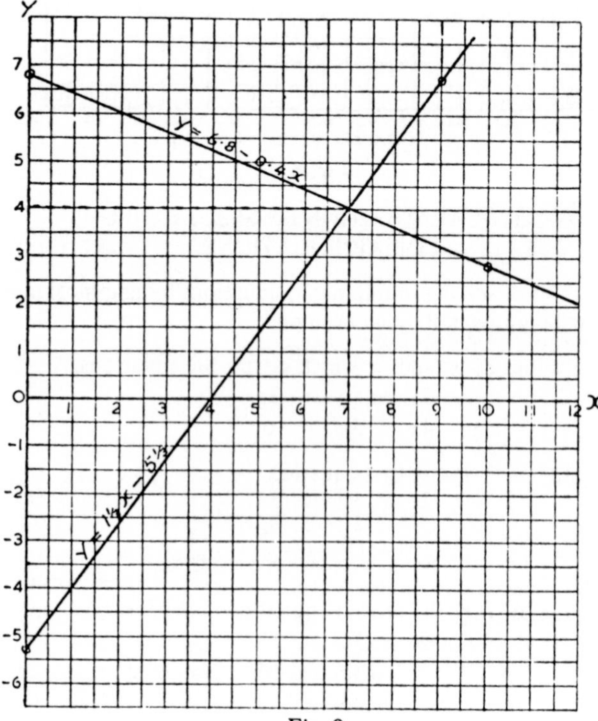

Fig. 9

GRAPHICAL SOLUTION OF QUADRATIC EQUATIONS

We have seen that graphs of equations containing x to the first power are all straight lines. Graphs of equations containing x to other powers, such as x^2 (quadratic equations) and x^3 (cubic equations) are curves and obviously more than two plotting points are necessary as a guide to the drawing of a curve; the more plotting points we have the more accurate can the curve be drawn.

As in previous cases the equation to be solved is first simplified, all terms are brought to one side in order of descending powers of the unknown, say x, leaving zero on the other side. Replace the nought by y, that is, let $y =$ the given expression and plot a graph for a series of values of x. Where this graph reads $y = 0$ gives the values of x which will satisfy the equation. The values of x when $y = 0$ are those points where the graph cuts the x axis, the values of y in proximity to these points therefore change from positive to negative or from negative to positive, the trial values of x for calculating the plotting points for the curve should therefore be chosen with this in mind. The following worked examples will clarify this explanation.

Example. Find, graphically, the values of x in the equation,

$$x^2 + 2x = 5{\cdot}5x - 1{\cdot}96$$

Simplify and bring all terms to one side,

$$x^2 - 3{\cdot}5x + 1{\cdot}96 = 0$$

$$\text{Let } y = x^2 - 3{\cdot}5x + 1{\cdot}96$$

Calculate values of y for selected values of x,

when $x = 0$	$y = 0 - 0 + 1{\cdot}96$	$= +1{\cdot}96$
,, $x = 1$	$y = 1 - 3{\cdot}5 + 1{\cdot}96$	$= -0{\cdot}54$
,, $x = 2$	$y = 4 - 7 + 1{\cdot}96$	$= -1{\cdot}04$
,, $x = 3$	$y = 9 - 10{\cdot}5 + 1{\cdot}96$	$= +0{\cdot}46$
,, $x = 4$	$y = 16 - 14 + 1{\cdot}96$	$= +3{\cdot}96$

We note that the value of y changes sign between $x = 0$ and $x = 1$, and again between $x = 2$ and $x = 3$, therefore the two values of x will be obtained from the graph at these two intersections of the x axis and hence there is no need to plot the graph beyond the limits of $x = 0$ and $x = +3$. The larger the scale of the graph the more accurate will be the reading of the values of x therefore the graph should be plotted to the largest scale possible on the paper.

Results of greater accuracy could be obtained by calculating values of y for a few points between $x = 0$ and $x = +1$, and also between $x = +2$ and $x = +3$ and drawing to a larger scale only those two parts of the curve which cross the x axis.

By drawing the graph as shown in Fig. 10 we read, when $y = 0$, $x = 0{\cdot}7$ and $2{\cdot}8$.

Therefore, $x = 0{\cdot}7$ or $2{\cdot}8$ Ans.

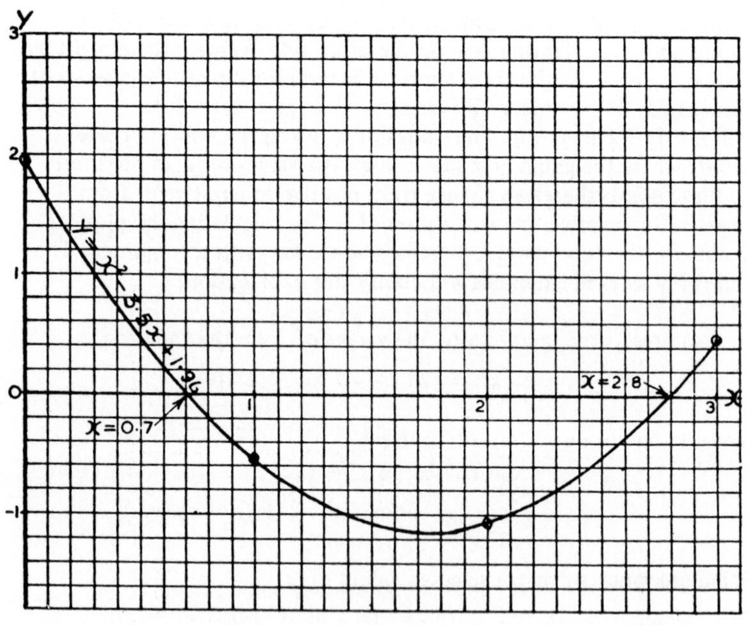

Fig. 10

Example. Solve, $x^2 - 0{\cdot}8x - 3{\cdot}84 = 0$

Let $y = x^2 - 0{\cdot}8x - 3{\cdot}84$

when $x = 0$,	$y = 0 - 0 - 3{\cdot}84$	$=$	$-3{\cdot}84$
,, $x = 1$,	$y = 1 - 0{\cdot}8 - 3{\cdot}84$	$=$	$-3{\cdot}64$
,, $x = 2$,	$y = 4 - 1{\cdot}6 - 3{\cdot}84$	$=$	$-1{\cdot}44$
,, $x = 3$,	$y = 9 - 2{\cdot}4 - 3{\cdot}84$	$=$	$+2{\cdot}76$

We note that the value of y changes sign from negative to positive between $x = 2$ and $x = 3$, therefore there is no need to proceed further in this direction. The other value where change of sign takes place, i.e. from positive to negative, must be when x is a negative quantity. Proceeding then in this direction,

when $x = -1, \quad y = 1 + 0{\cdot}8 - 3{\cdot}84 = -2{\cdot}04$
„ $x = -2, \quad y = 4 + 1{\cdot}6 - 3{\cdot}84 = +1{\cdot}76$

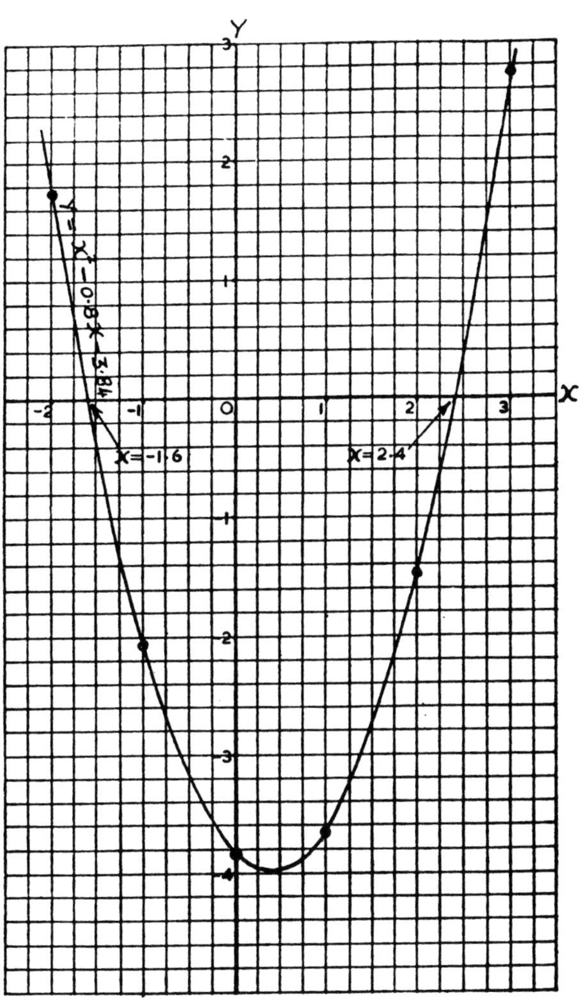

Fig. 11

Now that we have ascertained that $y = 0$ between $x = -1$ and -2, we have sufficient data to plot the graph. It may be of assistance to clear up the above by tabulating these results before attempting to plot them, thus,

x	-2	-1	0	$+1$	$+2$	$+3$
y	$+1{\cdot}76$	$-2{\cdot}04$	$-3{\cdot}84$	$-3{\cdot}64$	$-1{\cdot}44$	$+2{\cdot}76$

Plotting the graph as in Fig. 11 and reading the values of x when $y = 0$ we have,

$$x = 2{\cdot}4 \text{ or } -1{\cdot}6 \text{ Ans.}$$

Another method of solving quadratic equations, and one which can often be applied to solve cubic and more complicated equations is demonstrated by the following example.

To solve: $x^2 - 4x + 1\frac{3}{4} = 0$

This can be written

$$x^2 - (4x - 1\tfrac{3}{4}) = 0$$

Let y_1 represent x^2
and y_2 represent $(4x - 1\frac{3}{4})$

then, $x^2 - (4x - 1\frac{3}{4}) = 0$
is represented by, $y_1 - y_2 = 0$

Graph plotting points are found for each part:

$$y_1 = x^2 \quad \ldots \quad \ldots \quad \ldots \quad \ldots \quad \text{(i)}$$

when $x = 0$, $y_1 = 0$
,, $x = 1$, $y_1 = 1$
,, $x = 2$, $y_1 = 4$
,, $x = 3$, $y_1 = 9$
,, $x = 4$, $y_1 = 16$

$$y_2 = 4x - 1\tfrac{3}{4} \quad \ldots \quad \ldots \quad \ldots \quad \text{(ii)}$$

This is a straight line equation and two plotting points only are required.

when $x = 0$, $y_2 = 0 - 1\frac{3}{4} = -1\frac{3}{4}$
,, $x = 4$, $y_2 = 16 - 1\frac{3}{4} = 14\frac{1}{4}$

The curve of $y_1 = x^2$ and the straight line representing $y_2 = 4x - 1\frac{3}{4}$ are now plotted as shown in Fig. 12.

From the graph it is seen that the difference between the values of y_1 and y_2 is zero (i.e., $y_1 - y_2 = 0$) where the straight line intersects the curve.

$y_1 - y_2 = 0$ which means $x^2 - 4x + 1\frac{3}{4} = 0$ when $x = \frac{1}{2}$ and $x = 3\frac{1}{2}$

$$\text{therefore } x = \tfrac{1}{2} \text{ or } 3\tfrac{1}{2} \text{ Ans.}$$

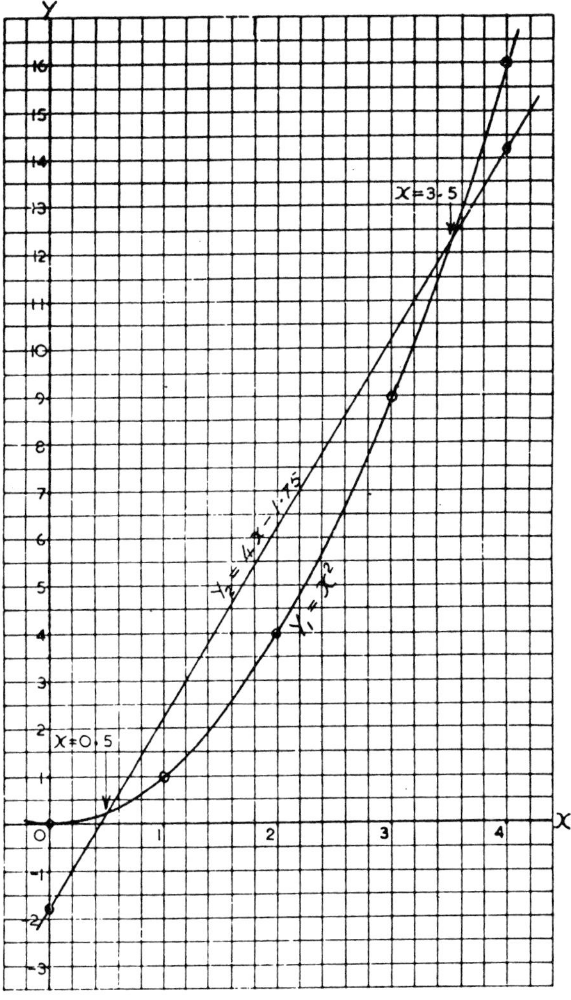

Fig. 12

The student should always check results by substituting them into the original equation to see if they are correct.

GRAPHICAL SOLUTION OF SIMULTANEOUS QUADRATICS

These are dealt with in a similar manner as previously explained with regard to simultaneous linear equations. Two quadratic equations could be involved, or one quadratic and one linear equation.

Example. Find the values of x and y which satisfy the equations,

$$y = 12 + 3x - 0{\cdot}5x^2$$
$$\text{and } y = 14 - 1{\cdot}25x$$

The extreme values of x to be taken for calculating the plotting points are chosen by making mental estimates with a view to the two graphs crossing.

Taking the linear equation first, being a straight line graph only two points are needed,

$$y = 14 - 1{\cdot}25x$$

when $x = 0$,	$y = 14 - 0$	$= 14$
,, $x = 10$,	$y = 14 - 12{\cdot}5$	$= 1{\cdot}5$

Taking the quadratic,

$$y = 12 + 3x - 0{\cdot}5x^2$$

when $x = 0$,	$y = 12 + 0 - 0$	$= 12$
,, $x = 1$,	$y = 12 + 3 - 0{\cdot}5$	$= 14{\cdot}5$
,, $x = 2$,	$y = 12 + 6 - 2$	$= 16$
,, $x = 3$,	$y = 12 + 9 - 4{\cdot}5$	$= 16{\cdot}5$
,, $x = 4$,	$y = 12 + 12 - 8$	$= 16$
,, $x = 5$,	$y = 12 + 15 - 12{\cdot}5$	$= 14{\cdot}5$
,, $x = 6$,	$y = 12 + 18 - 18$	$= 12$
,, $x = 7$,	$y = 12 + 21 - 24{\cdot}5$	$= 8{\cdot}5$
,, $x = 8$,	$y = 12 + 24 - 32$	$= 4$
,, $x = 9$,	$y = 12 + 27 - 40{\cdot}5$	$= -1{\cdot}5$

This appears to be sufficient.

Drawing the graphs as in Fig. 13 the points of intersection P_1 and P_2 produce the values of x and y, thus,

$$\left.\begin{array}{lll} x = 0{\cdot}5 & \text{and } y = 13{\cdot}375 \\ \text{or } x = 8 & \text{and } y = 4 \end{array}\right\} \text{Ans.}$$

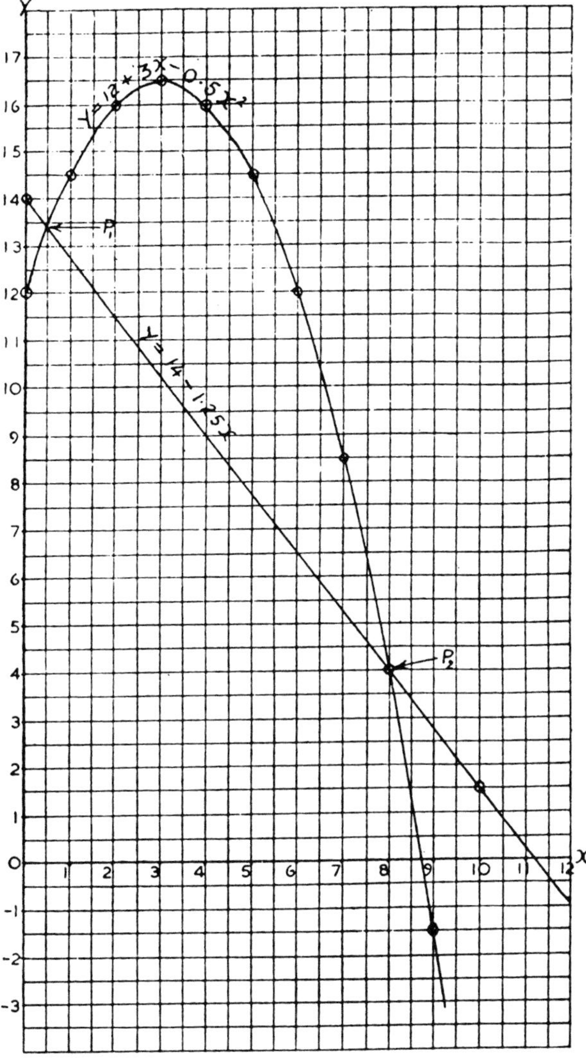

Fig. 13

DETERMINATION OF LAWS

One case has already been shown, that of determining the law connecting effort and load in a lifting machine, this relation was expressed by a straight line equation and the law found quite simply. However, in engineering many cases arise where the variation of one quantity with another are more complicated, most of them are represented by curves, but some of these may be reduced to a straight line from which the law can be found.

One typical example is that of a mass of a gas being expanded or compressed in a cylinder, the law connecting the variation of pressure as the volume of the gas is increased or decreased by the movement of the piston is:

$$pV^n = \text{a constant}$$

Writing this equation in log form:

$$\log p + n \log V = \log C$$

and transposing to express log p in terms of the other quantities:

$$\log p = \log C - n \log V$$

It is now reduced to a straight-line equation similar to $y = a + bx$

the variable		log p	taking the place of the variable				y
,,	,,	log V	,,	,,	,,	,,	x
,,	constant	log C	,,	,,	,,	constant	a
,,	,,	n	,,	,,	,,	,,	b

A graph of log p on a base of log V is now plotted and the value of n is determined in the same manner as the value of b would be found in the equation $y = a + bx$.

Example. The following ordinates and abscissae were measured from part of the expansion curve of an indicator diagram off an I.C. engine, where p is the pressure and V the volume of the gases in the cylinder. If the law of expansion can be expressed by $pV^n = C$, estimate the value of n.

p	28	24	16·6	13	9·7	6·8	4·7
V	0·6	0·7	0·9	1·1	1·4	1·8	2·4

Tabulating the values of p and V with their respective logarithms:

p	$\log p$	V	$\log V$	
28	1·4472	0·6	$\bar{1}$·7782	= −0·2218
24	1·3802	0·7	$\bar{1}$·8451	= −0·1549
16·6	1·2201	0·9	$\bar{1}$·9542	= −0·0458
13	1·1139	1·1	0·0414	
9·7	0·9868	1·4	0·1461	
6·8	0·8325	1·8	0·2553	
4·7	0·6721	2·4	0·3802	

When the values of V are less than unity, their logarithms have negative characteristics, to plot these they are expressed as all-negative values.

The graph is now plotted as shown in Fig. 14. Note that the lowest value of log p is 0·6721, the graph can be drawn to a larger scale by starting with a value of log p just a little lower than this, say 0·6, instead of commencing with zero origin.

Choosing two points on the graph,

$$n = \frac{\text{decrease of log } p}{\text{increase of log } V} = \frac{0{\cdot}65}{0{\cdot}5} = 1{\cdot}3$$

Therefore, $n = 1{\cdot}3$

Thus, $\log p = \log C - 1{\cdot}3 \log V$

Written in nominal form this is,

$$p = C \times V^{-1\cdot3}$$

or, as in the original setting,

$$pV^{1\cdot3} = C$$

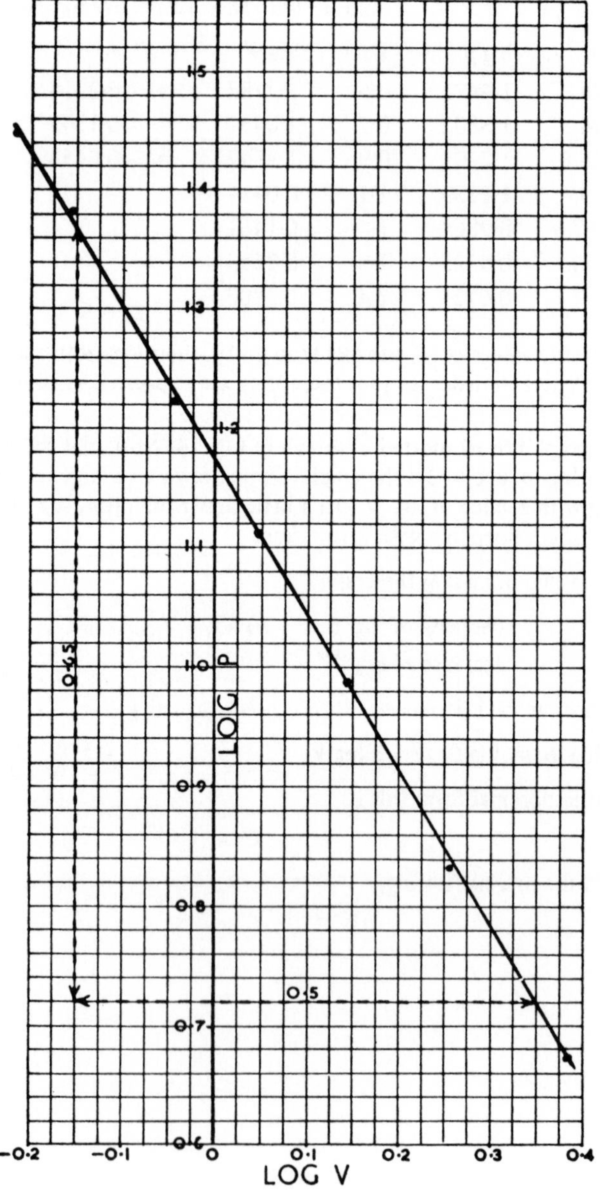

Fig. 14

TEST EXAMPLES 7

1. On a common base, plot graphs representing the equations,

(i)	$y = 2 + x$
(ii)	$y = 12 - 1{\cdot}5x$
(iii)	$y = -1 - 0{\cdot}5x$
(iv)	$y = -4 + 1{\cdot}25x$

all between the limits of $x = 0$ and $x = 12$

2. Find, by means of drawing a graph, the equation to the straight line which passes through the points (2, 4), (10, 16).

3. Draw a straight line through the two pairs of points (–2, 14·5), (8, –3) and from it derive the equation to the graph.

4. Plot a graph using the following values and find the law of the graph.

x	–2	–1	0	1	2	3
y	10	7	4	1	–2	–5

5. The following data were taken during an experiment on a small turbine where P represents the power developed and m the consumption of steam per hour. Assuming that the relationship between P and m can be represented by the straight line equation $m = a + bP$, draw a straight line as near as possible through the plotted points of the experimental results, estimate the values of a and b and hence the law connecting P and m.

P	20	25	30	35	40	45	50
m	220	265	315	365	410	455	505

6. Find, graphically, the values of x and y which satisfy the simultaneous equations,

$$3x + 5y = 23$$
$$\text{and } 5x - 2y = 12{\cdot}5$$

7. Find, graphically, the values of p and q in the simultaneous equations,

$$5p - 2q = 5{\cdot}6$$
$$\text{and } 2p - 3q = -4{\cdot}8$$

8. Find, graphically, the value of x in the equation,

$$x^2 - 5x + 5\tfrac{1}{4} = 0$$

9. Draw the graph of $y = 0{\cdot}5x^2 - 2x - 6$ between the values of $x = -4$ and $x = +8$.

   From the graph read the values of x in the following equations:

   (i) $0{\cdot}5x^2 - 2x - 6 = 0$
   (ii) $0{\cdot}5x^2 - 2x - 4 = 0$
   (iii) $0{\cdot}5x^2 - 2x - 1 = 0$
   (iv) $0{\cdot}5x^2 - 2x = 0$
   (v) $0{\cdot}5x^2 - 2x + 1 = 0$

10. On a common base draw graphs of $y_1 = x^2$, and $y_2 = 3{\cdot}5 + 2{\cdot}5x$, between the values of $x = -2$ and $x = +4$, find the values of x in the equation $x^2 - 2{\cdot}5x - 3{\cdot}5 = 0$.

11. Find, by graphical means, the values of x and y which satisfy the simultaneous equations,

$$y = 0{\cdot}4x^2 - 3x + 2$$
$$y = 1{\cdot}4x - 2$$

CHAPTER 8

MEASUREMENT OF ANGLES AND TRIGONOMETRIC RELATIONSHIPS

An *angle* is the corner of two joining lines and the magnitude of an angle is measured in either degrees or radians.

A degree is one three-hundred-and-sixtieth part of a circle, hence there are 360 degrees in a circle.

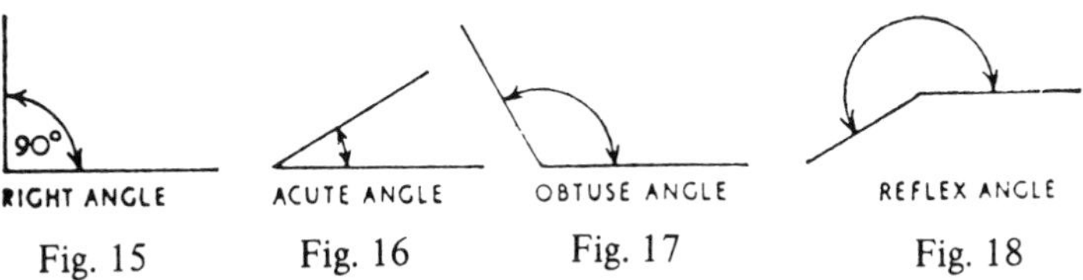

Fig. 15 Fig. 16 Fig. 17 Fig. 18

Fig. 15 shows a quarter of a circle which is 90 degrees and is termed a *right-angle*, an angle which is less than 90° (Fig. 16) is called an *acute* angle, greater than 90 but less than 180° (Fig. 17) is an *obtuse* angle, and greater than 180° (Fig. 18) is a *reflex* angle.

One sixtieth part of a degree is termed one *minute* and the sixtieth part of a minute is one *second*, thus,

60 seconds	=	1 minute
60 minutes	=	1 degree
360 degrees	=	1 circle.

Symbols are used to represent degrees, minutes and seconds. An angle of 35 degrees 23 minutes and 15 seconds is written $35^\circ\ 23'\ 15''$. In this work, accuracy to the nearest minute is all that is required.

A *radian* is the angle subtended by a circular arc of length equal to the radius.

In Fig. 19, the length of the arc A to B is equal to the radius OB and OA, and the enclosed angle AOB is one radian.

The length of an arc to subtend two radians is 2 × radius, in symbols:

If θ (theta) = number of radians
and r = radius
then, arc length = $\theta \times r$

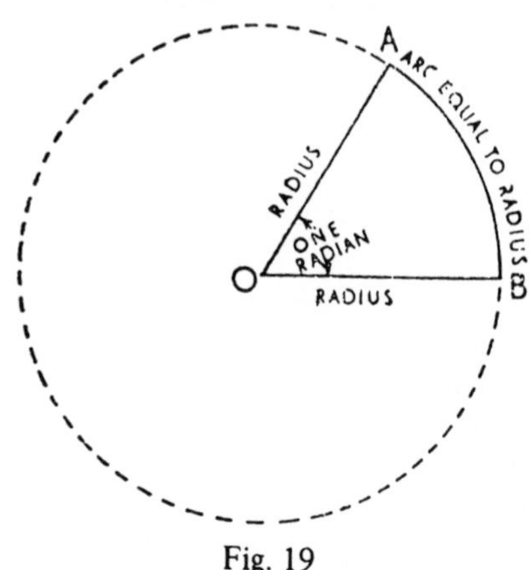

Fig. 19

Similarly, if a wheel of 0·3 metre radius turns through 4 radians in one second, then a point on the rim moves a linear distance of $4 \times 0{\cdot}3 = 1{\cdot}2$ metres in one second. In symbols:

If ω (omega) = angular velocity in radians per second,
v = linear velocity,
r = radius
then, $v = \omega r$

Circumference of a circle $= \pi \times$ diameter
$= 2\pi \times$ radius

Therefore one circle contains 2π radians

$$2\pi \text{ radians} = 360 \text{ degrees}$$

$$\therefore \text{ one radian} = \frac{360}{2\pi} \text{ degrees}$$

$$= 57{\cdot}3^\circ$$

The angular velocity of a rotating part of a mechanism is often more conveniently expressed in radians per second instead of in the practical units of revolutions per minute, to convert from one to the other:

$$\text{If } N = \text{revolutions per minute}$$
$$\text{and } \omega = \text{radians per second,}$$

$$\text{angular velocity} = N \text{ rev/min.}$$
$$= 2\pi N \text{ radians per minute}$$
$$= \frac{2\pi N}{60} \text{ radians per second}$$
$$\therefore \omega = \frac{2\pi N}{60}$$

Example. Express in (a) radians, (b) degrees, the angles subtended at the centre of a circle of 50 millimetres radius, by arc lengths of 50, 100, 125 and 140 millimetres respectively.

(i) Arc length of 50 mm:

$$\theta = \frac{\text{arc length}}{\text{radius}} = \frac{50}{50} = 1 \text{ radian}$$
$$= 57{\cdot}3 \text{ degrees}$$

(ii) Arc length of 100 mm:

$$\theta = \frac{100}{50} = 2 \text{ radians}$$
$$2 \times 57{\cdot}3 = 114{\cdot}6 \text{ degrees}$$

(iii) Arc length of 125 mm:

$$\theta = \frac{125}{50} = 2\tfrac{1}{2} \text{ radians}$$
$$2{\cdot}5 \times 57{\cdot}3 = 143{\cdot}25 \text{ degrees}$$

(iv) Arc length of 140 mm:

$$\theta = \frac{140}{50} = 2{\cdot}8 \text{ radians}$$
$$2{\cdot}8 \times 57{\cdot}3 = 160{\cdot}44 \text{ degrees}$$

Example. A flywheel of one metre diameter is rotating at 120 rev/min.

(i) express this in radians per second,

(ii) find the linear velocity, in metres per second, of a point on the rim.

$$\omega = \frac{120 \times 2\pi}{60}$$

$$= 12{\cdot}568 \text{ rad/s} \quad \ldots \quad \ldots \quad \text{(i)}$$

$$v = \omega r$$

$$= 12{\cdot}568 \times 0{\cdot}5$$

$$= 6{\cdot}284 \text{ m/s} \quad \ldots \quad \ldots \quad \ldots \quad \text{(ii)}$$

TRIGONOMETRIC RATIOS

A right angled triangle is one which contains an angle of 90°, the longest side (opposite the right angle) is termed the *hypotenuse*, the other two sides are termed the *opposite* and *adjacent* depending upon which of the other two angles are under consideration.

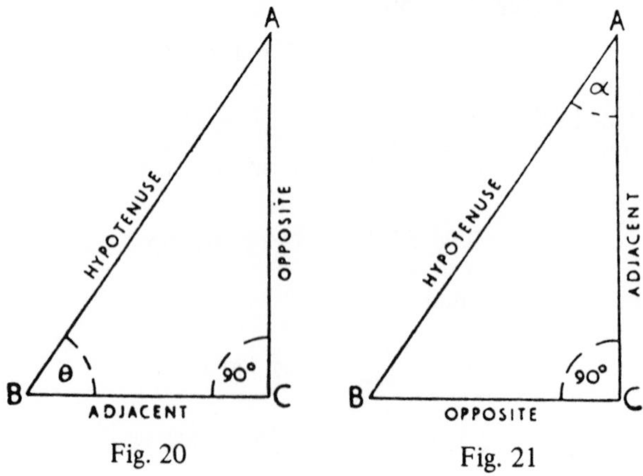

Fig. 20 Fig. 21

Considering angle θ (Fig. 20), side AC is opposite the angle and therefore referred to as such, the other side BC is the adjacent.

If angle α is being considered (Fig. 21), BC is the opposite and AC is the adjacent.

The ratios of the lengths of the sides of a right angled triangle are expressed by *sine* (abbreviated to sin), *cosine* (abbreviated cos) and *tangent* (abbreviated tan), as follows:

$$\text{sine of angle} = \frac{\text{opposite}}{\text{hypotenuse}}$$

$$\text{cosine of angle} = \frac{\text{adjacent}}{\text{hypotenuse}}$$

$$\text{tangent of angle} = \frac{\text{opposite}}{\text{adjacent}}$$

Referring to Fig. 20,

$$\sin\theta = \frac{AC}{AB}$$

$$\cos\theta = \frac{BC}{AB}$$

$$\tan\theta = \frac{AC}{BC}$$

The reciprocals of the above ratios are:

$$\text{cosecant } \theta\text{, abbreviated cosec } \theta = \frac{1}{\sin\theta} = \frac{AB}{AC}$$

$$\text{secant } \theta\text{, ,, sec } \theta = \frac{1}{\cos\theta} = \frac{AB}{BC}$$

$$\text{cotangent } \theta\text{, ,, cot } \theta = \frac{1}{\tan\theta} = \frac{BC}{AC}$$

Every angle has its own value of sine, cosine and tangent. As a preliminary, we can take the three most commonly used angles of 30, 45 and 60 degrees, draw triangles as in Figs. 22 and 23 by using protractor and ruler, measure the sides and calculate the ratios.

The sum of the angles in any triangle is 180 degrees, therefore if one angle of a right angled triangle is 30° the other must be 60° and if one angle is 45° the other must be 45°.

Referring to Fig. 22,

$$\sin 30 = \frac{100}{200} = 0{\cdot}5$$

$$\cos 30 = \frac{173{\cdot}2}{200} = 0{\cdot}866$$

$$\tan 30 = \frac{100}{173{\cdot}2} = 0{\cdot}5773$$

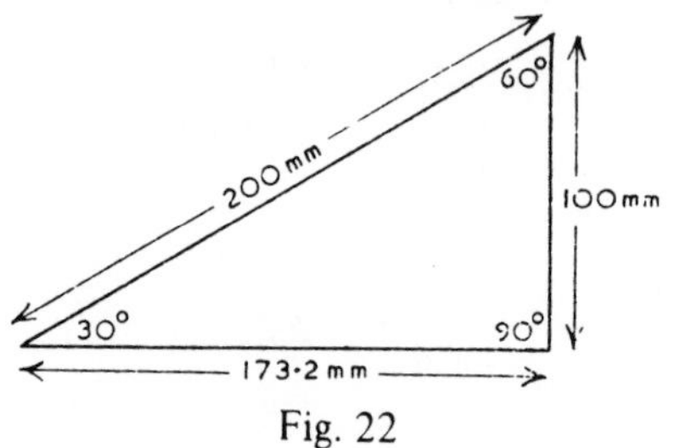

Fig. 22

Note that the sides are in the ratio, $1 : 2 : \sqrt{3}$

$$\sin 60 = \frac{173{\cdot}2}{200} = 0{\cdot}866$$

$$\cos 60 = \frac{100}{200} = 0{\cdot}5$$

$$\tan 60 = \frac{173{\cdot}2}{100} = 1{\cdot}732$$

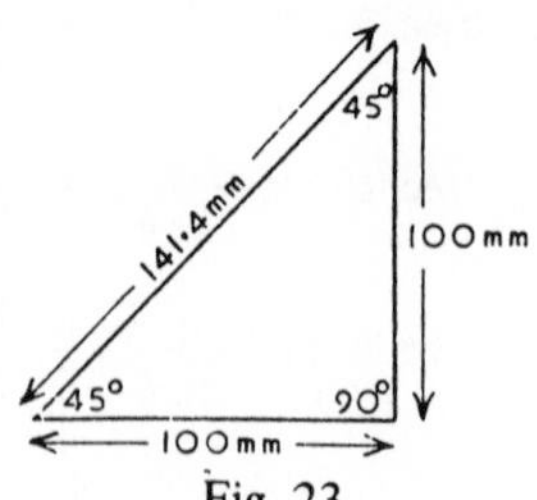

Fig. 23

Referring to Fig. 23,

$$\sin 45 = \frac{100}{141{\cdot}1} = 0{\cdot}7071$$

$$\cos 45 = \frac{100}{141{\cdot}1} = 0{\cdot}7071$$

$$\tan 45 = \frac{100}{100} = 1{\cdot}000$$

COMPLEMENTARY ANGLES

If the sum of any two angles is 90 degrees they are said to be *complementary angles.* 60° is the complementary angle to 30°, 50° is complementary to 40°, 70° is complementary to 20°, and so on. In all such cases *the sine of an angle is equal to the cosine of its complement.* The notes referring to Fig. 22 above show that the sine of 30° is the same as the cosine of 60°, and cos 30° is the same as sin 60°.

In general, referring to Fig. 24,

$$\sin\theta = \cos\alpha = \cos(90 - \theta)$$
$$\cos\theta = \sin\alpha = \sin(90 - \theta)$$
$$\tan\theta = \cot\alpha = \cot(90 - \theta)$$

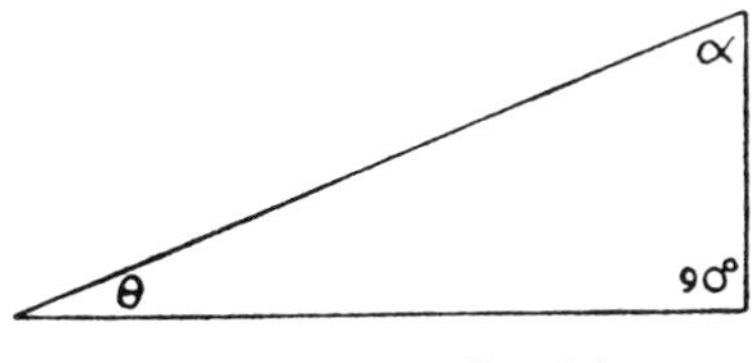

Fig. 24

SUPPLEMENTARY ANGLES

If the sum of two angles is 180 degrees they are said to be *supplementary* to each other. Thus 80° is the supplement of 100°, 53° is the supplementary angle of 127°, and so on.

THEOREM OF PYTHAGORAS

In a right angled triangle, *the square of the hypotenuse is equal to the sum of the squares of the other two sides.*

This may be shown by a small square of side c contained in a larger square of side $(a + b)$ arranged as in Fig. 25.

The area of a right angled triangle is $\dfrac{\text{base} \times \text{perpendicular height}}{2}$

In the triangles of sides a, b and c,

$$\text{area of one triangle} = \tfrac{1}{2}ab$$
$$\text{area of four triangles} = 4 \times \tfrac{1}{2}ab = 2ab$$
$$\text{area of large square} = (a + b)^2 = a^2 + 2ab + b^2$$
$$\text{area of small square} = c^2$$

$$\text{Area of small square} = \text{area of large square} - \text{area of 4 triangles}$$
$$c^2 = a^2 + 2ab + b^2 - 2ab$$
$$c^2 = a^2 + b^2$$

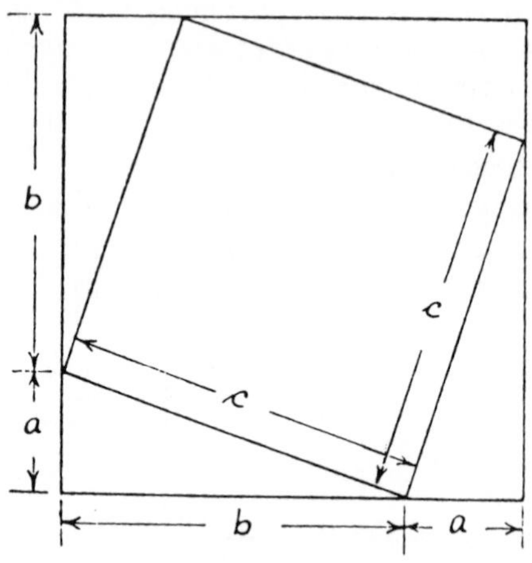

Fig. 25

The simplest example is that of a right angled triangle with a height of 3 and base 4,

$$(\text{hypotenuse})^2 = (\text{height})^2 + (\text{base})^2$$
$$\therefore \text{hypotenuse} = \sqrt{3^2 + 4^2}$$
$$= \sqrt{25} = 5$$

This is illustrated in Fig. 26.

RELATIONSHIPS BETWEEN SINE, COSINE AND TANGENT OF ANGLES

Referring to Fig. 20,

$$\sin\theta = \frac{AC}{AB} \text{ and } \cos\theta = \frac{BC}{AB}$$

$$\sin\theta \div \cos\theta = \frac{AC}{AB} \div \frac{BC}{AB} = \frac{AC}{AB} \times \frac{AB}{BC} = \frac{AC}{BC}$$

$$\text{and } \frac{AC}{BC} \text{ is } \tan\theta$$

$$\therefore \tan\theta = \frac{\sin\theta}{\cos\theta} \qquad \ldots \quad \ldots \quad \ldots \qquad \text{(i)}$$

$$(\text{hypotenuse})^2 = (\text{opposite})^2 + (\text{adjacent})^2$$
$$(AB)^2 = (AC)^2 + (BC)^2$$

Dividing every term by $(AB)^2$,

$$1 = \left(\frac{AC}{AB}\right)^2 + \left(\frac{BC}{AB}\right)^2$$

$$1 = \sin^2\theta + \cos^2\theta$$

usually written, $\sin^2\theta + \cos^2\theta = 1 \quad \ldots \quad \ldots \quad \ldots \quad \ldots \quad$ (ii)

The ratios of the reciprocals are also useful,

$$(AB)^2 = (AC)^2 + (BC)^2$$

Dividing every term by $(AC)^2$

$$\left(\frac{AB}{AC}\right)^2 = \left(\frac{AC}{AC}\right)^2 + \left(\frac{BC}{AC}\right)^2$$

$$\text{cosec}^2\theta = 1 + \cot^2\theta \quad \ldots \quad \ldots \quad \text{(iii)}$$

$$(AB)^2 = (AC)^2 + (BC)^2$$

Dividing every term by $(BC)^2$

$$\left(\frac{AB}{BC}\right)^2 = \left(\frac{AC}{BC}\right)^2 + \left(\frac{BC}{BC}\right)^2$$

$$\sec^2\theta = \tan^2\theta + 1 \quad \ldots \quad \ldots \quad \text{(iv)}$$

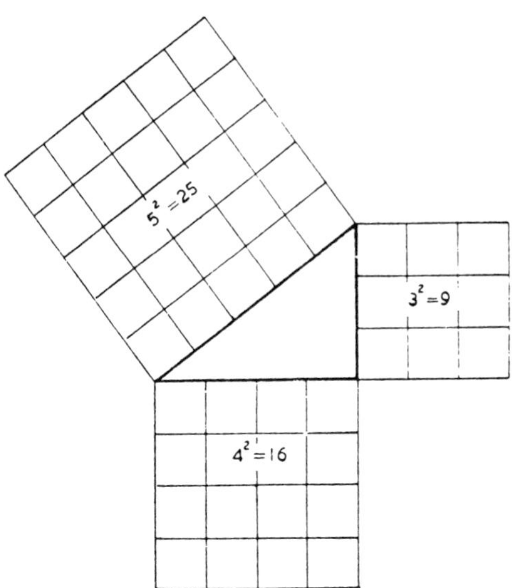

Fig. 26

Example. Without using tables, find the cosine and tangent of an angle whose sine is 0·8.

$$\begin{aligned}
\sin^2 \theta + \cos^2 \theta &= 1 \\
\cos^2 \theta &= 1 - \sin^2 \theta \\
\cos \theta &= \sqrt{1 - \sin^2 \theta} = \sqrt{1 - 0{\cdot}8^2} \\
&= \sqrt{1 - 0{\cdot}64} = \sqrt{0{\cdot}36} \\
&= 0{\cdot}6 \quad \dots \quad \dots \quad \dots \quad \dots \quad \text{(i)}
\end{aligned}$$

$$\begin{aligned}
\tan \theta &= \frac{\sin \theta}{\cos \theta} = \frac{0{\cdot}8}{0{\cdot}6} \\
&= 1{\cdot}333 \quad \dots \quad \dots \quad \dots \quad \text{(ii)}
\end{aligned}$$

Take care, $\sin \theta \times \sin \theta$ is written $\sin^2 \theta$ and not $\sin \theta^2$

$\cos \theta \times \cos \theta$,, $\cos^2 \theta$,, $\cos \theta^2$

and so on,

thus it is the *sine or cosine which is squared, and not the angle.*

IDENTITIES

An *identity* is an equation that is true for all values of the quantities involved in the equation, thus:

$$\begin{aligned}
x^2 - y^2 &= (x + y)(x - y) \\
(x + y)^2 &= x^2 + 2xy + y^2
\end{aligned}$$

are examples of identities because they are true for any values of x and y.

A *trigonometric identity* is an equation that is true for any angle, thus the four expressions derived in the previous paragraph:

$$\begin{aligned}
\sin \theta / \cos \theta &= \tan \theta \\
\sin^2 \theta + \cos^2 \theta &= 1 \\
\operatorname{cosec}^2 \theta &= 1 + \cot^2 \theta \\
\sec^2 \theta &= 1 + \tan^2 \theta
\end{aligned}$$

are all trigonometric identities because they are true for any value of angle θ, and many more can be proved from the knowledge of these fundamental identities and the trigonometric ratios. A few simple examples are shown below.

Prove that $\quad \operatorname{cosec} \theta + \cot \theta = \dfrac{1 + \cos \theta}{\sin \theta}$

Simplifying left hand side of the equation:

$$\operatorname{cosec}\theta + \cot\theta = \frac{1}{\sin\theta} + \frac{\cos\theta}{\sin\theta}$$

$$= \frac{1 + \cos\theta}{\sin\theta}$$

= right hand side of equation. Ans.

Prove that $\dfrac{\sec A - \cos A}{\sin A} = \tan A$

Simplifying left hand side of equation:

$$\frac{\sec A - \cos A}{\sin A} = \frac{\frac{1}{\cos A} - \cos A}{\sin A}$$

$$= \frac{\frac{1 - \cos^2 A}{\cos A}}{\sin A} = \frac{1 - \cos^2 A}{\cos A \,.\, \sin A}$$

$$= \frac{\sin^2 A}{\cos A \,.\, \sin A} = \frac{\sin A}{\cos A}$$

$$= \tan A$$

= right hand side of equation. Ans.

Prove that $\sin\theta = \dfrac{\tan\theta}{\sqrt{1 + \tan^2\theta}}$

Simplifying right hand side of equation:

$$\frac{\tan\theta}{\sqrt{1 + \tan^2\theta}} = \frac{\tan\theta}{\sqrt{\sec^2\theta}} = \frac{\tan\theta}{\sec\theta}$$

$$= \frac{\sin\theta}{\cos\theta} \times \frac{\cos\theta}{1} = \sin\theta$$

= l.h. side of equation. Ans.

READING TRIGONOMETRIC TABLES

The values of the sine, cosine and tangent for angles 0 to 90° are included in the usual Four-Figure Mathematical Tables which also contain logarithms. The student should now have a set of these tables at hand to refer to while studying the following.

SINE TABLES. These are read in the same manner as logarithms. Turning to the page of Natural Sines we see degrees (usually in heavy type) in the left hand margin, followed by a number of vertical columns for minutes from 0 to 54 by intervals of 6 minutes: the right hand margin contains the differences for each minute from 1 to 5.

Note that as the angle increases from 0 to 90 degrees, the sine increases from 0 to 1. All values of sines are therefore decimal fractions except for 90° where the sine is of maximum value, i.e., unity.

Example (i), to find the sine of 23°.

Glance down the left hand margin for 23 degrees, read the sine of this angle in the column headed 0 minutes.

$$\text{Sine } 23^\circ = 0{\cdot}3907$$

Example (ii), to find the sine of 57° 18′.

First find 57 degrees in the left hand margin then read horizontally along into the column headed 18 minutes.

$$\text{Sine } 57^\circ\ 18' = 0{\cdot}8415$$

Example (iii), to find the sine of 19° 44′.

Read horizontally along the level of 19 degrees into the column headed 42 minutes, this reads 3371; the difference between 44′ and 42′ is 2′, therefore read further along into the right hand margin on the same horizontal line and in the column under the heading of 2 minutes, here we see the figure 5; add 5 to 3371 to get 3376.

$$\text{Sine } 19^\circ\ 44' = 0{\cdot}3376$$

COSINE TABLES. Turn to the page of Natural Cosines and note that as the angle increases, the value of the cosine gets smaller. The maximum value of the cosine is unity for an angle of 0 degrees, decreasing to zero for an angle of 90 degrees.

These tables are read similar to sine tables except that the mean difference figures in the right hand margin are *subtracted* instead of added.

Example, to find the cosine of 71° 34′.

Along the level of 71 degrees to the column headed 30 minutes, read 3173; for the extra 4 minutes look into the right hand margin along the same horizontal level, into the column headed 4 minutes, we read 11. *Subtract* 11 from 3173 to get 3162.

$$\text{Cosine } 71^\circ\ 34' = 0{\cdot}3162$$

TANGENT TABLES. Tangents are read in the same way as sines, the figures in the mean differences margin being added. Note that the value of the tangent increases from zero for 0° to infinity for 90°, below 45° the tangents are decimal fractions, the tangent of 45° is unity; above 45° the tangent is greater than unity and care must be taken not to forget to include the figure preceding the decimal which is usually given only in the first column (headed 0 minutes).

Example, to find the tangent of 48° 44′.

On the horizontal level of 48 degrees note the figure preceding the decimal in the first (0 minutes) column, this is 1; now into the column headed 42 minutes read 1383, further along into the margin in the column for the extra 2 minutes read 13, add 13 to 1383 to get 1396.

Tangent 48° 44′ = 1·1396

In some sets of tables the values are given in columns of 10 minutes intervals, i.e., 0, 10, 20, etc., up to 50, instead of 6 minute intervals as described above, the method of reading these tables is similar and will be readily followed.

The student should now carefully check the following readings.

GIVEN ANGLE	TO FIND SINE	COSINE	TANGENT
3° 45′	0·0654	0·9978	0·0656
29° 2′	0·4853	0·8743	0·5551
46° 35′	0·7264	0·6873	1·0569
72° 15′	0·9524	0·3049	3·1242

To find the angles of given sines, cosines or tangents is the reverse process of the above.

Example (i), to find the angle whose sine is 0·8136.

Look for the nearest figure below 8136 in the sine tables, this is 813 on the level of 54 degrees and in the column headed 24 minutes. The difference between 8136 and 8131 is 5, now look along the same horizontal level into the margin for 5 (or nearest figure to 5), this is in the 3 minutes column, therefore add 3 minutes to the 24. Hence the angle is 54° 27′.

Example (ii), to find the angle whose cosine is 0·2609.

The nearest figure below 2609 in the cosine tables is 2605 which is for an angle of 74° 54′. The difference between 2609 and 2605 is 4, the nearest to 4 in the margin is 3 in the column headed 1 minute, *subtract* 1 minute from 54, hence the angle is 74° 53′.

ANGLES GREATER THAN 90 DEGREES

The usual tables give the sine, cosine and tangent for angles only up to 90 degrees, but we require the values for any angle up to 360 degrees.

Figs. 27b, 27c and 27d show circles of 360° with a radius arm swept around through given angles, in all cases the angle between the arm and the horizontal diameter has the same numerical values of sine, cosine and tangent as the given angle. The student will find it helpful to make the appropriate sketch when dealing with an angle over 90°. For angles between 0 and 90° all the ratios have positive values; for angles between 90 and 180° only the sine is positive, the cosine and tangent are negative for angles between 270 and 360° only the cosine is positive, the sine and tangent are negative. As an aid to remember whether the ratios are positive or negative, divide a circle into its four quadrants and memorise the positive values only, thus, All, Sin, Tan, Cos, respectively, as illustrated in Fig. 27a, the remainder being negative values.

The following examples will clarify the above statements.

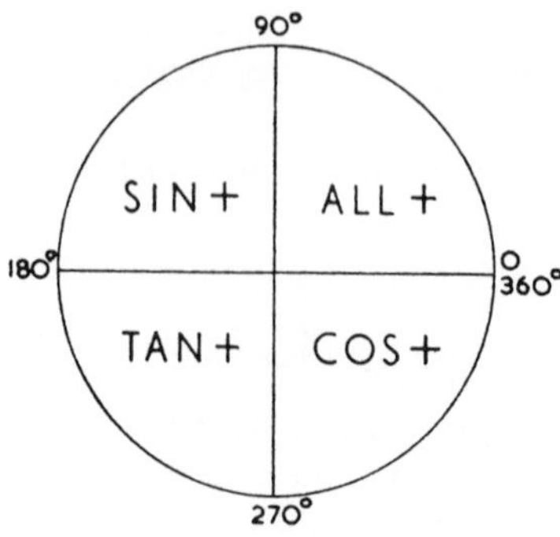

Fig. 27a

ANGLES BETWEEN 90 AND 180°. Subtract the angle from 180° and read the values of the resultant angle from the tables, these are numerically the same as for the given angle but, although the sine of the given angle is positive, its cosine and tangent have negative values. See Figs. 27a and 27b.

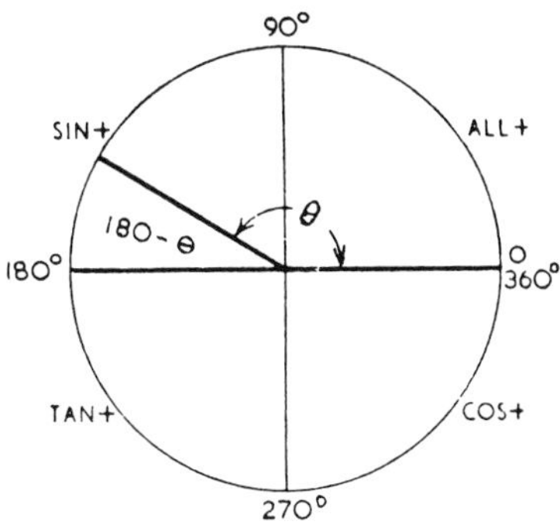

Fig. 27b

Example.

$$
\begin{aligned}
\text{Sin } 124^\circ\, 23' &= \sin(180^\circ - 124^\circ\, 23') \\
&= \sin 55^\circ\, 37' \\
&= 0{\cdot}8253 \\
\cos 124^\circ\, 23' &= -\cos(180^\circ - 124^\circ\, 23') \\
&= -\cos 55^\circ\, 37' \\
&= -0{\cdot}5648 \\
\tan 124^\circ\, 23' &= -\tan(180^\circ - 124^\circ\, 23) \\
&= -\tan 55^\circ\, 37' \\
&= -1{\cdot}4614
\end{aligned}
$$

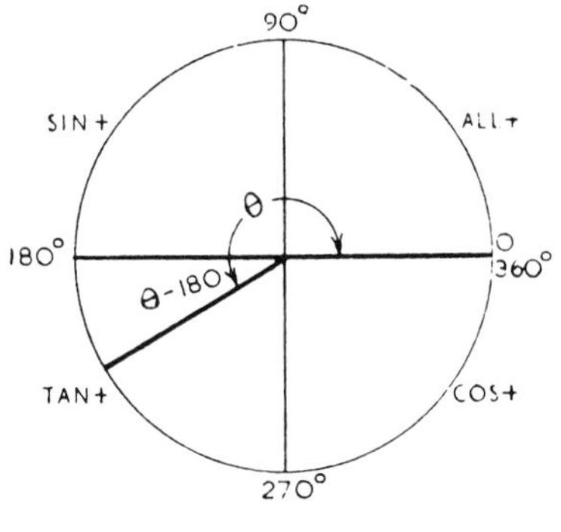

Fig. 27c

ANGLES BETWEEN 180 AND 270°. Subtract 180° from the angle and read the values of the resultant angle. In this case only the tangent of the given angle is positive, the sine and cosine are negative. See Figs. 27a and 27c.

Example.

$$\sin 203°\,14' = -\sin(203°\,14' - 180°)$$
$$= -\sin 23°\,14'$$
$$= -0{\cdot}3944$$
$$\cos 203°\,14' = -\cos(203°\,14' - 180°)$$
$$= -\cos 23°\,14'$$
$$= -0{\cdot}9189$$
$$\tan 203°\,14' = \tan(203°\,14' - 180°)$$
$$= \tan 23°\,14'$$
$$= 0{\cdot}4293$$

ANGLES BETWEEN 270 AND 360°. Subtract the angle from 360° and read the values of the resultant angle. The cosine is now the only positive value, the sine and tangent are negative. See Figs. 27a and 27d.

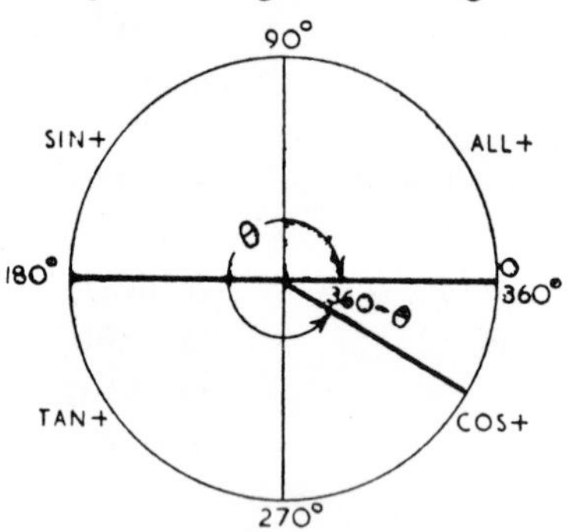

Fig. 27d

Example.

$$\sin 304°\,44' = -\sin(360° - 304°\,44')$$
$$= -\sin 55°\,16'$$
$$= -0{\cdot}8218$$
$$\cos 304°\,44' = \cos(360° - 304°\,44')$$
$$= \cos 55°\,16'$$
$$= 0{\cdot}5697$$
$$\tan 304°\,44' = -\tan(360° - 304°\,44')$$
$$= -\tan 55°\,16'$$
$$= -1{\cdot}4424$$

Now check the following readings.

ANGLE	SINE	COSINE	TANGENT
171° 18′	0·1513	–0·9885	–0·1530
225° 50′	–0·7173	–0·6968	1·0295
347° 27′	–0·2173	0·9761	–0·2226

We see from the foregoing that when it is required to find the angle corresponding to a given sine, cosine or tangent, unless we previously know in which quadrant the angle lies, there is more than one possible solution.

Example (i), given that the sine of an angle is 0·3636, to find the angle.

The sine being of positive value for all angles from 0 to 180 degrees, the angle can lie in the first or second quadrant. From the tables we read 0·3636 is the sine of 21° 19′, but it is also the sine of (180° – 21° 19′) which is 158° 41′.

Example (ii), given that the cosine of an angle is –0·3352, to find the angle.

The cosine is negative for all angles from 90 to 270 degrees therefore the angle can be in the second or third quadrants. From the tables we read 0·3352 is the cosine of 70° 25′, therefore,

–0·3352 is the cosine of (180° – 70° 25′)
= 109° 35′
or 180° + 70° 25′
= 250° 25′

Example (iii), given that the tangent of an angle is 2·4876, to find the angle.

The tangent is positive for angles between 0 and 90 degrees, and also between 180 and 270 degrees, therefore the angle can be in the first or third quadrants. From the tables we read 2·4876 is the tangent of 68° 6′, therefore,

2·4876 is the tangent of 68° 6′,
or (180° + 68° 6′)
= 248° 6′

It is useful to note the values of the following very commonly used angles.

ANGLE	SINE	COSINE	TANGENT
0	0	1	0
30	$\frac{1}{2}$ or 0·5	$\frac{\sqrt{3}}{2}$ or 0·866	$\frac{1}{\sqrt{3}}$ or 0·5774
45	$\frac{1}{\sqrt{2}}$ or 0·7071	$\frac{1}{\sqrt{2}}$ or 0·7071	1
60	$\frac{\sqrt{3}}{2}$ or 0·866	$\frac{1}{2}$ or 0·5	$\sqrt{3}$ or 1·732
90	1	0	$\propto$
120	$\frac{\sqrt{3}}{2}$ or 0·866	$-\frac{1}{2}$ or $-0·5$	$-\sqrt{3}$ or $-1·732$
135	$\frac{1}{\sqrt{2}}$ or 0·7071	$-\frac{1}{\sqrt{2}}$ or $-0·7071$	-1
150	$\frac{1}{2}$ or 0·5	$-\frac{\sqrt{3}}{2}$ or $-0·866$	$-\frac{1}{\sqrt{3}}$ or $-0·5774$
180	0	-1	0

The student is advised to continue the table for angles up to 360°.

We see from the above that the values of the sine and cosine are between zero and ±1, never numerically greater than unity. The tangent varies from zero to ± infinity.

It is sometimes useful to express some angles in radians instead of degrees. There are 2π radians in a full circle of 360°, therefore

$$360° \text{ may be expressed as } 2\pi \text{ radians}$$
$$180° \text{ may be expressed as } \pi \text{ radians}$$
$$90° \text{ may be expressed as } \pi/2 \text{ radians}$$

and so on.

GRAPHICAL REPRESENTATION

A little more study of the previous circle diagrams, and also Fig. 28, will show that if the radius arm of the circle is unity, then the vertical height above the horizontal axis represents the value of sine θ when θ is the angle between the arm and the horizontal. Since the cosine of an angle is equal to the sine of its complement, the vertical height above the horizontal will represent the value of cosine θ if θ is taken as the angle between the arm and the *vertical* axis. Thus, although graphs of

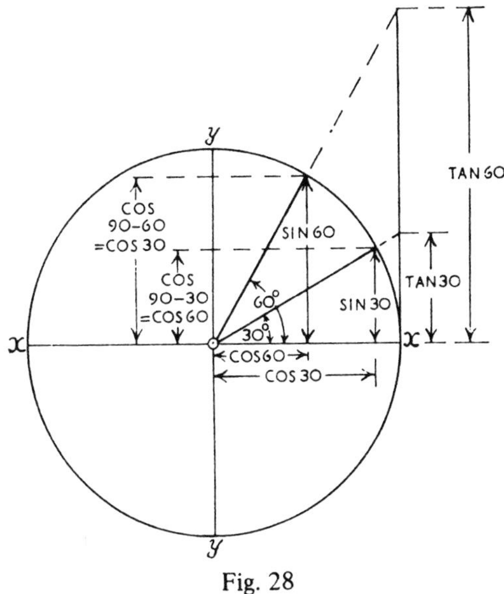

Fig. 28

sine and cosine can be plotted by taking values from the tables, a quicker and more convenient method is to project their values from a circle diagram as shown in Fig. 29.

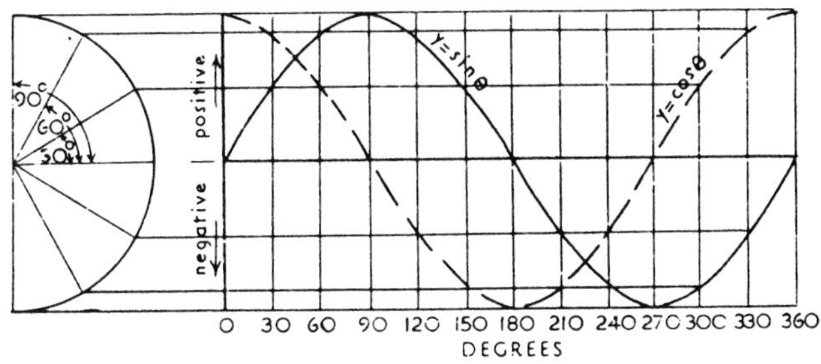

Fig. 29

It will be seen that, with a proper understanding of the positive and negative values, a half-circle only is required from which to project all points from 0 to 360°.

Further, the tangent of an angle being opposite ÷ adjacent, then the vertical height (opposite) = adjacent × tan θ. Hence if the base line Ox is made constant at unity, the vertical height of the radius-arm projected as shown in Fig. 28 will represent tan θ. The graph of y = tan θ is shown in Fig. 30.

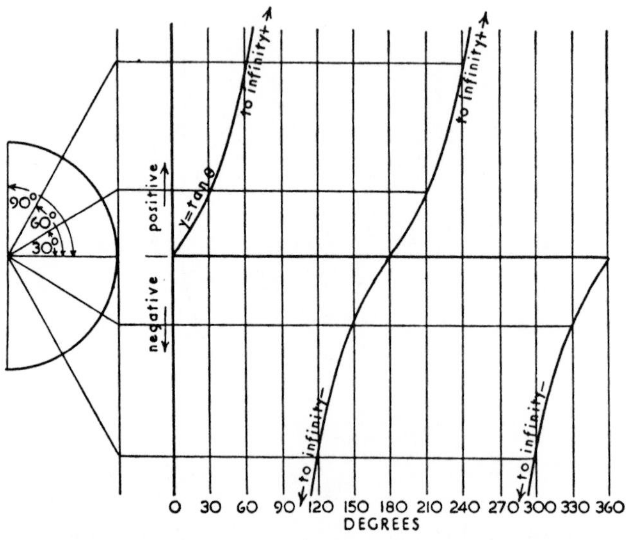

Fig. 30

LATITUDE AND LONGITUDE

The shape of the earth is approximately spherical, a straight line joining the North and South Poles and passing through the centre of the earth is called the Polar Axis. The earth rotates about its polar axis in the direction from West to East, that is, anticlockwise looking down on to the North Pole, and makes one complete revolution once in (approximately) every twenty-four hours.

Parallels of Latitude are circles around the earth, perpendicular to the polar axis. The parallel of latitude midway between the two poles is the Equator and this is the standard from which other parallels of latitude are measured, North and South of it, from 0 to 90 degrees either way.

Meridians of Longitude are circles around the earth passing through the two poles and cutting the equator at right angles. The Greenwich

Meridian is the meridian of longitude which passes through Greenwich and this is taken as the standard from which other meridians of longitude are measured, East and West of it, from 0 to 180 degrees.

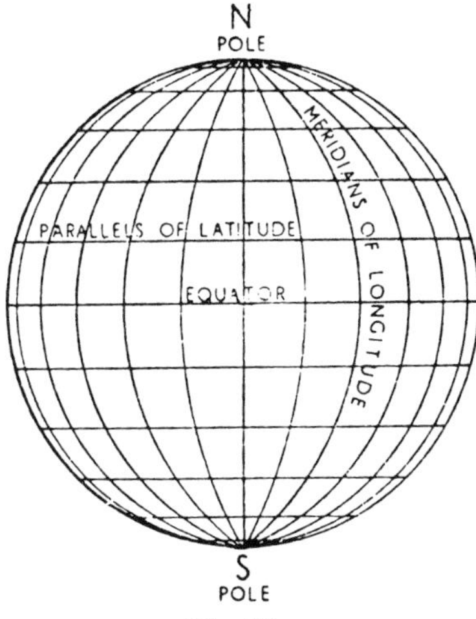

Fig. 31a

Thus any position on the earth's surface can be defined by its latitude and its longitude. Referring to Fig. 31b, let NGLS be the Greenwich Meridian. Angle QOP is the degrees latitude of point P, North of the equator. Angle LOQ (which is the same as GMP) is the degrees longitude of P, East of Greenwich.

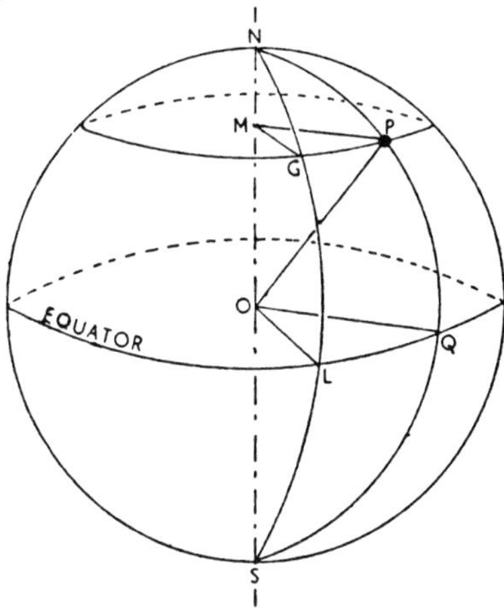

Fig. 31b

Since the earth rotates in the direction West to East and the sun being a fixed body, any point on the earth's surface East of another will pass a direct line of the sun first. Hence at places East of Greenwich the time is ahead of the time at Greenwich, and at places West of Greenwich the time is behind.

Taking the earth to rotate once in 24 hours:

$$360 \text{ degrees} = 24 \text{ hours}$$
$$15 \text{ degrees} = 1 \text{ hour}$$
$$1 \text{ degree} = \tfrac{1}{15} \text{ hour} = 4 \text{ minutes.}$$

As an example, the longitude of Karachi is 67 degrees East, therefore the difference in time between Greenwich and Karachi is 67×4 minutes $=$ 4 hours 28 minutes, and if the clocks were set by the sun, it would be 4-28 p.m. in Karachi when it is 12 noon at Greenwich.

It is obvious that different parts of one country should not have different times by the clock to other parts of the same country, modifications are therefore made. The time by the sun at Greenwich is referred to as Greenwich Mean Time (G.M.T.), all clocks in Britain are set to G.M.T. in the winter, and put forward one hour in the summer.

THE NAUTICAL MILE is the length of an arc on the earth's surface subtending an angle of one minute ($\frac{1}{60}$ degree) at the centre of the earth.

Taking the circumference of the earth as 40 000 kilometres:

$$360^\circ \text{ around surface} = 40\,000 \text{ km}$$

$$1 \text{ minute} \quad ,, \quad ,, \quad = \frac{40\,000}{360 \times 60}$$

$$= 1{\cdot}852 \text{ km}$$

This is the international nautical mile which is the nautical unit of distance.

Ship's speeds are expressed in nautical miles per hour, which are termed *knots*, one international knot being equal to 1·852 kilometres per hour.

COMPASS BEARINGS

A direction expressed with regard to the points of the compass is termed a *bearing* and it is usual to give a bearing in one of the two following methods.

(i) Stating the angle in degrees less than 90° from either North or South (whichever of the two is the nearer direction), the angle thus lies within one of the four quadrants and is therefore called a *quadrant bearing.*

(ii) Stating the angle in degrees measured clockwise from North and expressing this in three figures, using noughts for hundred, tens or units when there are none. This is called a *three-figure bearing.*

To illustrate these terms, Fig. 32 shows the direction of movement of a ship in relation to the four cardinal points of the compass.

Fig. 32a is the quadrant bearing and may be stated as 83° East of South, or South 83° East which is abbreviated to S 83° E.

Fig. 32b shows the same direction expressed as a three-figure bearing, 097°.

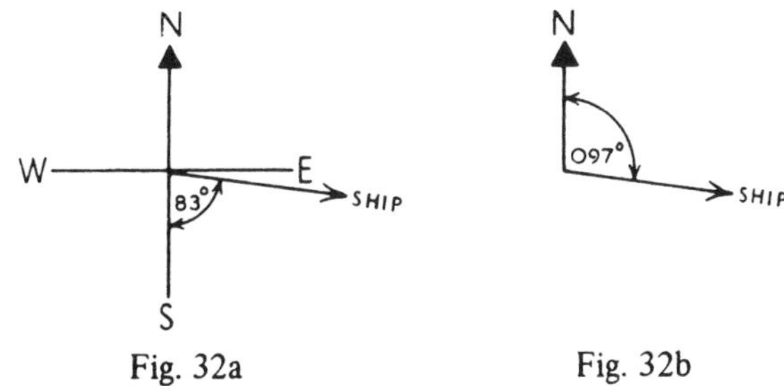

Fig. 32a Fig. 32b

TEST EXAMPLES 8

1. Express the following angles in radians,

   114° 36′, 143° 15′, 186° 13·5′, 286° 30′.

2. Express in (*a*) radians, (*b*) degrees, the angles subtended by arc lengths of 10, 21·25, 27·5 and 30·4 metres respectively, on a circle of 10 metres diameter.

3. The following are engine speeds in revolutions per minute, express each in radians per second,

   84, 120, 350, 3000.

4. The following are speeds in radians per second, convert each to revolutions per minute,

   9·25, 12·5, 37·25, 500.

5. A disc flywheel is rotating at a speed of 10·52 radians per second. Calculate the linear velocity, in metres per second, of points on the wheel at radii of 100, 250 and 500 millimetres respectively.

6. Find the linear velocity in metres per second of the rim of a flywheel of 2 metres diameter when rotating at 125 revolutions per minute.

7. In a right angled triangle ABC, length AC is 36 mm, length BC is 27 mm, and the angle at C is 90 degrees. Calculate the length of the hypotenuse AB and the sine, cosine and tangent of the angle at B.

8. Find, without the use of tables, the sine and tangent of an angle whose cosine is 0·4924.

9. Tabulate the sines and cosines of angles between 0° and 360° at intervals of 30° and draw graphs representing these values on a common base of angles.

10. Write down the sine, cosine and tangent of the following angles,

$$10^\circ\,33' \qquad 46^\circ\,55' \qquad 150^\circ\,47' \qquad 201^\circ\,21' \qquad 287^\circ\,14'$$

11. The following refer to angles between 0 and 360°,

(*a*) find the angles whose sines are:
0·3783, --0·7005,

(*b*) find the angles whose cosines are:
0·9687, –0·8769,

(*c*) find the angles whose tangents are:
0·2010, –3·2006.

12. If $\theta = 80^\circ$, find the values of:

$$\sin\theta, \quad \cos\theta, \quad \sin 2\theta, \quad \cos 2\theta, \quad \sin^2\theta, \quad \cos^2\theta$$

13. Find the values of θ between 0° and 360° which satisfy the equation,

$$\cos\theta - \sin^2\theta = 0$$

14. If $a = 2\sin\theta$, and $b = 5\cos\theta$, find the values of θ for angles between 0° and 180° when:

$$\frac{a^2}{2} = \frac{2b^2}{5} - 4$$

15. Prove that $\operatorname{cosec}^2 A + \sec^2 A = \operatorname{cosec}^2 A.\sec^2 A$

16. Prove that $\dfrac{1 - \sin A}{1 + \sin A} = (\sec A - \tan A)^2$

17. Prove that:

$$(\sec A - \cos A)(\operatorname{cosec} A - \sin A) = \frac{1}{\tan A + \cot A}$$

18. The velocity of the piston in a reciprocating engine is given by:

$$v = \omega r \left(\sin\theta + \frac{\sin 2\theta}{2n}\right) \text{ metres per second}$$

and the acceleration is given by:

$$a = \omega^2 r \left(\cos\theta + \frac{\cos 2\theta}{n}\right) \text{ metres per (second)}^2$$

where ω = angular velocity of crank in radians per second,
r = throw of crank ($\frac{1}{2}$ stroke) in metres,
θ = angle of crank past top centre, in degrees,
n = ratio of connecting rod length to crank length.

Calculate (i) the velocity, (ii) the acceleration, of the piston of an engine of 1 metre stroke, connecting rod length 2 metres, at the instant the crank is 80° past top centre and running at 150 revolutions per minute.

CHAPTER 9

GEOMETRY

The construction and rules given here have been selected as those which are found to be the most useful to engineers. The best method of learning these is for the student to take a pencil, ruler, pair of compasses and protractor and perform each example himself.

To erect a perpendicular from a given point on a given line. Referring to Fig. 33 let *xy* be the line and O the point at which the perpendicular is to be erected. With centre O and compasses set to any convenient radius, cut *xy* at both sides of O at A and B so that OA is equal to OB. With centres A and B describe two arcs of equal radius to intersect at C. Join CO. This line CO is perpendicular to *xy*.

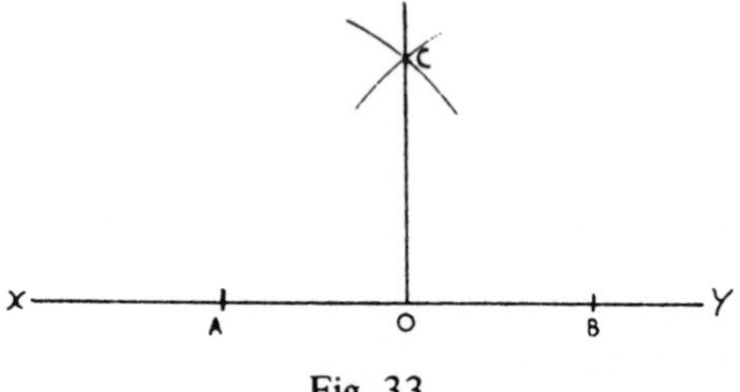

Fig. 33

Note that "perpendicular" means "at right angles", for example in Fig. 33 CO is perpendicular to *xy* because angles *x*OC and *y*OC are 90 degrees. In Fig. 34 CD is perpendicular to AB, and in Fig. 35 LM is perpendicular to PQ.

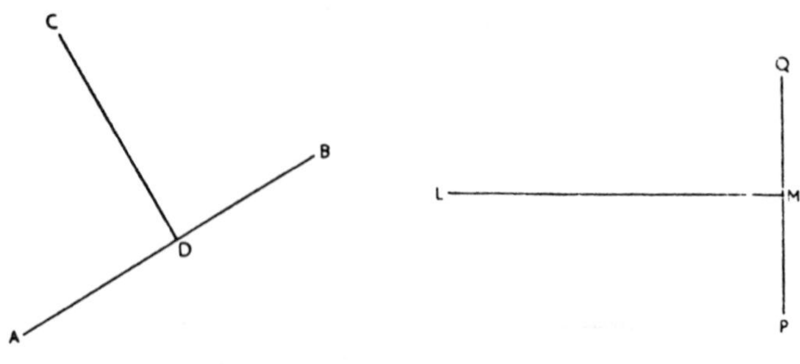

Fig. 34

Fig. 35

To bisect a given line.

Set the compasses to a radius a little more than half the length of the line. With centres x and y at each end of the line (Fig. 36) draw two arcs of equal radius, join the intersections of the arcs by drawing a line AB through them, this line cuts through the centre of xy and is perpendicular to it.

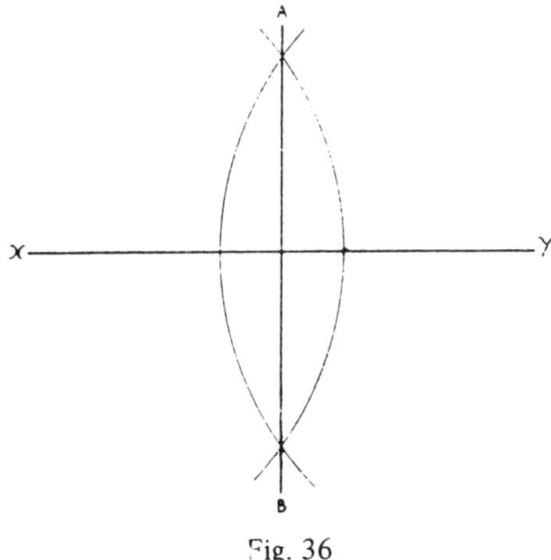

Fig. 36

To bisect a given angle.

Referring to Fig. 37, with centre O describe an arc of any convenient radius cutting the two legs of the angle at A and B. With centres A and B and the compasses set to a radius a little greater than half the length of the arc AB, describe two arcs of equal radius to intersect each other at C. Join OC and this bisects the angle AOB into two equal angles, AOC and BOC, each being half of AOB.

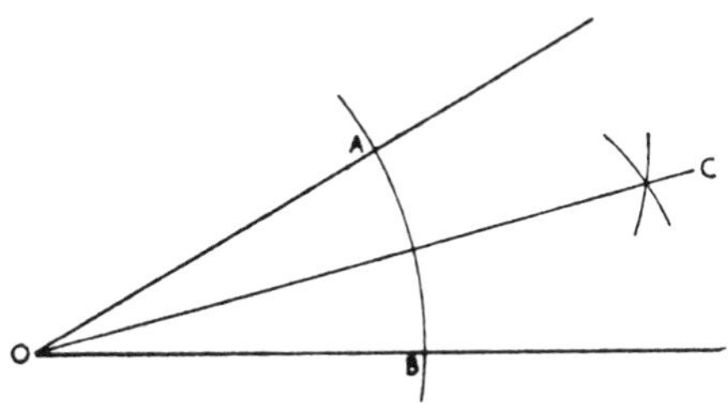

Fig. 37

EXAMPLES ON THE CONSTRUCTION OF TRIANGLES

Example (i), to construct a triangle of sides 60, 50 and 40 mm respectively, and measure the three angles.

Referring to Fig. 38, draw AB equal to 60 mm. Set compasses to 50 mm radius and, with centre A, describe an arc somewhere about where the other angle is likely to be. With centre B, describe an arc of 40 mm radius to cut the first arc. Join the intersection of the arcs, C, with A and B.

The angles can now be measured by means of a protractor and should read, as near as can be read with a protractor:

$$A = 41^\circ \quad B = 56^\circ \quad C = 83^\circ$$

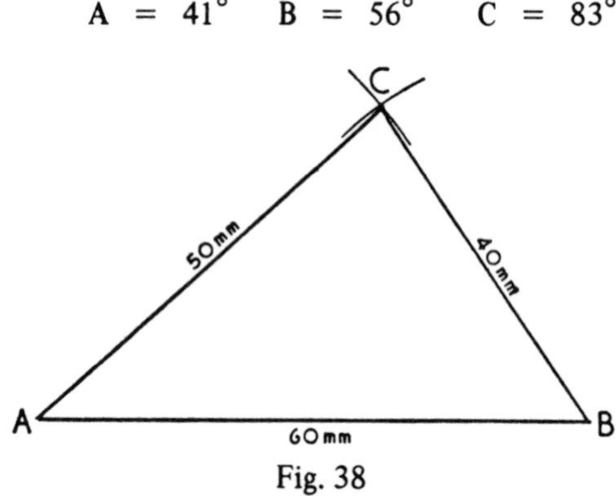

Fig. 38

Example (ii), to construct a triangle, two sides of which are 65 and 45 mm respectively and the angle between these sides being 35 degrees, and to measure the remaining side and angles.

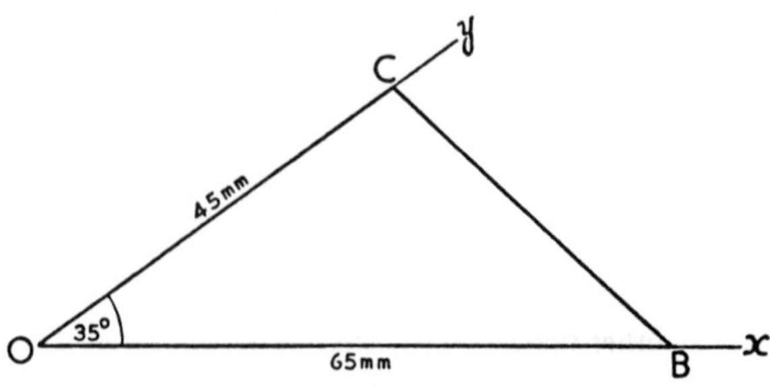

Fig. 39

Referring to Fig. 39 draw a line Ox, set the protractor with its centre at O, mark off 35 degrees at y, and draw the line Oy. Mark off 65 mm along Ox from O (point B) and 45 mm along Oy from O (point C). Join B and C.

Measure the length of the side BC and the angles at B and C which should read:

$$BC = 38 \text{ mm} \quad B = 42\tfrac{1}{2}^\circ \quad C = 102\tfrac{1}{2}^\circ$$

Example (iii), to construct a triangle of sides 64 and 48 mm and an angle of 42 degrees opposite the 48 mm side, and to measure the remaining side and angles.

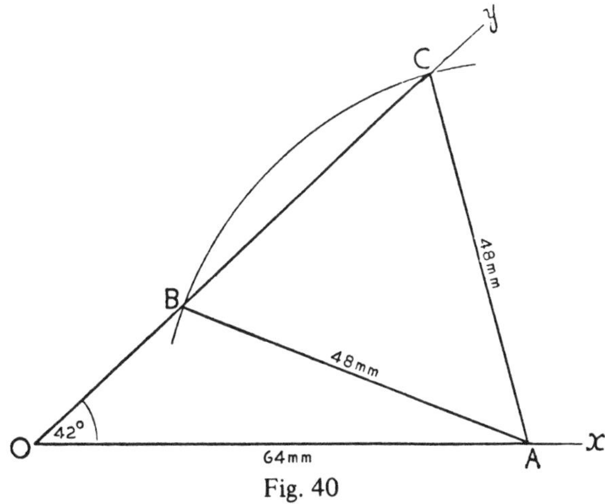

Fig. 40

Referring to Fig. 40, draw the base Ox, measure 42 degrees and draw Oy. Mark A at 64 mm from O. With centre A and compasses set to a radius of 48 mm, describe an arc to cut Oy, we find here that the arc will cut Oy at two points B and C. Join AB and we have one triangle OAB which satisfies the description. Join AC and now we have another triangle OAC which also satisfies the description. Hence there are two possible solutions to this example. Problems of this nature often occur in our work.

Measuring the remaining side and angles we get:

For one triangle,

Remaining side, OB = 26 mm
Remaining angles, OAB = 21° and OBA = 117°

For the other triangle,

Remaining side, OC = 69 mm
Remaining angles, OAC = 75° and OCA = 63°

SOME IMPORTANT FACTS

The opposite angles of the intersection of two straight lines are equal. See Fig. 41 and note also that $\alpha = 180 - \theta$.

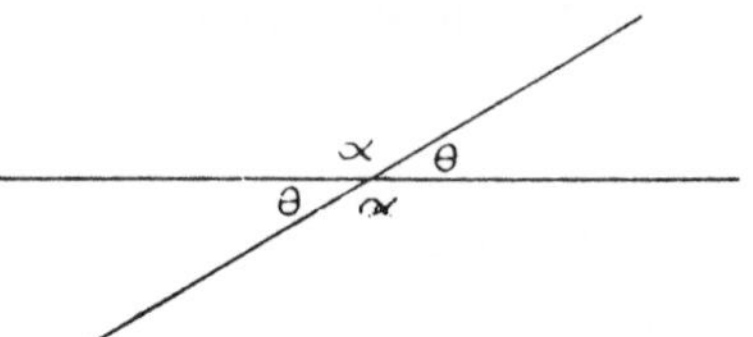

Fig. 41

The alternate angles produced by a straight line cut of a pair of parallel lines are equal to each other. See Fig. 42. When the student makes this diagram he should also measure all the other angles and insert them.

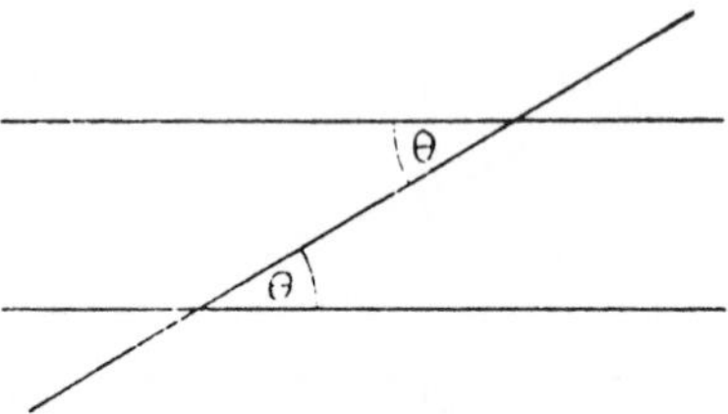

Fig. 42

The exterior angle formed by extending one side of a triangle is equal to the sum of the two opposite internal angles. Fig. 43 illustrates this. The student is advised to extend other sides of his triangle and test the truth of this for all possible exterior angles.

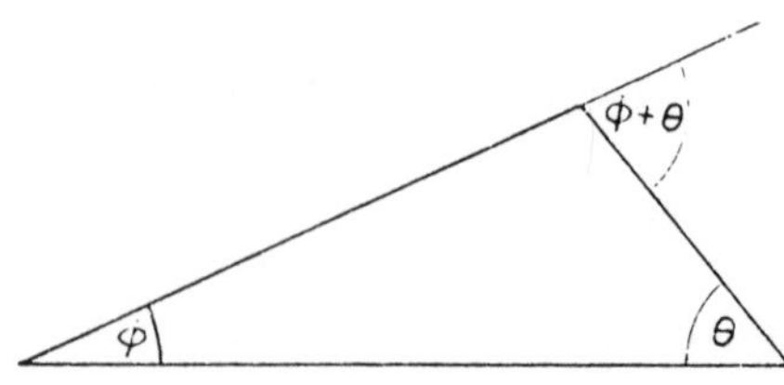

Fig. 43

If an angle between two sides of a triangle is bisected and the bisecting line produced to meet the opposite side, the opposite side is divided into two parts having lengths of the same ratio as the two sides which form the bisected angle. Thus, referring to Fig. 44, angle B is bisected and side b is divided into two parts x and y, and the ratio x to y is equal to the ratio a to c.

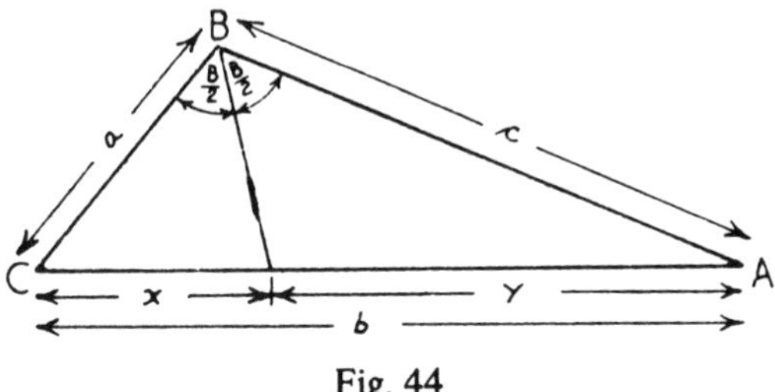

Fig. 44

TWO NON-PARALLEL CHORDS OF A CIRCLE

Referring to Fig. 45, the triangles CAB and CDB have the common base CB which is a chord of the circle, therefore the angles at the apices of these triangles where they touch the circumference of the circle are equal (see Fig. 49). Angle CAB = Angle CDB. Also, at the intersection of the two straight lines AB and CD, angle COA = angle BOD (see Fig. 41). Now considering triangles COA and BOD, since two angles of each of these are equal, the remaining angle of each must be equal, therefore angle ACO = angle DBO.

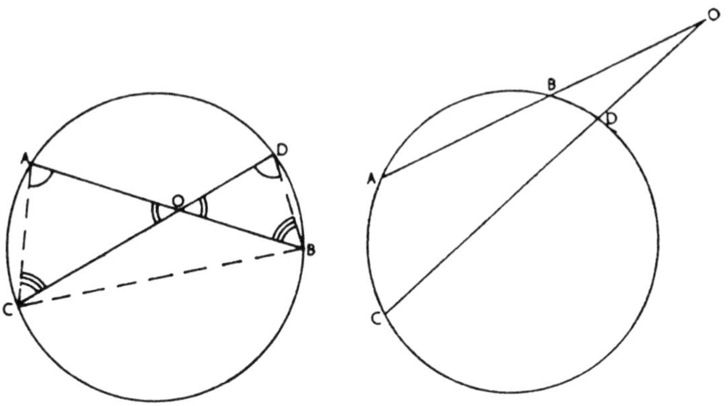

Fig. 45 Fig. 46

In such similar triangles, the sides opposite the corresponding angles must be in the same ratio, hence

$$\frac{AO}{CO} = \frac{DO}{BO}$$

$$\therefore AO \times BO = CO \times DO$$

where A and B are the points where the two ends of the chord AB touch the circle, C and D the points where the two ends of another chord CD touch the circle, and O is the point of intersection of these two chords.

This rule of crossed chords is also true when the intersection of the chords occurs outside the circle as illustrated in Fig. 46.

A useful application of the case where the chords intersect inside the circle is the estimation of the diameter of a circular vessel when a rule of sufficient length is not available. Thus, one chord is made the diameter itself and this cuts another chord perpendicular to it as in Fig. 47. As an example, let AB be the diameter of a boiler shell, a 1 metre straight-edge is laid horizontally across to form the chord CD, the vertical distance between bottom of shell and straight-edge is measured to be 100 mm, this is BO, then:

$$\begin{aligned} AO \times BO &= CO \times DO \\ AO \times 100 &= 500 \times 500 \\ AO &= 2500 \text{ mm} \\ \text{Diameter of shell} = AB &= AO + BO \\ &= 2500 + 100 \\ &= 2600 \text{ mm or } 2{\cdot}6 \text{ m} \end{aligned}$$

An interesting example of the case where the chords meet outside the circle is to estimate the distance of observation from a given point above the earth's surface to the horizon. Fig. 48 shows a circle representing the earth, AB is one chord which is the diameter, say 12 750 km, this is extended to the observation point O, let this be 24 m above the earth's surface. The other chord is imagined to be so small that it becomes tangential to the circle, i.e., the length of CD is nil and points C and D are at one. The extension of this chord meets the other at O. CO (or DO) is the distance from the point of observation to the horizon.

$$\begin{aligned} 24 \text{ metres} &= 0{\cdot}024 \text{ kilometre} \\ AO \times BO &= CO \times DO \\ (12\,750{\cdot}024) \times (0{\cdot}024) &= CO \times DO \end{aligned}$$

Adding 0·024 to 12 750 makes no difference in four-figure accuracy, i.e., 12 750·024 can be taken as 12 750.

$$CO \times DO = (CO)^2 \text{ because CO is equal to DO}$$

$$\therefore 12\,750 \times 0{\cdot}024 = (CO)^2$$

$$CO = \sqrt{12\,750 \times 0{\cdot}024}$$

$$= 17{\cdot}49 \text{ kilometres}$$

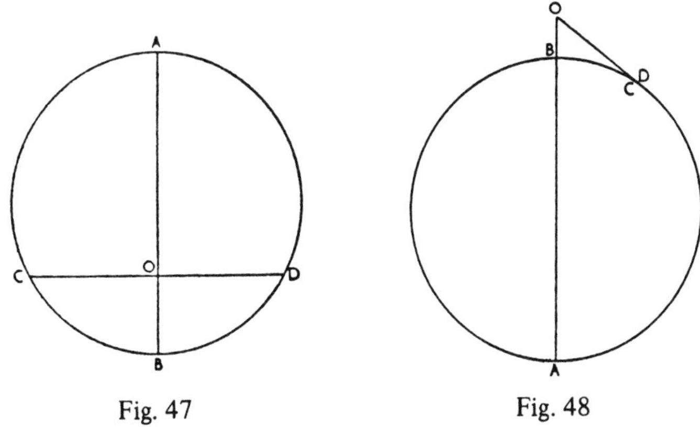

Fig. 47 Fig. 48

CYCLIC TRIANGLE ON COMMON BASES

The apex angles of all triangles at the circumference within the same segment of a circle and having the same chord as a base, are equal. Stated briefly: Angles in the same segment of a circle are equal. See Fig. 49.

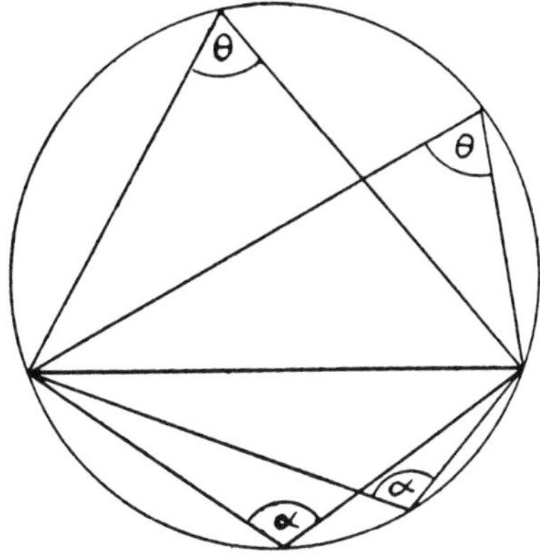

Fig. 49

If the base of the triangle is the diameter of the circle, then the apex angle at the circumference is 90 degrees. Stated briefly: The angle in a semi-circle is a right angle. See Fig. 50.

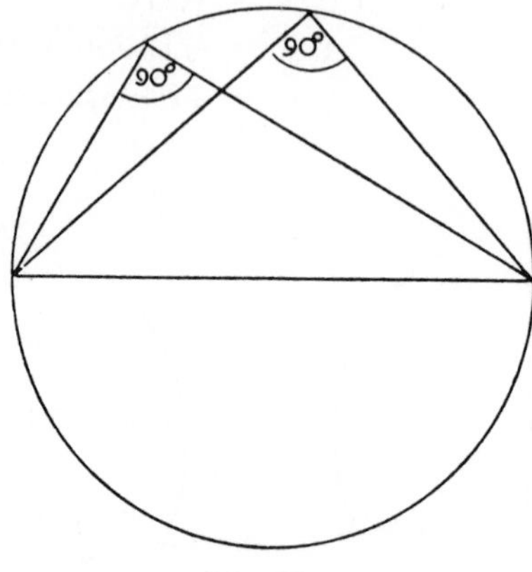

Fig. 50

If two triangles be drawn within the same segment of a circle and having one chord as their common base, the apex of one being at the centre of the circle and the apex of the other touching the circumference, the angle at the centre is twice the angle at the circumference. Briefly: The angle at the centre of a circle is double the angle at the circumference for triangles in the same segment standing on the same chord. See Fig. 51.

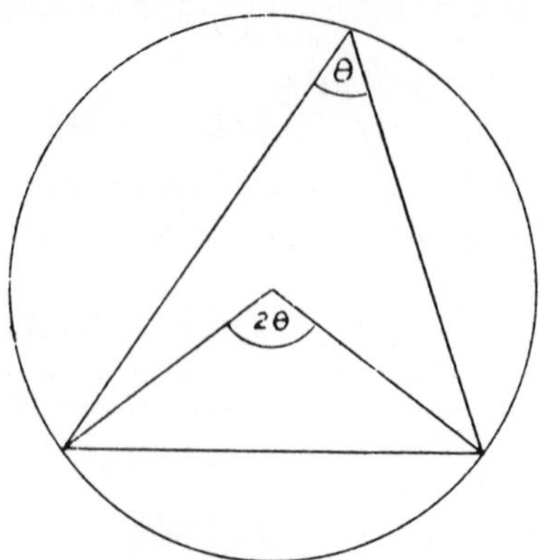

Fig. 51

The sum of either pair of opposite angles in a quadrilateral (i.e., a four-sided figure) inscribed in a circle is equal to 180 degrees. This is illustrated in Fig. 52 with measured angles. Briefly: The opposite angles of a cyclic quadrilateral are supplementary. It can also be seen that any exterior angle produced by extending one side, is equal to the opposite interior angle.

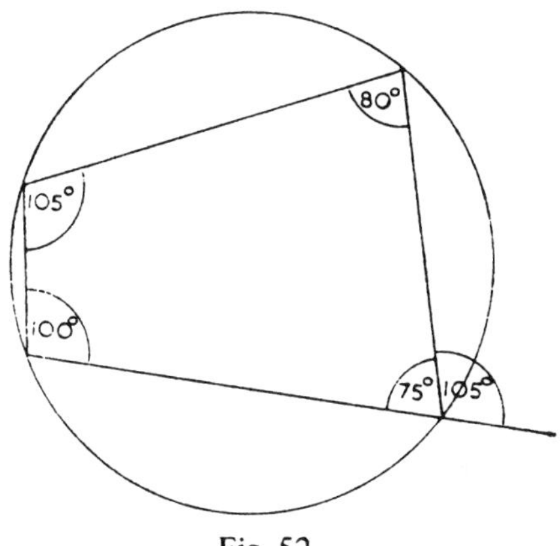

Fig. 52

As an exercise on the last few cases, draw a circle as large as possible depending upon the size of paper and compasses at hand. With Fig. 53 as reference, draw a chord AB, construct one triangle on this chord with its apex at the circle centre C, construct another two triangles on the same chord and in the same segment with their apices touching the circumference, such as at D and E.

Measure angles ACB, ADB and AEB.

ADB should be equal to AEB.

ACB should be double ADB.

With the same chord AB as a base, draw another two triangles in the other segment with their apices touching the circumference, such as at F and G.

Measure angles AFB and AGB.

AFB should be equal to AGB.

There are a number of cyclic quadrilaterals here, ADBF, ADBG, AEBF and AEBG.

Measure any remaining angles in these quadrilaterals and check that all opposite angles are supplementary.

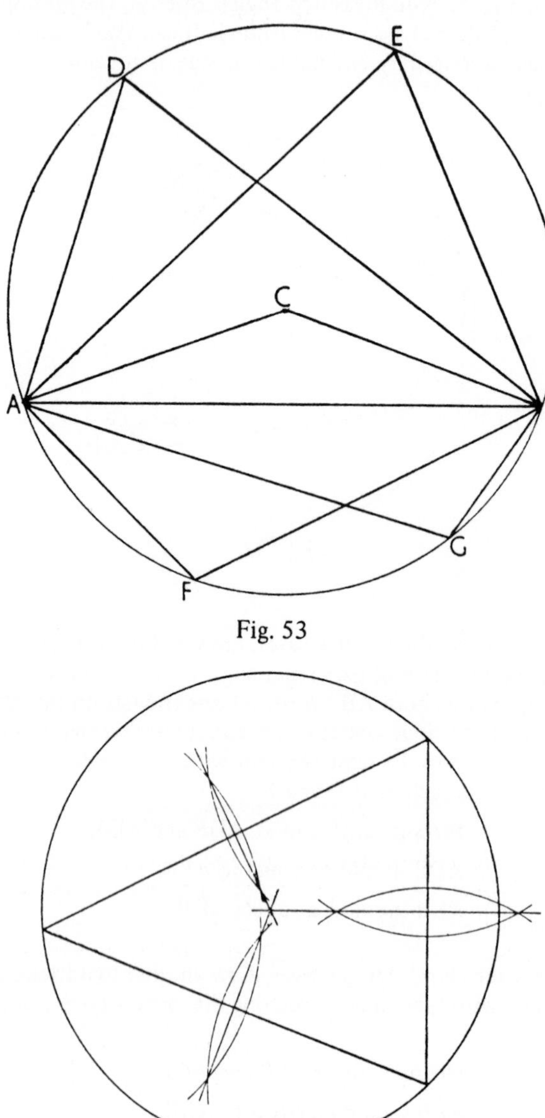

Fig. 53

Fig. 54

CIRCUMSCRIBED CIRCLE

If a triangle is drawn inside a circle with its three corners touching the circumference, the perpendicular bisectors of the three sides meet at the centre of the circle. See Fig. 54.

INSCRIBED CIRCLE

If a triangle is drawn around a circle such that the three sides of the triangle just touch the circumference of the circle, the bisectors of the three angles meet at the centre of the circle. See Fig. 55.

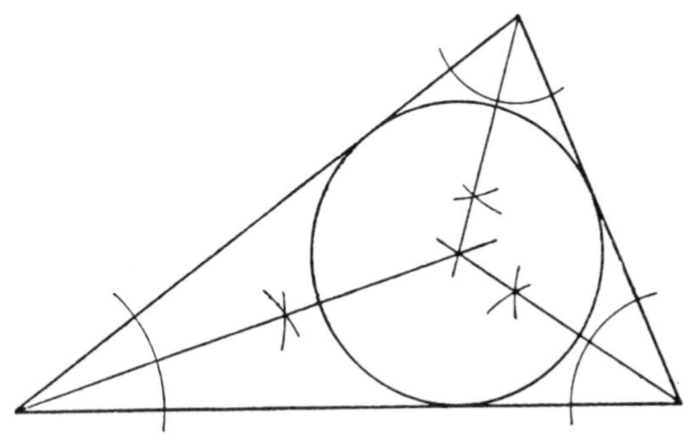

Fig. 55

SIMILAR TRIANGLES

Triangles are *similar* when they are equiangular (i.e., contain the same angles). It follows that their corresponding sides are in the same proportion and the shape of the triangles are the same.

To ascertain whether two triangles are similar, we can apply any of the following tests to see if:

(i) they are equiangular,

(ii) their corresponding sides are in proportion,

(iii) one angle in each triangle are equal and the sides containing these angles are in proportion.

Fig. 56 shows some pairs of similar triangles, in each pair triangle ABC is similar to triangle XYZ.

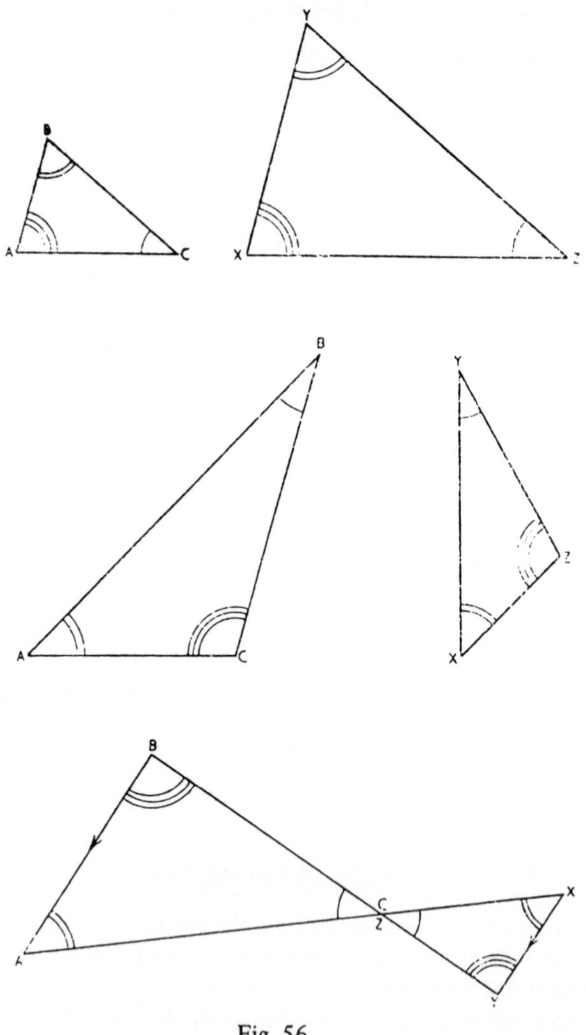

Fig. 56

A given triangle can be divided into two or more similar triangles by drawing one or more lines parallel to one of the sides. When lines are drawn parallel to one side, they divide the other two sides into parts of the same ratio. This is illustrated in Fig. 57, the usual method of indicating parallel lines by arrows is used.

ABC, XYZ and PQR are similar triangles and their sides are in the same ratio.

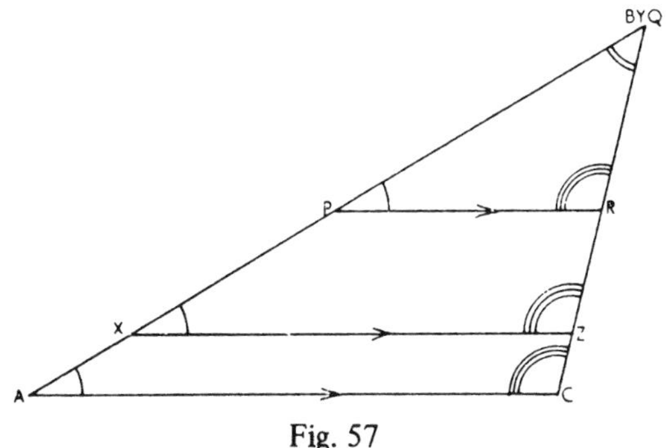

Fig. 57

Congruent triangles have the same shape and *size*, and thus contain the same area. Fig. 58 illustrates some congruent triangles.

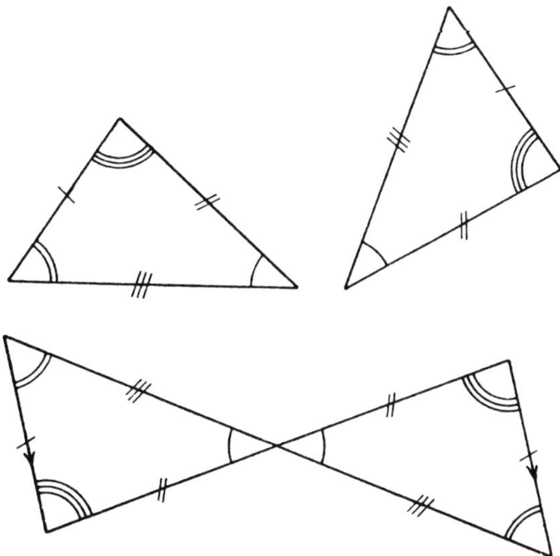

Fig. 58

TEST EXAMPLES 9

1. The bottom end of a derrick 11 m long is hinged at the base of a vertical mast. A topping lift 6 m long connects the top end of the derrick to a point on the mast at 7 m above the base. Draw the triangular framework of this jib crane to scale, measure the three internal angles and also the perpendicular distance from the base of the mast to a wire rope hanging vertically from the crane head.

2. Construct a triangle, two sides of which are 90 and 70 mm respectively and the angle included between these sides being 44 degrees. Measure the remaining side and angles.

3. Construct a triangle, ABC, such that AB = 48 mm, BC = 39 mm and the angle at A = 50°, measure the remaining side and angles.

4. A $\frac{1}{2}$ m straight-edge is laid horizontally inside an air vessel, at right angles to the longitudinal axis, and the distance from the centre of the straight-edge to the bottom of the vessel is measured to be 125 mm. Calculate the diameter of the air vessel.

5. The width of the water-level in a cylindrical vessel lying with its longitudinal axis horizontally, is 1·86 m and the inside diameter of the shell is 1·92 m. Calculate the depth of the water from water-level to bottom of vessel.

6. A trammel gauge supplied by the makers of an engine as a measurement of the diameter of a cylinder when new, was exactly 750 mm. The cylinder is now gauged and, with one end of the trammel held against the cylinder wall, there is 78 mm chord travel at the other end. Find the wear (increase in diameter).

7. Find the distance from an observation point 30 m above sea level, to the horizon, assuming the earth is 12 750 km diameter.

8. Construct a triangle of sides 56, 48 and 40 mm long respectively, bisect each side to find the centre of a circle which will pass through the three points of the triangle, draw the circle and state its radius.

9. Construct a triangle of sides 68, 58 and 42 mm long respectively, bisect the angles to find the centre of a circle which will touch the three sides of the triangle, draw the circle and state its radius.

CHAPTER 10

SOLUTION OF TRIANGLES BY CALCULATION

To solve a triangle means to find the other sides and angles of the triangle remaining from the data given. Of the three sides and three angles of any triangle, at least one side must be given with two other quantities; the two other quantities may be the other two sides, or one other side and one angle, or two of the angles.

In the last chapter it was shown how triangles could be solved by drawing them to scale and measuring the unkown quantities, this method has its limitations with regard to accuracy and we will now see how the required quantities can be calculated. Still however, a sketch of the triangle is necessary to understand the problem clearly and the student will find that to draw the triangle to scale takes very little longer than making a free-hand sketch, therefore he is well advised to make a scale drawing, especially in the early stages of learning, and thus produce a graphical check on his calculations.

RIGHT ANGLED TRIANGLES

The simplest of triangles is the right angled triangle. It was shown in Chapter 8 that the sides and the ratio of any two of them are given special names. Thus, Fig. 59 is a right angled triangle because one of its

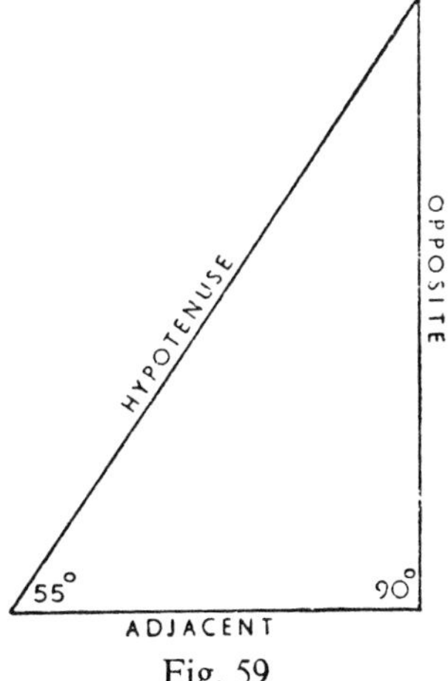

Fig. 59

angles is 90 degrees. The side opposite the right angle is the longest side and called the hypotenuse, the other two sides are called the opposite or the adjacent depending upon which angle is being dealt with. In Fig. 59 one of the angles (other than the right angle) is given, therefore the side opposite this angle is referred to as the opposite and the other side is the adjacent.

The ratio of the sides are:

$$\text{Sine} = \frac{\text{opposite}}{\text{hypotenuse}}$$

$$\text{Cosine} = \frac{\text{adjacent}}{\text{hypotenuse}}$$

$$\text{Tangent} = \frac{\text{opposite}}{\text{adjacent}}$$

After the notes on right angled triangles in Chapter 8 it should now only be necessary to show a few worked examples relating to them.

Example. One of the angles in a right angled triangle is 59° 42′ and the hypotenuse is 55 mm. Find the remaining angle and sides.

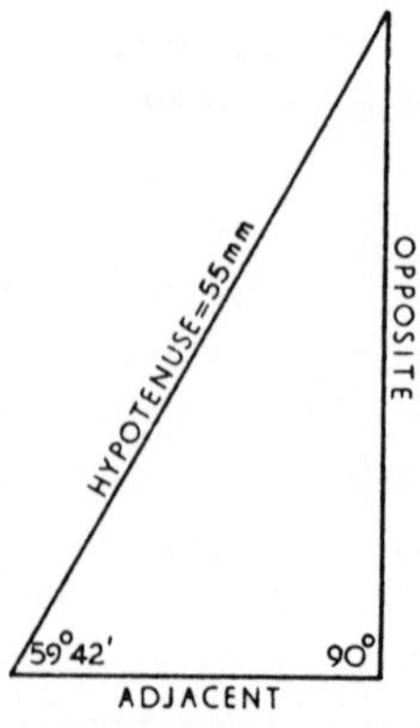

Fig. 60

Referring to Fig. 60

$$\frac{\text{opposite}}{\text{hypotenuse}} = \sin 59^\circ\ 42'$$

$$\begin{aligned} \text{opposite} &= \text{hypotenuse} \times \sin 59^\circ\ 42' \\ &= 55 \times 0{\cdot}8634 \\ &= 47{\cdot}49 \text{ mm} \end{aligned}$$

The remaining side can now be found by the theorem of Pythagoras:

$$\begin{aligned} (\text{hypotenuse})^2 &= (\text{opposite})^2 + (\text{adjacent})^2 \\ \therefore \text{adjacent} &= \sqrt{55^2 - 47{\cdot}49^2} \\ &= 27{\cdot}73 \text{ mm} \end{aligned}$$

or by using the cosine ratio:

$$\frac{\text{adjacent}}{\text{hypotenuse}} = \cos 59^\circ\ 42'$$

$$\begin{aligned} \text{adjacent} &= 55 \times \cos 59^\circ\ 42' \\ &= 27{\cdot}74 \text{ mm} \end{aligned}$$

There may be a very small difference in the fourth figure due to different tables, such small inaccuracies are negligible.

Since the three angles of any triangle add up to 180 degrees, the third angle is found by subtracting the sum of the other two from 180.

$$\begin{aligned} \text{Third angle} &= 180^\circ - (90^\circ + 59^\circ\ 42') \\ &= 90^\circ - 59^\circ\ 42' \\ &= 30^\circ\ 18' \end{aligned}$$

Example. A light on a cliff is viewed from a position at sea, the angle of elevation being 32° 15′. At a point 100 m further away from the cliff and in direct line with the first observation point, the light is viewed again and the angle of elevation is now 20° 44′. Find (*a*) the height of the light above sea level, and (*b*) the horizontal distance from the second observation point to the light.

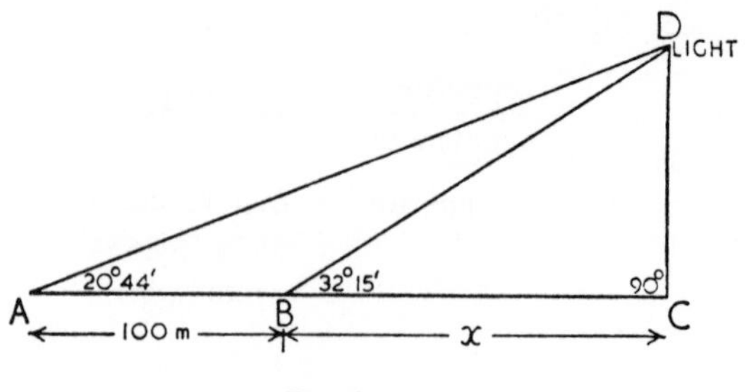

Fig. 61

Referring to Fig. 61, $\frac{CD}{x} = \tan 32° 15'$

$\therefore CD = x \times 0{\cdot}6309$ ... ... (i)

$\frac{CD}{100 + x} = \tan 20° 44'$

$\therefore CD = (100 + x) \times 0{\cdot}3786$

$CD = 37{\cdot}86 + 0{\cdot}3786x$ ... (ii)

From (i) and (ii),

$$0{\cdot}6309x = 37{\cdot}86 + 0{\cdot}3786x$$
$$0{\cdot}2523x = 37{\cdot}86$$
$$x = 150{\cdot}1 \text{ m}$$
$$AC = 100 + 150{\cdot}1$$
$$= 250{\cdot}1 \text{ m} \quad \text{Ans. } (b)$$

From (i),

$$CD = x \times 0{\cdot}6309$$
$$= 150{\cdot}1 \times 0{\cdot}6309$$
$$= 94{\cdot}69 \text{ m} \quad \text{Ans. } (a)$$

TRIANGLES OTHER THAN RIGHT ANGLED

If the triangle does not contain one angle of 90 degrees, the sine, cosine and tangent ratios of the sides cannot be applied as in the above right angled triangles. Other rules are employed which will now be explained.

The commonest method of notation of triangles is to let capital letters represent the angles and their corresponding small letters represent the sides opposite to these angles. Thus in Fig. 62 the three corners are lettered A, B and C, this means that the three angles at these corners are represented by these letters. The side opposite angle A is represented by its lower case a, the side opposite angle B is represented by b, and the side opposite angle C is represented by c. A perpendiuclar erected from the base of the triangle to its apex is usually denoted by h, this perpendicular divides the triangle into two right angled triangles which can often provide a solution by the knowledge of right angled triangles only.

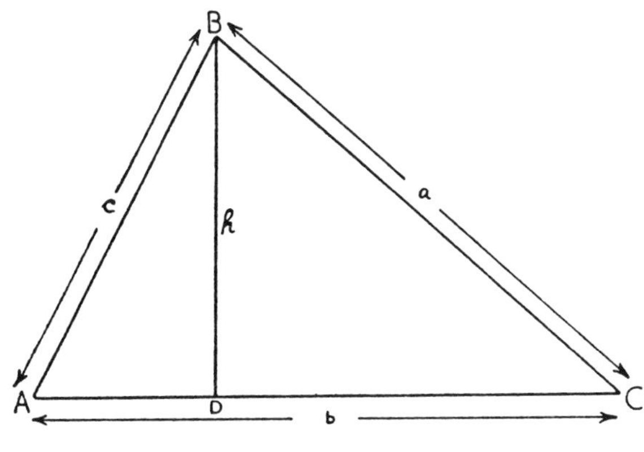

Fig. 62

SINE RULE

Referring to Fig. 62, and considering the right angled triangular part BCD,

$$\frac{h}{a} = \sin C \quad \therefore h = a \sin C$$

Considering the right angled triangular part ABD,

$$\frac{h}{c} = \sin A \quad \therefore h = c \sin A$$

hence,

$$a \sin C = c \sin A$$

or,

$$\frac{a}{\sin A} = \frac{c}{\sin C} \quad \ldots \quad \ldots \quad \ldots \quad \text{(i)}$$

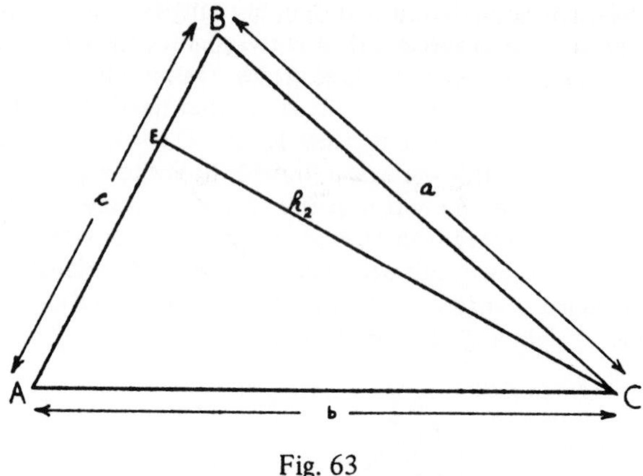

Fig. 63

Fig. 63 represents the same triangle as Fig. 62, but with a perpendicular h_2 from side c to angle C.

Considering the right angled triangular part BCE,

$$\frac{h_2}{a} = \sin B \quad \therefore \; h_2 = a \sin B$$

Considering the right angled triangular part ACE,

$$\frac{h_2}{b} = \sin A \quad \therefore \; h_2 = b \sin A$$

hence,

$$a \sin B = b \sin A$$

or,

$$\frac{a}{\sin A} = \frac{b}{\sin B} \qquad \ldots \quad \ldots \quad \ldots \qquad \text{(ii)}$$

From (i) and (ii) we have:

$$\frac{a}{\sin A} = \frac{b}{\sin B} = \frac{c}{\sin C}$$

This is the SINE RULE which is very important and most useful in trigonometrical calculations. Note that the length of any side is proportional to the sine of the angle opposite that side, the longest side is opposite the largest angle, and the shortest side is opposite the smallest angle.

Example. The vertical post of a jib crane is 10 m long. The angle between jib and post is 40 degrees, and between jib and tie the angle is 45 degrees. Find the length of the tie and the length of the jib.

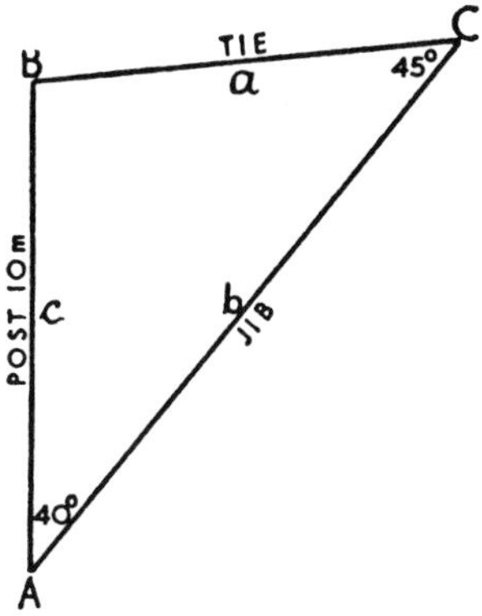

Fig. 64

Referring to Fig. 64,

$$\frac{a}{\sin A} = \frac{c}{\sin C}$$

$$a = \frac{c \sin A}{\sin C}$$

$$= \frac{10 \times 0{\cdot}6428}{0{\cdot}7071} = 9{\cdot}088 \text{ m}$$

$$B = 180 - (40 + 45) = 95^\circ$$

$$\frac{b}{\sin B} = \frac{c}{\sin C}$$

$$b = \frac{c \sin B}{\sin C}$$

$$= \frac{10 \times 0{\cdot}9962}{0{\cdot}7071} = 14{\cdot}09 \text{ m}$$

Length of tie = 9·088 m
,, ,, jib = **14·09 m** Ans.

Example. Two ships leave the same port at the same time on courses which diverge at 29 degrees. When one ship had travelled 40 nautical miles the two ships were then 21 nautical miles apart, find how far the other ship had travelled from port.

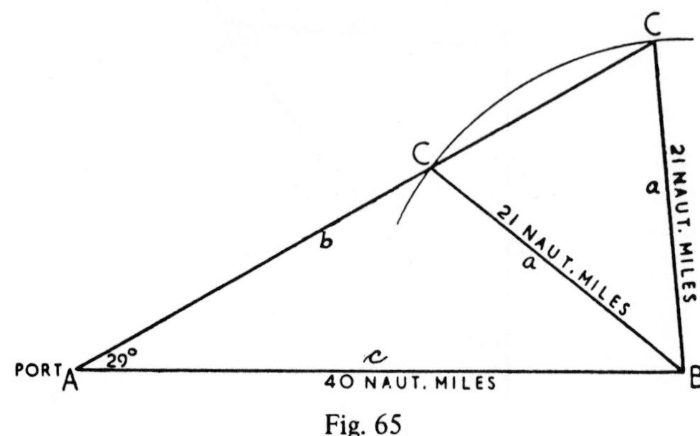

Fig. 65

Referring to Fig. 65,

$$\frac{a}{\sin A} = \frac{c}{\sin C}$$

$$\sin C = \frac{c \sin A}{a}$$

$$= \frac{40 \times 0{\cdot}4848}{21} = 0{\cdot}9234$$

From tables, 0·9234 is the sine of 67° 26′
but it is also the sine of 180° – 67° 26′ = 112° 34′

Therefore there are two possible answers to this problem, as illustrated in Fig. 65.

If $C = 67° 26'$:

$$B = 180° - (29° + 67° 26') = 83° 34'$$

$$\frac{a}{\sin A} = \frac{b}{\sin B}$$

$$b = \frac{a \sin B}{\sin A}$$

$$= \frac{21 \times 0{\cdot}9937}{0{\cdot}4848} = 43{\cdot}04 \qquad \text{(i)}$$

If $C = 112^\circ 34'$

$$B = 180^\circ - (29^\circ + 112^\circ 34') = 38^\circ 26'$$

$$b = \frac{a \sin B}{\sin A}$$

$$= \frac{21 \times 0{\cdot}6216}{0{\cdot}4848} = 26{\cdot}93 \qquad \text{(ii)}$$

For the two ships to be 21 nautical miles apart and one having travelled 40 nautical miles, the other ship has travelled:

either 43·04 or 26·93 nautical miles. Ans.

COSINE RULE

Inspection of the Sine Rule will show that it would not solve a triangle when two sides and the included angle between these sides were the only given quantities, nor if the given quantities were the three sides without any angles. Such cases can be solved by the Cosine Rule which will now be explained.

Referring to Fig. 62 and considering the right angled triangular part ABD,

let AD be represented by x, then by the Theorem of Pythagoras,

$$h^2 = c^2 - x^2 \quad \ldots \quad \ldots \quad \ldots \qquad \text{(i)}$$

Now considering the right angled triangular part BCD,

$$\text{DC} = b - x$$
$$h^2 = a^2 - (b - x)^2$$
$$= a^2 - (b^2 - 2bx + x^2)$$
$$= a^2 - b^2 + 2bx - x^2 \qquad \text{(ii)}$$

From (i) and (ii)

$$a^2 - b^2 + 2bx - x^2 = c^2 - x^2$$
$$a^2 = b^2 + c^2 - 2bx$$

but $x = c \times \cos A$, therefore,

$$a^2 = b^2 + c^2 - 2bc \cos A$$

This is the COSINE RULE and equally important as the sine rule for solving triangles. It will be obvious that this can be written in either of the three forms:

$$a^2 = b^2 + c^2 - 2bc \cos A$$
$$b^2 = a^2 + c^2 - 2ac \cos B$$
$$c^2 = a^2 + b^2 - 2ab \cos C$$

Note that the last term is a negative value, if the angle involved is between 90 and 180 degrees, its cosine is negative, and $-2bc$ multiplied by a minus quantity will produce a positive term.

Transposing the cosine rule to find one angle given the three sides:

$$a^2 = b^2 + c^2 - 2bc \cos A$$
$$2bc \cos A = b^2 + c^2 - a^2$$
$$\cos A = \frac{b^2 + c^2 - a^2}{2bc}$$

this can also be written:

$$\cos B = \frac{a^2 + c^2 - b^2}{2ac}$$
$$\text{or, } \cos C = \frac{a^2 + b^2 - c^2}{2ab}$$

Example. A ship steaming due North at 16 knots runs into a 5 knot current moving North-East. Find the resultant speed and direction of the ship.

Ref. Fig. 66,

$$\text{Angle } A = 90 + 45 = 135°$$
$$a^2 = b^2 + c^2 - 2bc \cos A$$
$$= 16^2 + 5^2 - 2 \times 16 \times 5 \times \cos 135°$$
$$= 256 + 25 - 160 \times (-0{\cdot}7071)$$
$$= 256 + 25 + 113{\cdot}1$$
$$a = \sqrt{394{\cdot}1}$$
$$= 19{\cdot}85 \text{ knots}$$
$$\frac{a}{\sin A} = \frac{c}{\sin C}$$
$$\sin C = \frac{c \times \sin A}{a}$$
$$= \frac{5 \times 0{\cdot}7071}{19{\cdot}85} = 0{\cdot}1781$$
$$C = 10° \ 15'$$

Resultant speed = 19·85 knots
,, direction = 10° 15′ East of North } Ans.

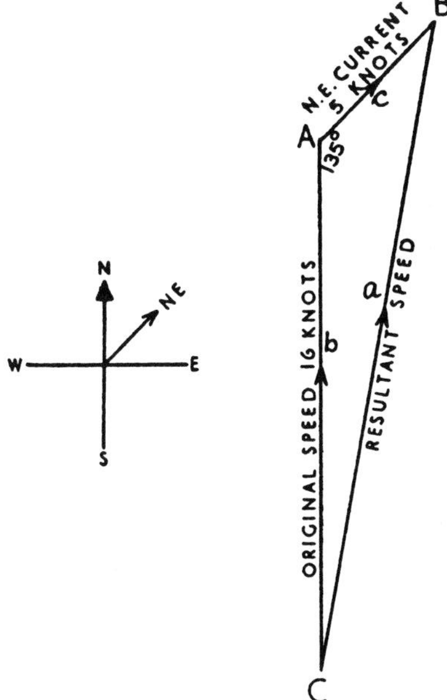

Fig. 66

Example. The three sides of a triangle measure 80, 60 and 40 mm respectively. Find the angles.

Working in centimetres:

$$\cos A = \frac{b^2 + c^2 - a^2}{2bc}$$

$$= \frac{4^2 + 8^2 - 6^2}{2 \times 4 \times 8} = 0{\cdot}6875$$

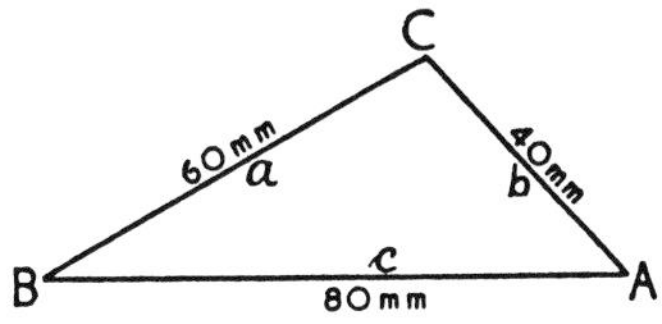

Fig. 67

$$\text{Angle } A = 46° \ 34'$$

$$\frac{a}{\sin A} = \frac{b}{\sin B}$$

$$\sin B = \frac{b \sin A}{a}$$

$$= \frac{4 \times 0{\cdot}7262}{6} = 0{\cdot}4841$$

$$\text{Angle } B = 28° \ 57'$$

$$\text{Angle } C = 180° - (46° \ 34' + 28° \ 57')$$

$$= 180° - 75° \ 31'$$

$$= 104° \ 29'$$

The three angles are:

46° 34′, 28° 57′ and 104° 29′ Ans.

Summarising the formulae given so far for solving triangles, the sine, cosine and tangent ratios of sides are used for right angled triangles, and either the sine rule or cosine rule for other triangles. The sine rule is easier to calculate than the cosine rule and is therefore used wherever possible, the two cases where the sine rule cannot be applied and the cosine rule becomes necessary are when the only data available is:

(i) two sides and the angle between these sides,

(ii) the three sides.

AREAS OF TRIANGLES

It can be seen from any of the previous illustrations of triangles, or from Figs. 68 and 69 that the area of a triangle is half of the area of its circumscribing rectangle, therefore,

$$\text{Area of triangle} = \tfrac{1}{2}(\text{base} \times \text{perpendicular height})$$

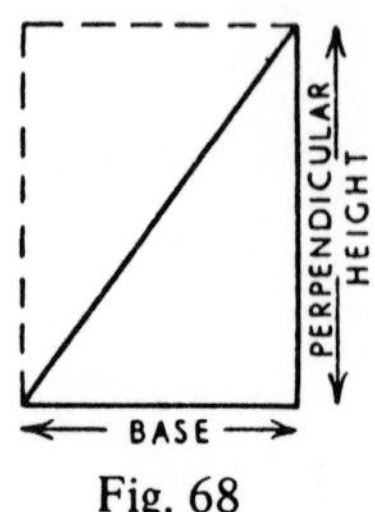

Fig. 68

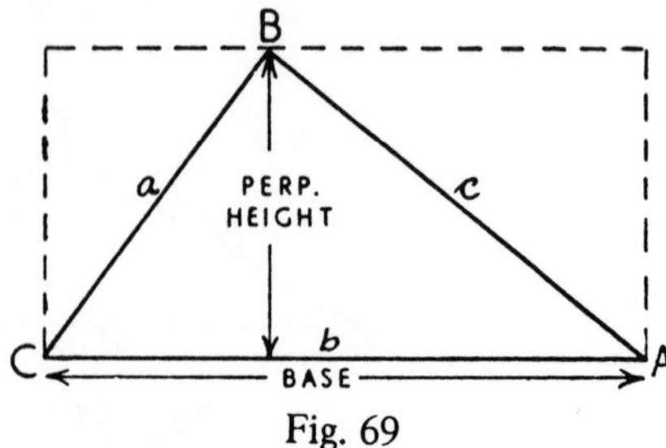

Fig. 69

Referring to Fig. 69,

$$\text{Perpendicular height} = a \times \sin C$$
$$\text{Area} = \tfrac{1}{2}(b \times a \sin C)$$
$$\therefore \text{Area} = \frac{ab \sin C}{2}$$

Note this formula, in words it is "half the product of two sides and the sine of the angle between these sides". An easy formula to apply and a very useful one.

Another important formula for finding the area of a triangle is one which calculates the area directly from the three sides of the triangle:

$$\text{Area} = \sqrt{s(s-a)(s-b)(s-c)}$$

where a, b and c = the three lengths of the sides respectively,

and s = semisum of sides,

$$= \tfrac{1}{2}(a + b + c)$$

Example. The three sides of a triangle measure 80, 100 and 140 mm respectively, find the area enclosed.

Working in centimetres:

$$s = \tfrac{1}{2}(a + b + c)$$
$$= \tfrac{1}{2}(8 + 10 + 14)$$
$$= 16$$
$$s - a = 16 - 8 = 8$$
$$s - b = 16 - 10 = 6$$
$$s - c = 16 - 14 = 2$$
$$\text{Area} = \sqrt{s(s-a)(s-b)(s-c)}$$
$$= \sqrt{16 \times 8 \times 6 \times 2} = \sqrt{1536}$$
$$= 39{\cdot}2\ \text{cm}^2 \quad \text{Ans.}$$

$$= 39{\cdot}2 \times 10^2\ \text{mm}^2 \text{ or } 3{\cdot}92 \times 10^3\ \text{mm}^2 \quad \text{Ans.}$$

EQUILATERAL TRIANGLE

An equilateral triangle is one which has three equal sides and three equal angles. Each angle is therefore 60 degrees (Fig. 70).

$$\text{Perpendicular height} = \text{side} \times \sin 60^\circ$$
$$= \text{side} \times 0{\cdot}866$$
$$\text{Area} = \tfrac{1}{2}(\text{base} \times \text{perp. ht.})$$
$$= \tfrac{1}{2}(\text{side} \times \text{side} \times 0{\cdot}866)$$
$$= 0{\cdot}433\ \text{side}^2$$

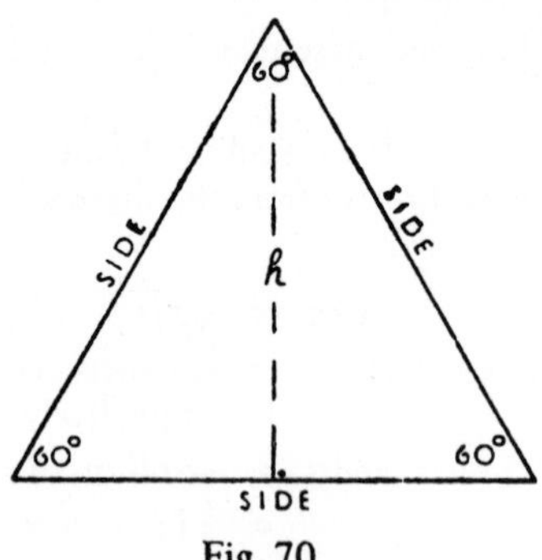

Fig. 70

ISOSCELES TRIANGLE

An isosceles triangle is one which had two equal sides and two equal angles.

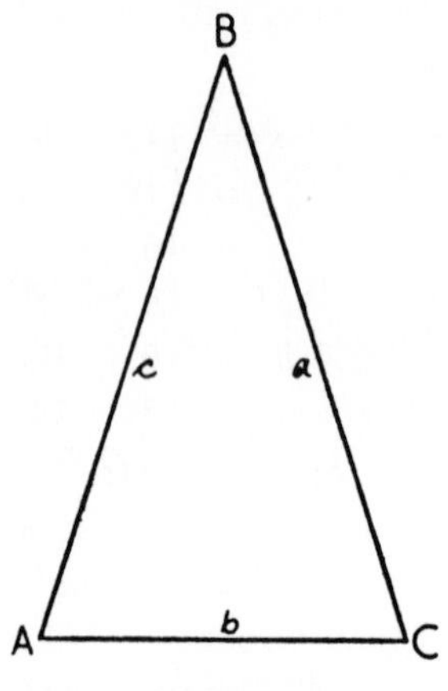

Fig. 71

In Fig. 71 angles A and C are equal, therefore the lengths of the sides a and c are equal.

$$A = C = \tfrac{1}{2}(180 - B)$$

It is thus an easy matter to calculate the perpendicular height or other data required to find the area by any of the general formulae.

CIRCUMSCRIBED CIRCLE

To find the radius of a circle whose circumference will pass through the three angular points of a triangle. This construction was shown in Chapter 9 (Fig. 54). Let the given triangle be ABC as shown in Fig. 72.

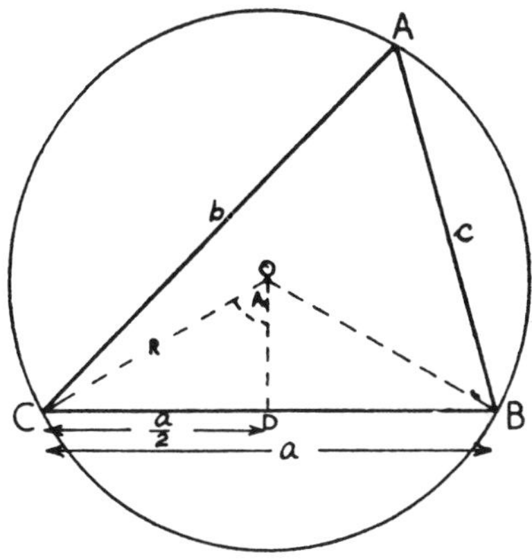

Fig. 72

Since the angle at the centre is twice the angle at the apex, that is, angle COB $= 2A$, then angle COD $= A$.

$$\frac{a}{2} \div R = \sin A$$

$$\frac{a}{2R} = \sin A$$

$$\therefore 2R = \frac{a}{\sin A}$$

Therefore, by sine rule,

$$2R = \frac{a}{\sin A} = \frac{b}{\sin B} = \frac{c}{\sin C} \quad \ldots \quad \ldots \quad \text{(i)}$$

$$\text{Also, area of triangle} = \tfrac{1}{2}ab \sin C$$

$$\text{Substituting the value of } \sin C = \frac{c}{2R} \text{ we have}$$

$$\text{Area of triangle} = \tfrac{1}{2} \times a \times b \times \frac{c}{2R}$$

$$\therefore R = \frac{abc}{4 \times \text{area of triangle}} \quad \ldots \quad \text{(ii)}$$

INSCRIBED CIRCLE

To find the radius of a circle inscribed within a triangle, this construction was shown in Chapter 9 (Fig. 55).

Let R = radius of the circle

Let a, b and c = lengths of the sides of the triangle.

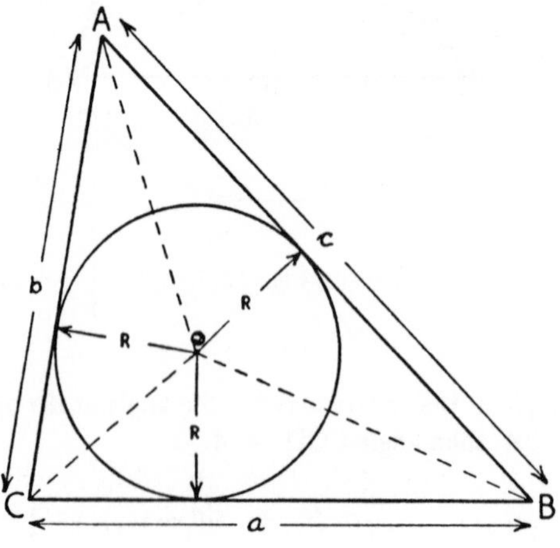

Fig. 73

Dividing the triangle ABC into three small triangles BOC, COA and AOB as shown in Fig. 73, the radius of the circle R is the perpendicular height in each of these.

Finding the area of each triangle by the rule, area $= \frac{1}{2}$ (base $\times$ perp. height) and then adding together we have,

$$\text{Area of BOC} = \tfrac{1}{2} \times a \times R$$
$$\text{,, COA} = \tfrac{1}{2} \times b \times R$$
$$\text{,, AOB} = \tfrac{1}{2} \times c \times R$$
$$\therefore \text{Area of whole triangle ABC} = \tfrac{1}{2} R(a + b + c)$$
$$\therefore R = \frac{2 \times \text{area of triangle}}{a + b + c}$$
$$\text{or, } R = \frac{2 \times \text{area of triangle}}{\text{perimeter of triangle}}$$

COMPOUND ANGLES

In some problems it is often convenient to express the relationship between the trigonometrical ratios of a compound angle and those of the two single component angles.

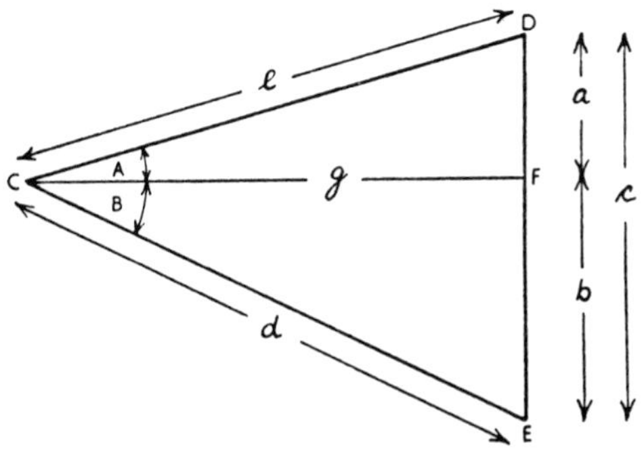

Fig. 74

Referring to Fig. 74,

$$\text{Area of CED} = \text{area of CFD} + \text{area of CFE}$$
$$\tfrac{1}{2}\,[ed \sin (A + B)] = \tfrac{1}{2}\,(ge \sin A) + \tfrac{1}{2}\,(gd \sin B)$$

Dividing throughout by $\frac{1}{2}ed$,

$$\sin (A + B) = \frac{g}{d} \sin A + \frac{g}{e} \sin B$$
$$= \cos B \sin A + \cos A \sin B$$

Usually written,

$$\sin(A + B) = \sin A \cos B + \cos A \sin B \qquad \text{(i)}$$

By a similar process it can be shown that,

$$\sin(A - B) = \sin A \cos B - \cos A \sin B \qquad \text{(ii)}$$

Again referring to Fig. 74 and applying the cosine rule,

$$\cos C = \frac{e^2 + d^2 - c^2}{2ed}$$

$$\cos(A + B) = \frac{e^2 + d^2 - (a + b)^2}{2ed}$$

$$= \frac{e^2 + d^2 - a^2 - 2ab - b^2}{2ed}$$

but, by Pythagoras, $e^2 - a^2 = g^2$
and $d^2 - b^2 = g^2$

$$\therefore \cos(A + B) = \frac{2g^2 - 2ab}{2ed}$$

$$= \frac{g^2}{ed} - \frac{ab}{ed}$$

$$= \frac{g}{e} \times \frac{g}{d} - \frac{a}{e} \times \frac{b}{d}$$

$$= \cos A \times \cos B - \sin A \times \sin B$$

Usually written,

$$\cos(A + B) = \cos A \cos B - \sin A \sin B \qquad \text{(iii)}$$

By a similar kind of process it can also be shown that,

$$\cos(A - B) = \cos A \cos B + \sin A \sin B \qquad \text{(iv)}$$

The tangent of an angle being the sine divided by its cosine, the relationship for the tangent of a compound angle can be obtained by dividing $\sin(A + B)$ by $\cos(A + B)$,

$$\tan(A + B) = \frac{\sin(A + B)}{\cos(A + B)}$$

$$= \frac{\sin A \cos B + \cos A \sin B}{\cos A \cos B - \sin A \sin B}$$

Dividing all terms, top and bottom, by $\cos A \cos B$,

$$= \frac{\dfrac{\sin A \cos B}{\cos A \cos B} + \dfrac{\cos A \sin B}{\cos A \cos B}}{\dfrac{\cos A \cos B}{\cos A \cos B} - \dfrac{\sin A \sin B}{\cos A \cos B}}$$

$$= \frac{\tan A + \tan B}{1 - \tan A \tan B} \qquad \ldots \qquad \text{(v)}$$

and by a similar process,

$$\tan(A - B) = \frac{\tan A - \tan B}{1 + \tan A \tan B} \qquad \ldots \qquad \text{(vi)}$$

Collecting the above formulae we have:

$$\sin(A \pm B) = \sin A \cos B \pm \cos A \sin B$$
$$\cos(A \pm B) = \cos A \cos B \mp \sin A \sin B$$
$$\tan(A \pm B) = \frac{\tan A \pm \tan B}{1 \mp \tan A \tan B}$$

DOUBLE ANGLES

Formulae expressing the relationship between the ratios of angles and double the angles are useful and can be obtained from those above by letting $A = B$, thus,

$$\sin(A + B) = \sin A \cos B + \cos A \sin B$$

$B = A$, therefore,

$$\sin 2A = \sin A \cos A + \cos A \sin A$$
$$\sin 2A = 2 \sin A \cos A \qquad \ldots \qquad \ldots \qquad \text{(i)}$$

$$\cos(A + B) = \cos A \cos B - \sin A \sin B$$

$B = A$, therefore,

$$\cos 2A = \cos A \cos A - \sin A \sin A$$
$$\cos 2A = \cos^2 A - \sin^2 A \qquad \ldots \qquad \text{(iia)}$$

Also, since $\sin^2 A + \cos^2 A = 1$ (see Chapter 8)

then, $\cos^2 A = 1 - \sin^2 A$

and, $\sin^2 A = 1 - \cos^2 A$

Substituting for $\cos^2 A$ and $\sin^2 A$ in turn into (iia):

$$\begin{aligned}\cos 2A &= \cos^2 A - \sin^2 A \\ &= (1 - \sin^2 A) - \sin^2 A \\ \cos 2A &= 1 - 2\sin^2 A \quad \ldots \quad \ldots \quad \text{(iib)}\end{aligned}$$

and,

$$\begin{aligned}\cos 2A &= \cos^2 A - \sin^2 A \\ &= \cos^2 A - (1 - \cos^2 A) \\ &= \cos^2 A - 1 + \cos^2 A \\ \cos 2A &= 2\cos^2 A - 1 \quad \ldots \quad \ldots \quad \text{(iic)}\end{aligned}$$

$$\tan (A + B) = \frac{\tan A + \tan B}{1 - \tan A \tan B}$$

$B = A$, therefore,

$$\tan 2A = \frac{\tan A + \tan A}{1 - \tan A \tan A}$$

$$\tan 2A = \frac{2\tan A}{1 - \tan^2 A} \quad \ldots \quad \ldots \quad \text{(iii)}$$

TEST EXAMPLES 10

1. The top of a vertical mast is viewed from a position at 15 m from its base on a level ground and the angle of elevation measured to be 45° 34′. Find the height of the mast.

2. A boat is sighted from a point on a cliff 95 m above the sea, the angle of depression of the line of view being 14° 25′. Find (i) the horizontal distance from cliff to boat, and (ii) the distance from observation point to boat.

3. From a ship steaming due East at 18 knots a lighthouse is sighted and appeared to be in the direction 58° 32′ East of North from the ship. Half an hour later, the lighthouse appears due North of the ship, what is now the distance between ship and lighthouse?

4. A point A is 880 m North and 440 m East of a point B. Another point C is 220 m North-West of B. Find the distance from A to C and the relative direction of C from A.

5. A right angled triangular plate has one angle of 28° 37′ and the length of the hypotenuse is 120 mm. Find the lengths of the other two sides and the area of the triangle.

6. The top of a flagstaff is viewed from a point on a level ground at some distance from the foot of the staff, and the angle of elevation is 48° 30′. At another point 10 m further away from the foot, the staff top is viewed again and this time the angle of elevation is 37° 38′. Find the distance from the foot of staff to the first observation point and also the height of the flagstaff.

7. The length of the sides of a cube is 60 mm. Find the length of the diagonal across the face and the length of the cross diagonal from one corner to the opposite corner passing through the centre of the cube.

8. A ship steaming at $17\frac{1}{2}$ knots due East, runs into a current moving 30° West of North, and the resultant speed of the ship was 16 knots. Find the probable speed of the current and direction of the ship.

9. A ship steaming due South at 17 knots runs into a 4 knot current running 50° East of North. Find the resultant speed and direction of the ship.

10. The length of the vertical post of a jib crane is 15 m. The angle between jib and post is 35° 30′ and between tie and post the angle is 105° 30′. Calculate the lengths of the jib and tie.

11. In a triangle ABC, angle A is 60 degrees, length a is 30 mm and length b is 27 mm. Find angles B and C and length c.

12. In a reciprocating engine of 800 mm stroke, the connecting rod is 1600 mm long. Find how far the crosshead has moved from the top of its stroke when the crank is 35 degrees past its top dead centre.

13. The fuel valve of a diesel engine closes when the piston has moved one-tenth of its stroke from the top. If the length of the connecting rod is twice the length of the stroke, find the angular position of the crank from top dead centre when the valve closes.

14. The length of the connecting rod of a 250 mm stroke reciprocating engine is 480 mm. Find the length of a perpendicular drawn from the connecting rod to the centre of the shaft when the crank is 116 degrees past top dead centre.

15. Find the lengths of the sides a and b of a triangle ABC wherein angle A = 38° 25′, angle B = 84° 14′, and side c = 35 mm.

16. Two ships leave the same port at the same time. One ship sails at 14 knots in a direction 20° 12′ East of South, the other at 16 knots due West. Find the distance between the ships after two hours.

17. Two ships approach a port, their courses converging at an angle of 23 degrees. At a certain time one ship is twice as far from port as the other and their distance apart is 32 nautical miles; find how far each ship is from port.

18. Two ships are 50 nautical miles apart and steaming towards the same port at the same speed of 18 knots, on courses which converge at 73° 39′. If one ship will arrive in port half-an-hour before the other, find their distances from port.

19. Three sides of a triangle measure 16·4, 10·2 and 9·8 m respectively. Find the angles.

20. Two sides of a triangle measure 6·5 and 7·5 m respectively and the angle included between these two sides is 46° 51′. Find the area of the triangle in square metres.

21. Find the area in square centimetres of a triangular plate of sides 71, 42 and 53 mm long respectively.

22. The area of an equilateral triangle is 57·27 cm². Find the length of its sides.

23. From, $\sin^2\theta + \cos^2\theta = 1$

and, $\cos 2\theta = \cos^2\theta - \sin^2\theta$

Prove that (i) $\cos 2\theta = 1 - 2\sin^2\theta$

(ii) $\cos 2\theta = 2\cos^2\theta - 1$

and, (iii) find the value of $(1 + \cot^2\theta)(1 - \cos 2\theta)$

24. Given that, $\sin 2\theta = 2\sin\theta\cos\theta$

and, $\cos 2\theta = \cos^2\theta - \sin^2\theta$

prove:

(i) $$\sin 2\theta = \frac{2\tan\theta}{1 + \tan^2\theta}$$

(ii) $$\cos 2\theta = \frac{1 - \tan^2\theta}{1 + \tan^2\theta}$$

and (iii) find the value of θ between 0 and 90 degrees which satifies the equation:

$$\sin 2\theta + 2\cos 2\theta = 1$$

25. Prove, $\sin 3\theta = 3\sin\theta - 4\sin^3\theta$

CHAPTER 11

MENSURATION OF AREAS

The SI unit of length is the metre, therefore, since area is the result of the product of two dimensions measured in similar units, the SI unit of area is the "square metre" (abbreviation m^2).

Other recommended units of length are the kilometre ($1\ km = 10^3\ m$) and the millimetre ($1\ mm = 10^{-3}\ m$), these conform to the special recommendations of using prefixes representing 10 to a power which is a multiple of 3, therefore other accepted units of area are the square kilometre [$1\ km^2 = (10^3)^2\ m^2 = 10^6\ m^2$] and the square millimetre [$1\ mm^2 = (10^{-3})^2\ m^2 = 10^{-6}\ m^2$].

However, the large difference of 10^6 in the size of these units of area justifies the use of an intermediate unit between the square metre and the square millimetre in engineering calculations. The square centimetre (cm^2) is a very practical size and commonly used.

Note the scale of areas:

$$1\ cm = 10\ mm \quad \therefore\ 1\ cm^2 = 10^2\ mm^2$$
$$1\ m = 10^2\ cm \quad \therefore\ 1\ m^2 = 10^4\ cm^2$$
$$1\ m = 10^3\ mm \quad \therefore\ 1\ m^2 = 10^6\ mm^2$$

Note also

$$1\ cm^2 = 10^{-4}\ m^2$$
$$1\ mm^2 = 10^{-6}\ m^2 = 10^{-2}\ cm^2$$

Most simple regular shaped figures will already be familiar to most students at this stage therefore they are given here briefly as revision.

A PARALLELOGRAM is a four-sided figure whose opposite sides are parallel and equal in length to each other. It therefore follows that opposite angles are equal, one pair of opposite angles being obtuse, and the other pair acute and supplementary to the obtuse angles. It may be considered as a rectangular framework leaning over to one side as in Fig. 75 wherein it can be seen that the outer triangular area (shown dotted) at one end is equal to the inner triangular area at the other end. Hence the area of the parallelogram is equal to that of a rectangle of the same base and same perpendicular height. Also, if a diagonal is drawn from one corner to the opposite corner, it will bisect the parallelogram into two equal triangles, the area of each being half that of the parallelogram.

$$\text{Area of Parallelogram} = \text{base} \times \text{perp. height.}$$

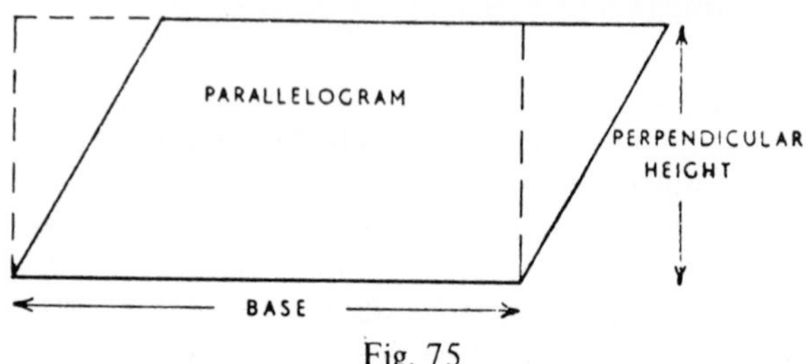

Fig. 75

A RHOMBUS is a special kind of parallelogram. It is a diamond-shaped four-sided figure with all sides of equal length and opposite sides parallel to each other. See Fig. 76. The diagonals of a rhombus are perpendicular to each other and their intersection form right angles; the diagonals bisect each other and each bisects its corner angles. The area of a rhombus is half of the product of its diagonals.

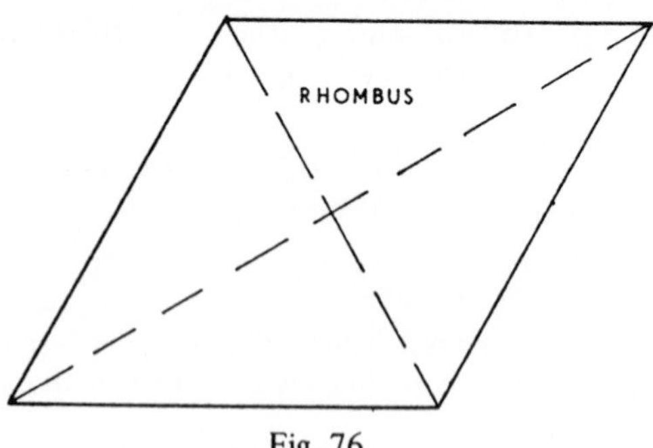

Fig. 76

A TRAPEZIUM is a four-sided figure, two sides only of which are parallel. Referring to Fig. 77, if a and b are the respective lengths of the two parallel sides and h is the perpendicular height, then,

$$\begin{aligned}\text{Area of Trapezium} &= \text{average length} \times \text{perp. height}\\ &= \tfrac{1}{2}(a + b) \times h\end{aligned}$$

The equivalent rectangle of the trapezium is shown in dotted lines.

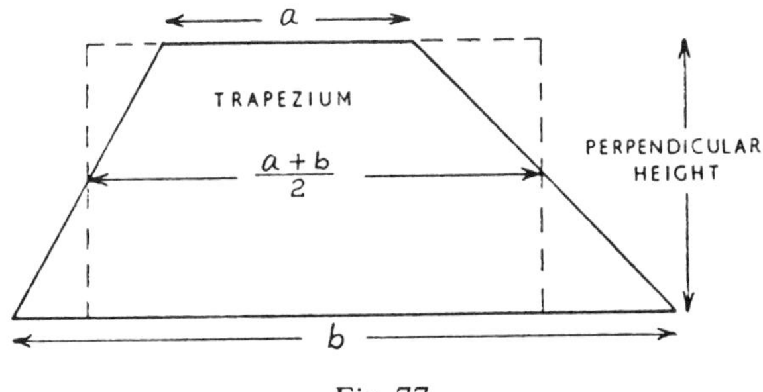

Fig. 77

The area of any four-sided figure, such as shown in Fig. 78, wherein all sides are of different length and all angles different, can be found by dividing the figure into two triangles by a diagonal, calculating the area of each triangle and adding them together.

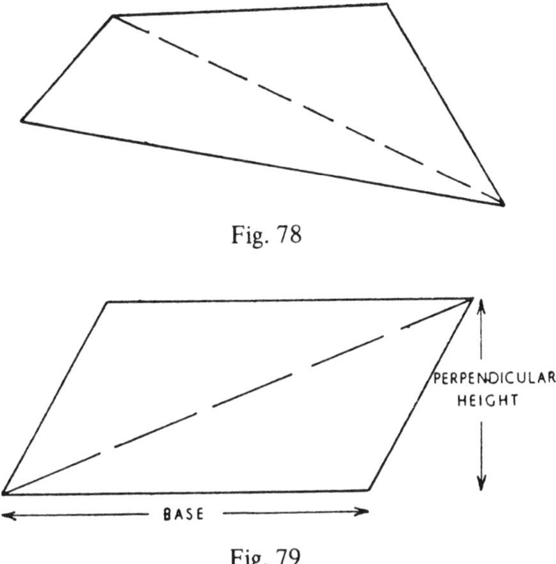

Fig. 78

Fig. 79

TRIANGLES

A diagonal drawn joining two opposite corners of a parallelogram divides the figure into two *equal* triangles, therefore the area of a triangle is half that of a parallelogram of the same base and perpendicular height. This and other rules for the areas of triangles were given in the previous chapter, thus,

$$\text{Area of triangle} = \tfrac{1}{2}(\text{base} \times \text{perp. height})$$
$$= \tfrac{1}{2}(ab \sin C)$$
$$= \sqrt{s(s - a)(s - b)(s - c)}$$

where a, b and c are the lengths of the three sides and s is their semisum.

POLYGONS

A Polygon is a figure bounded by more than four straight sides. A *regular* polygon has all its sides equal in length and all its angles equal, any regular polygon is therefore made up of as many equal triangles as there are sides and the method of finding the area of a polygon is to first find the area of one triangle and then multiply by the number of triangles which constitute the polygon.

Special names are given to some regular polygons as follows,

A Five-sided figure is called a Pentagon
,, Six ,, ,, ,, Hexagon
,, Seven ,, ,, ,, Heptagon
An Eight ,, ,, ,, an Octagon
A Nine ,, ,, ,, a Nonagon
,, Ten ,, ,, ,, Decagon

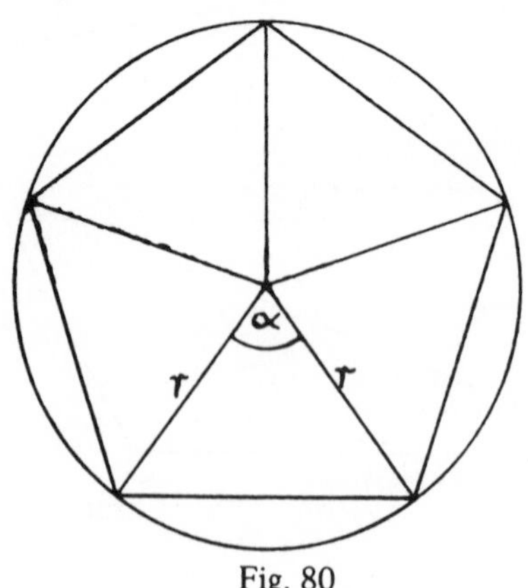

Fig. 80

For any polygon, if n = number of sides, then,

$$\text{number of triangles} = n$$

$$\text{Apex angle of each triangle} = \alpha = \frac{360}{n}$$

If r = radius of the circumscribing circle, then r is also the slant height of each triangle. Note that all the triangles are isosceles, see figures 80 to 82.

$$\text{Area of each triangle} = \tfrac{1}{2}(ab \sin C)$$
$$= \tfrac{1}{2}r^2 \sin \alpha$$
$$\therefore \text{Area of polygon} = \tfrac{1}{2}nr^2 \sin \alpha$$

It is, however, also convenient to express the area of the more common polygons in terms of the length of the 'sides' or 'flats'.

The HEXAGON is the most important polygon from an engineer's point of view because most nuts and bolt heads are hexagonal.

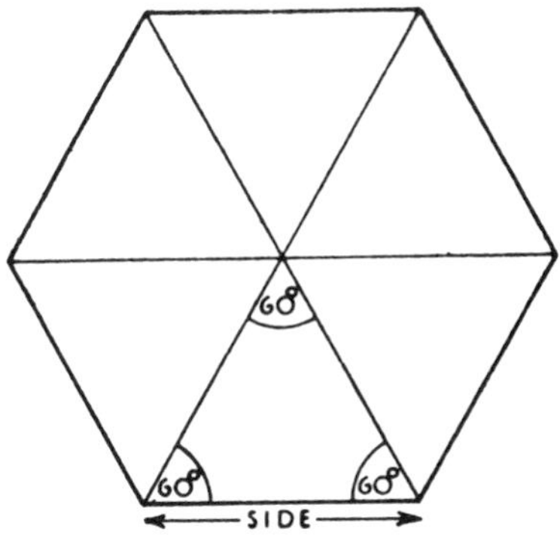

Fig. 81

As there are six sides to the hexagon, it is composed of six triangles, the apex of each triangle is 360 ÷ 6 = 60°

The base angles are each $\tfrac{1}{2}$(180 − 60) = 60°

Therefore the triangles must be equilateral (see Fig. 81).

$$\text{The area of an equilateral triangle} = 0{\cdot}433 \text{ side}^2 \text{ (previously shown)}$$
$$\therefore \text{Area of Hexagon} = 6 \times 0{\cdot}433 \text{ side}^2$$
$$= 2{\cdot}598 \text{ side}^2$$

The OCTAGON is another section which crops up quite often in engineering. Being composed of eight isosceles triangles, the apex angle of each is 360 ÷ 8 = 45° (see Fig. 82).

The base angles are each $\frac{1}{2}(180 - 45) = 67\frac{1}{2}°$

$$\text{Perp. height of triangle} = \tfrac{1}{2}\text{ side} \times \tan 67\tfrac{1}{2}°$$

$$\begin{aligned}\text{Area ,, ,,} &= \tfrac{1}{2}(\text{base} \times \text{perp. height}) \\ &= \tfrac{1}{2} \times \text{base} \times \tfrac{1}{2} \times \text{side} \times \tan 67\tfrac{1}{2} \\ &= \tfrac{1}{2} \times \text{side} \times \tfrac{1}{2} \times \text{side} \times \tan 67\tfrac{1}{2}°. \\ &= 0{\cdot}60355\text{ side}^2\end{aligned}$$

$$\begin{aligned}\text{Area of Octagon} &= 8 \times 0{\cdot}60355\text{ side}^2 \\ &= 4{\cdot}8284\text{ side}^2\end{aligned}$$

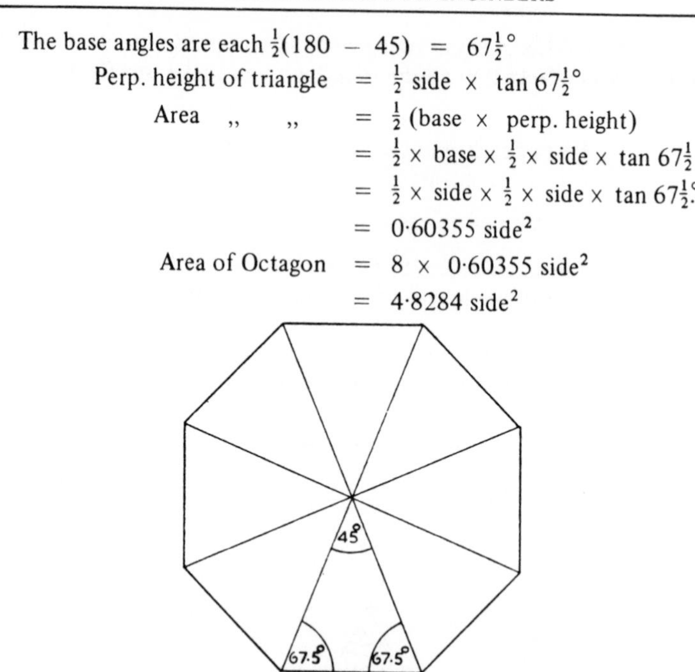

Fig. 82

THE CIRCLE

The Circumference is the outer rim of the circle, an Arc is part of the Circumference, other common terms are illustrated in Fig. 83.

$$\begin{aligned}\text{Circumference} &= \pi \times \text{diameter} = \pi d \\ &= 2\pi \times \text{radius} = 2\pi r \\ \text{where } \pi &= 3{\cdot}142 \text{ to nearest four figures.}\end{aligned}$$

In many cases π may be conveniently taken as the nearest vulgar fraction, which is $3\frac{1}{7}$ or $\frac{22}{7}$.

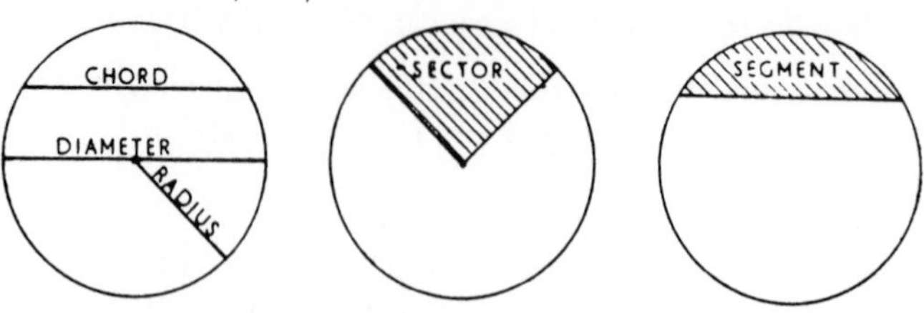

Fig. 83

The Area of a circle can be considered as being made up of a great number of small triangles, like a polygon of an infinite number of sides,

$$r = \text{height of each triangle}$$

$$\text{Area of each triangle} = \frac{r \times \text{base}}{2}$$

$$\text{Area of circle} = \text{sum of areas of all triangles}$$

$$= \text{sum of} \left\{ \frac{r \times \text{base}}{2} \right\}$$

$\frac{r}{2}$ is a common multiplier,

$$\therefore \text{Area of circle} = \frac{r}{2} \times \text{sum of bases}$$

but the sum of the bases constitute the whole circumference $= 2\pi r$

$$\therefore \text{Area of circle} = \frac{r}{2} \times 2\pi r = \pi r^2$$

$$\text{Also, since } r = \frac{d}{2}$$

$$\text{Area of circle} = \frac{\pi}{4} d^2 = 0{\cdot}7854 d^2$$

ANNULUS OR CIRCULAR RING

$$\text{Area of Annulus} = \text{area of outer circle} - \text{area of inner circle}$$

$$= \pi R^2 - \pi r^2$$

$$= \pi(R^2 - r^2)$$

$$\text{or } \frac{\pi}{4}(D^2 - d^2)$$

$$\text{or } 0{\cdot}7854(D^2 - d^2)$$

Note that the factors of $R^2 - r^2$ are $(R + r)(R - r)$
and ,, ,, ,, $D^2 - d^2$ are $(D + d)(D - d)$

Use of the factors make calculations of annular areas much quicker and simpler.

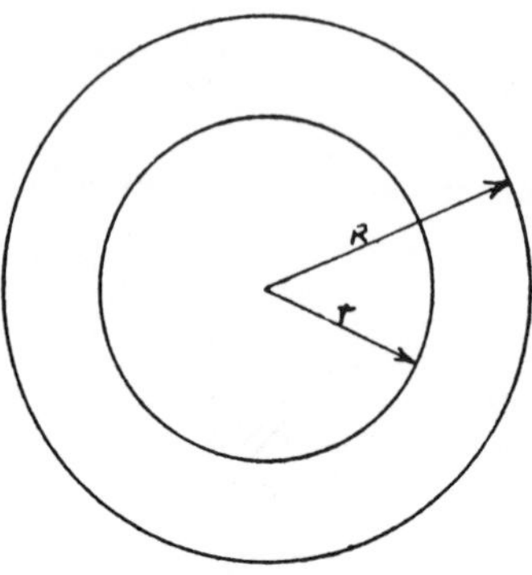

Fig. 84

Example. Find the effective under-face area of a reciprocating pump piston 41·5 mm diameter if the piston rod diameter is 8·5 mm.

$$\begin{aligned}\text{Area of Annulus} &= 0{\cdot}7854\,(D^2 - d^2)\\ &= 0{\cdot}7854\,(D + d)(D - d)\\ &= 0{\cdot}7854\,(41{\cdot}5 + 8{\cdot}5)(41{\cdot}5 - 8{\cdot}5)\\ &= 0{\cdot}7854 \times 50 \times 33\\ &= 1296\ \text{mm}^2 \text{ or } 12{\cdot}96\ \text{cm}^2 \quad \text{Ans.}\end{aligned}$$

SECTOR OF A CIRCLE

A Sector of a circle is shown in Fig. 85. As in the case of the whole circle we can consider the sector as being made up of a number of small triangles:

$$\begin{aligned}\text{Area of Sector} &= \frac{r}{2} \times \text{sum of bases}\\ &= \frac{r}{2} \times \text{length of arc}\end{aligned}$$

alternatively, if θ is the angle at the centre, in *degrees*,

$$\text{area of circle of } 360^\circ = \pi r^2$$

$$\text{,, sector ,, } 1^\circ = \frac{\pi r^2}{360}$$

$$\text{,, sector ,, } \theta^\circ = \frac{\theta}{360} \times \pi r^2$$

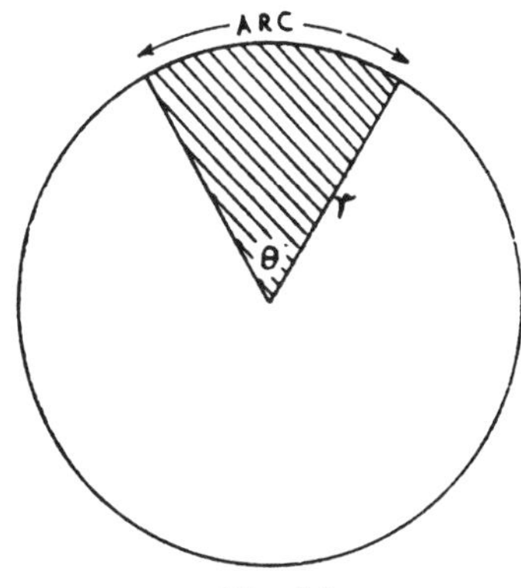

Fig. 85

Further, if the angle θ at the centre be measured in *radians*,

$$\text{area of circle } (= 2\pi \text{ radians}) = \pi r^2$$

$$\text{,, sector of 1 radian} = \frac{\pi r^2}{2\pi} = \frac{r^2}{2}$$

$$\text{,, sector of } \theta \text{ radians} = \frac{\theta r^2}{2}$$

SEGMENT OF A CIRCLE

Referring to Fig. 86,

$$\text{Area of Segment} = \text{area of sector} - \text{area of triangle}$$

$$\text{area of sector (shown above)} = \frac{\theta r^2}{2}$$

where θ is the centre angle in *radians*.

$$\text{Area of triangle} = \frac{ab \sin C}{2} \text{ in this case, } a = r,$$

$b = r$, and C is the centre angle θ in *degrees.*

$$\therefore \text{Area of triangle} = \frac{r^2 \sin \theta}{2}$$

$$\text{Area of segment} = \text{area of sector} - \text{area of triangle}$$

$$= \frac{\theta r^2}{2} - \frac{r^2 \sin \theta}{2}$$

$$= \frac{r^2}{2}[\theta - \sin \theta]$$

Note carefully that in this apparently simple formula, the first θ in the brackets is the angle expressed in *radians*, whereas the second θ is in *degrees* so that its sine can be read from the tables.

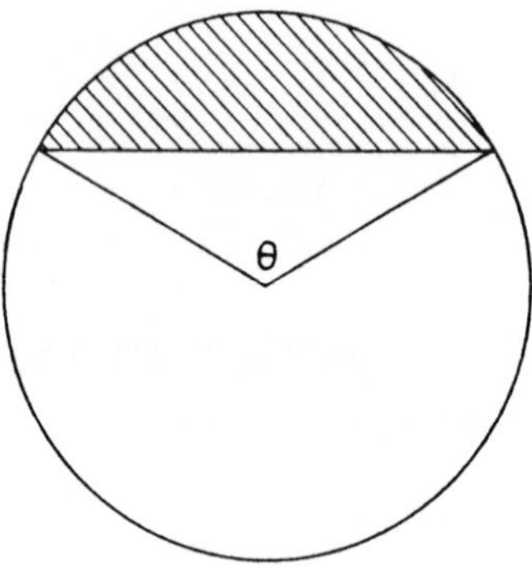

Fig. 86

Example. Calculate the area of a segment which subtends an angle of 150 degrees at the centre of a circle 150 mm diameter.

$$\theta \text{ in radians} = \frac{150}{57{\cdot}3} \text{ or } \frac{150 \times 2\pi}{360}$$

$$= 2{\cdot}618 \text{ radians}$$

$$\sin 150^\circ = \sin(180 - 150) = \sin 30^\circ = 0{\cdot}5$$

$$\text{Area of Segment} = \frac{r^2}{2}[\theta - \sin\theta]$$

$$= \frac{75^2}{2}[2{\cdot}618 - 0{\cdot}5]$$

$$= \frac{75^2 \times 2{\cdot}118}{2}$$

$$= 5957 \text{ mm}^2 \text{ or } 59{\cdot}57 \text{ cm}^2 \quad \text{Ans.}$$

ELLIPSE

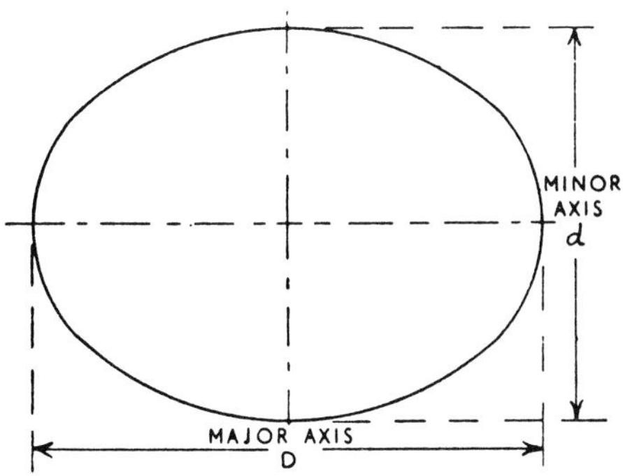

Fig. 87

If D = major axis, and d = minor axis of an ellipse,

$$\text{Circumference} = \pi \times \text{average diameter (approx.)}$$

$$= \pi\left(\frac{D + d}{2}\right)$$

$$\text{Area} = \frac{\pi}{4} \times D \times d$$

SURFACE OF CYLINDER

If we imagine the shell of a cylinder being un-rolled as in Fig. 88 it can be seen that,

$$\text{curved surface area of a cylinder} = \pi d \times h$$

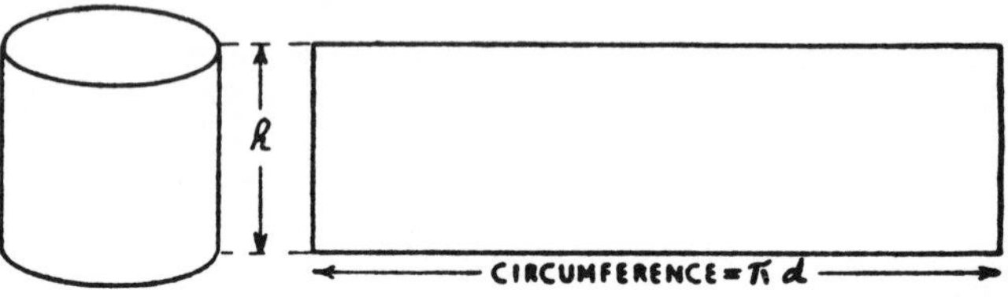

Fig. 88

SURFACE OF SPHERE

The curved surface area of a sphere is equal to the curved surface area of its circumscribing cylinder, that is, a cylinder of equal diameter and height.

The curved surface area of a segment of the sphere or any such sliced portion, is equal to the curved surface area of the corresponding slice off the circumscribing cylinder.

In all cases, referring to Fig. 89,

$$\text{Curved Surface Area} = \pi d \times h$$

$$\text{For the whole sphere, } h = d,$$

$$\text{Curved Surface Area of Sphere} = \pi d^2$$

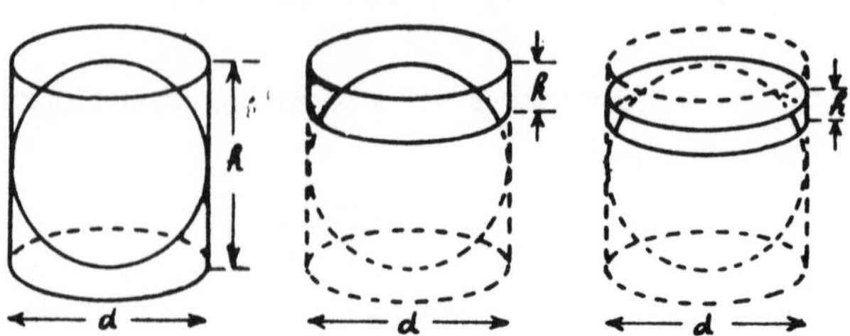

Fig. 89

THEOREM OF PAPPUS OR GULDINUS

This theorem is one of the most useful to employ in finding areas and volumes of objects of circular shapes, or it the area or volume is known it can be used to determine the position of the centre of gravity of many sections. With regard to areas the theorem is stated thus:

If a line, lying wholly on one side of a fixed axis, be rotated about that axis in its own plane, it will sweep out a surface area equal to the length of the line multiplied by the distance its centre of gravity moves.

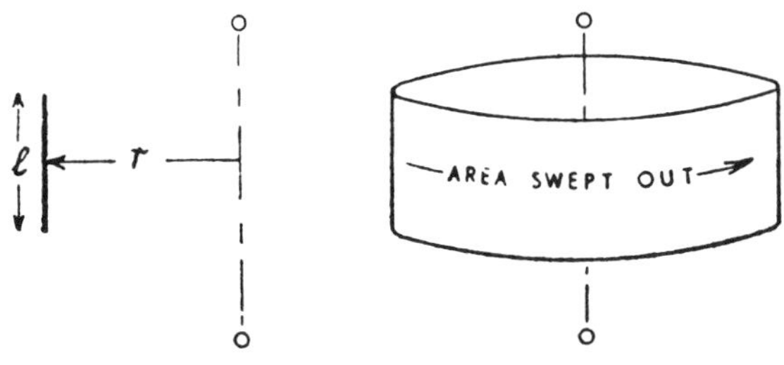

Fig. 90

Consider a 'line' such as a straight piece of wire of length l, positioned at r from an axis parallel to the wire, if the wire is moved around through a complete circle about the axis $o\ o$ as centre, a surface area like a thin cylinder will be swept out. The centre of gravity of the piece of wire is at its mid-length, the distance moved by the centre of gravity in one revolution is the circumference $= 2\pi r$, therefore:

$$\begin{aligned}\text{Area swept out} &= \text{length of line} \times \text{distance c.g. moves} \\ &= l \times 2\pi r \\ &= 2\pi rl\end{aligned}$$

SURFACE OF CONE

Now let the line be inclined with one end touching the fixed axis, as in Fig. 91. If this line is rotated through one complete revolution the area swept out is that of the curved surface area of a cone. The centre of gravity of the line is at its mid-length, let this be x from $o\ o$.

$$\text{Area swept out} = \text{length of line} \times \text{distance c.g. moves}$$
$$= l \times 2\pi x$$

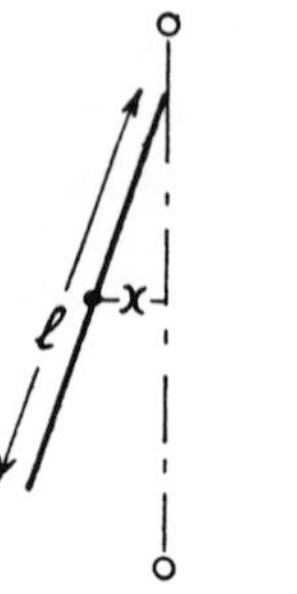

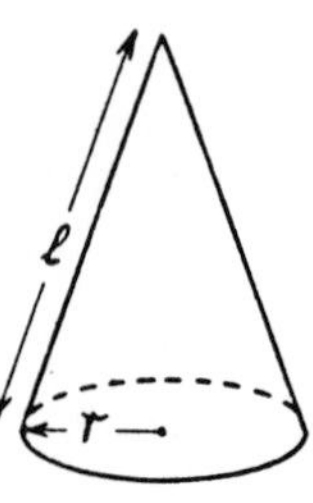

Fig. 91

If the radius of the base of the swept out cone be represented by r, it can be seen that x is half of r, therefore:

$$\text{Curved surface area of cone} = l \times 2\pi \times \frac{r}{2}$$

$$= \pi r l$$

l being the slant height of the cone.

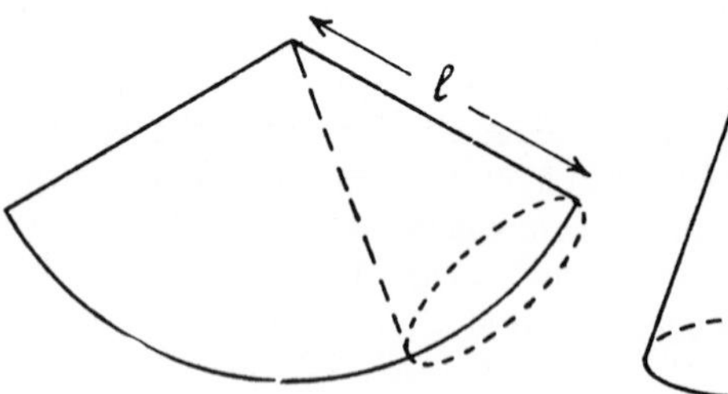

Fig. 92

As an alternative to the above, referring to Fig. 92, if we imagine a paper cone being unrolled, the curved surface area develops into a flat sector of a circle. The slant height of the cone, l, forms the radius of this sector, and the circumference of the base of the cone (which is $2\pi r$) forms the arc of the sector.

We have previously expressed the area of a sector of a circle as $r/2 \times$ length of arc, therefore if this is now expressed in terms of the dimensions of the corresponding cone, we get:

$$\text{Curved surface area of cone} = \text{area of sector}$$
$$= \frac{l}{2} \times 2\pi r$$
$$= \pi r l \text{ (as before).}$$

SURFACE OF FRUSTUM OF CONE

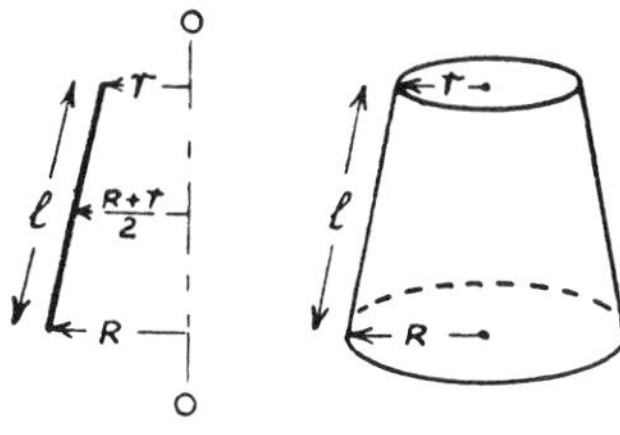

Fig. 93

A Frustum of a cone is the part left after a portion of the top of the cone has been sliced off. Referring to Fig. 93, the line to be rotated is equivalent to the slant height of the frustum, $= l$. If the radii of the frustum at bottom and top be represented by R and r respectively, these are the distances of the bottom and top ends of the line, respectively, from the axis. The centre of gravity of the line is at its mid-length, the position of this point from the axis is the mean of R and r which is $\frac{1}{2}(R + r)$, hence,

$$\text{Area swept out} = \text{length of line} \times \text{distance c.g. moves}$$
$$\text{Curved surface area of frustum} = l \times 2\pi \times \tfrac{1}{2}(R + r)$$
$$= \pi l(R + r)$$

SURFACE OF A CIRCULAR RING OF CIRCULAR SECTION

If we now consider a line as a piece of wire bent around into a circle of radius r, the centre of this circle being at R from the axis (Fig. 94),

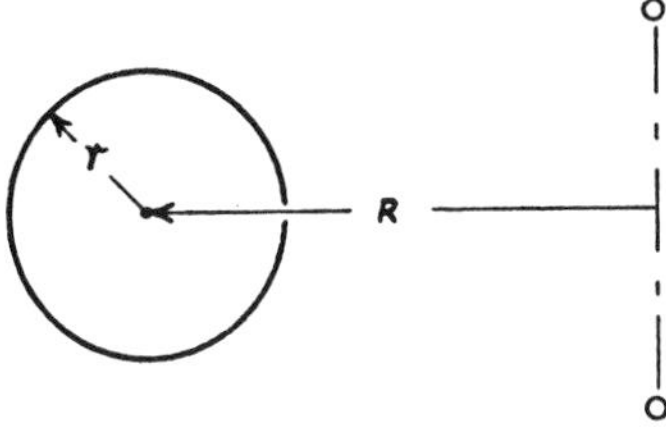

Fig. 94

when this circle is swept through a complete revolution about the axis oo, the area swept out is the curved surface area of a ring. A practical example of this is the surface area of an anchor ring – that area which is to be painted.

$$\text{Area swept out} = \text{length of line} \times \text{distance c.g. moves}$$

$$\text{Curved surface area of ring} = 2\pi r \times 2\pi R$$
$$= 4\pi^2 Rr$$

SURFACE OF A CIRCULAR RING OF ELLIPTICAL SECTION

If the line is an ellipse of major axis D and minor axis d, the centre being at R from the axis, the area swept out in one revolution is the curved surface area similar to that of a circular lifebuoy of elliptical section.

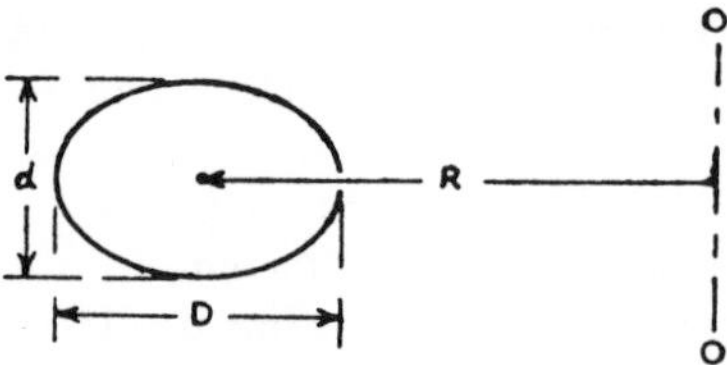

Fig. 95

$$\text{Area swept out} = \text{length of line} \times \text{distance c.g. moves}$$

$$\text{Curved surface of lifebuoy} = \pi \left(\frac{D + d}{2}\right) \times 2\pi R$$
$$= \pi^2 R (D + d)$$

From the foregoing examples the importance of this Theorem will be appreciated. Not only can curved surface areas be easily calculated, but if the area is known in the first place, the centre of gravity of the line could be determined, this is demonstrated in the example:

To find the centre of gravity of a piece of wire bent into the shape of a semi-circle of diameter d.

If this semi-circular piece of wire is swept through one revolution about a fixed axis on its own diameter (Fig. 96) the area swept out will be the surface area of a sphere, this has already been shown to be πd^2.

Let the centre of gravity be at x from the axis, that is, the position of a knife-edge for the wire to rest upon and be balanced.

$$\text{Area swept out} = \text{length of line} \times \text{distance c.g. moves}$$
$$\pi d^2 = \tfrac{1}{2} \times \pi d \times 2\pi x$$

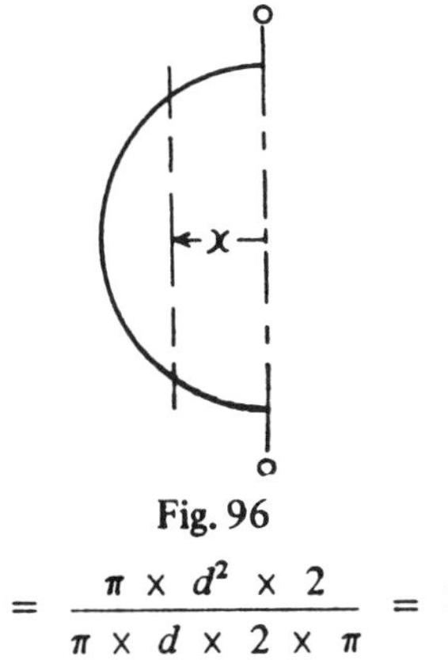

Fig. 96

$$x = \frac{\pi \times d^2 \times 2}{\pi \times d \times 2 \times \pi} = \frac{d}{\pi}$$

Hence the centre of gravity of this piece of wire lies on a line parallel to the diameter, at d/π from the diameter.

SIMILAR FIGURES

Areas of similar figures vary as the square of their corresponding linear dimensions.

Similar figures mean that they are of the same shape and proportions, although their sizes are different.

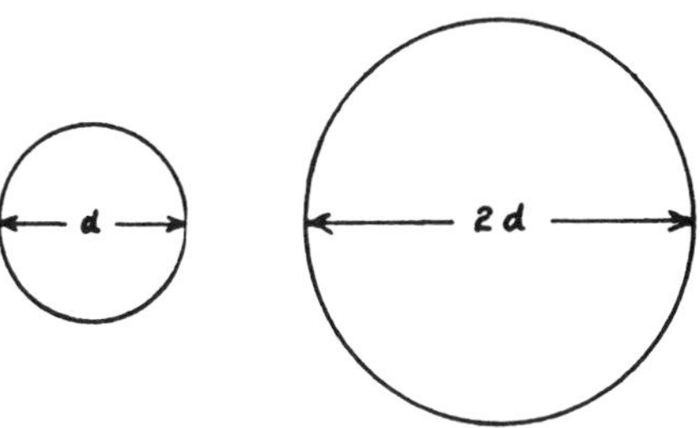

Fig. 97

Consider two circles, one of diameter d, the other twice as big in diameter $= 2d$, as in Fig. 97.

$$\text{Area of small circle} = 0{\cdot}7854d^2$$
$$\text{Area of large circle} = 0{\cdot}7854 \times (2d)^2$$

Therefore the ratio of their areas is,

$$d^2 : (2d)^2$$
$$= d^2 : 2^2 \times d^2$$
$$= 1 : 2^2$$

Thus, if the diameter of the larger circle is twice the diameter of the smaller circle, all linear dimensions (i.e., radius, circumference) will be twice as much, but the area is 2^2 which is *four* times the area of the small circle.

If the diameter of one circle is three times that of another, the radius and circumference of the larger is three times the radius and circumference of the smaller, but the area is $3^2 =$ *nine* times as much.

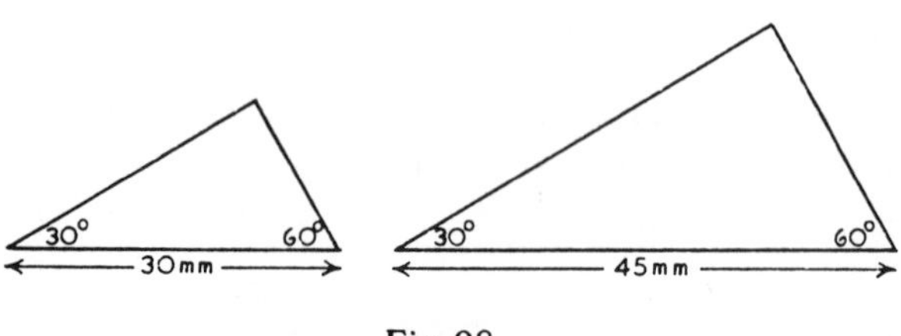

Fig. 98

Fig. 98 shows two similar triangles, the ratio of the base dimensions is 30 : 45 which is 1 : 1·5, hence the base of the larger triangle is $1\frac{1}{2}$ times greater than the base of the smaller.

Therefore all linear dimensions of the larger triangle are $1\frac{1}{2}$ times greater than the corresponding linear dimension of the smaller. The ratio of their areas is proportional to the square of their corresponding dimensions thus:

$$\text{Ratio of dimensions} = 1 : 1{\cdot}5$$
$$\therefore \;\; \text{,,} \;\; \text{areas} = 1^2 : 1{\cdot}5^2$$
$$= 1 : 2{\cdot}25$$

Hence the area of the larger triangle is $2\frac{1}{4}$ times the area of the smaller.

Example. A hexagonal plate is cut out of a sheet. Due to an error in marking off, the sides were all made 10% longer than intended, find the percentage error in area.

$$\begin{aligned}
\text{Ratio correct dimensions to wrong} &= 100 : 110 \\
&= 1 : 1{\cdot}1 \\
\text{Ratio correct area to wrong} &= 1^2 : 1{\cdot}1^2 \\
&= 1 : 1{\cdot}21 \\
&= 100 : 121
\end{aligned}$$

$\therefore$ Area is 21% too large. Ans.

IRREGULAR FIGURES

SIMPSON'S FIRST RULE This method of finding the area of an irregular figure is one of the most useful to marine engineers and naval architects. Briefly it is stated:

"To the sum of the first and last ordinates, add four times the even ordinates and twice the odd ordinates, multiply this sum by one-third the common interval and the result is the area of the figure."

An odd number of ordinates, equally spaced, must be used for this rule. Step by step, the procedure is as follows, referring to Fig. 99,

1. Divide the given figure into an even number of equally spaced parts, this gives an odd number of ordinates.
2. Measure the ordinates and the common distance between them.
3. Add together: the first ordinate, the last ordinate, four times the even ordinates and twice the odd ordinates.
4. Multiply the above sum by one-third of the common distance between the ordinates.

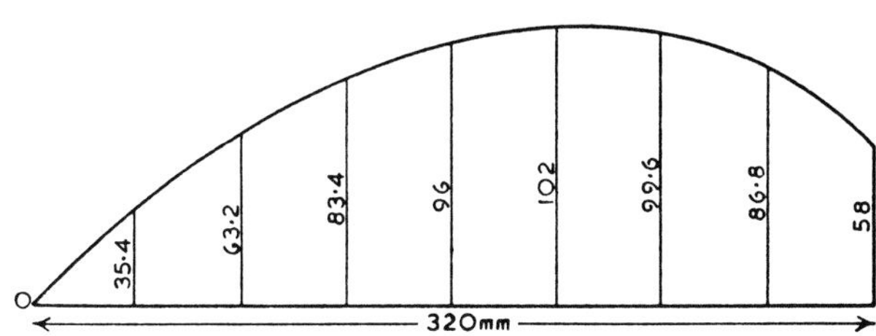

Fig. 99

Example. A flat plate is shaped as shown in Fig. 99, the dimensions being in millimetres, find its area by Simpson's rule.

Working in centimetres and setting out in tabulated form:

Ordinates	Simpson's Multipliers	Products
0	1	0
3·54	4	14·16
6·32	2	12·64
8·34	4	33·36
9·6	2	19·20
10·2	4	40·80
9·96	2	19·92
8·68	4	34·72
5·8	1	5·80
	Sum =	180·60

$$\text{Common interval} = \text{length} \div \text{number of spaces}$$
$$= 320 \div 8$$
$$= 40 \text{ mm} = 4 \text{ cm}$$

$$\text{Area} = \frac{180{\cdot}6 \times 4}{3}$$
$$= 240{\cdot}8 \text{ cm}^2 \quad \text{Ans.}$$

From the explanation of Simpson's Rule, the multipliers of the ordinates are:

For 3 ordinates	1, 4, 1
,, 5 ,,	1, 4, 2, 4, 1
,, 7 ,,	1, 4, 2, 4, 2, 4, 1
,, 9 ,,	1, 4, 2, 4, 2, 4, 2, 4, 1

and so on.

This rule is often expressed in formula fashion thus:

$$\text{Area} = \frac{h}{3}(a + 4b + 2c + 4d + e)$$

h being the common interval,
a, b, c, etc., being the ordinates.

The student, however, is advised to set out the work in tabulated form as in the example shown because, in work to follow, Simpson's rule is applied in finding first and second moments of irregular shapes and this is most neatly done by extending the table as for areas.

The mean (average) height of an irregular figure can be obtained by dividing the area by its length, or can be found direct by Simpson's rule by dividing the sum of the products of the ordinates and their multipliers, by the total of multipliers.

In the foregoing example the two methods of obtaining the mean height would be:

(i)

$$\text{Area} = 240{\cdot}8 \text{ cm}^2$$

$$\text{Length} = 32 \text{ cm}$$

$$\text{Mean height} = 240{\cdot}8 \div 32 = 7{\cdot}525 \text{ cm} = 75{\cdot}25 \text{ mm}$$

(ii)

$$\text{Sum of products} = 180{\cdot}6$$

$$\text{Sum of multipliers} = 1 + 4 + 2 + 4 + 2 + 4 + 2 + 4 + 1$$

$$= 24$$

$$\text{Mean height} = 180{\cdot}6 \div 24 = 7{\cdot}525 \text{ cm}$$

Example. The ordinates measured athwartships across a ship at her load water line are: 0·2, 9, 15·5, 20, 21·5, 20·5, 18·5, 12·5 and 1·3 metres respectively, and the length is 180 metres. Find the water plane area.

Ordinates	Simpson's Multipliers	Products
0·2	1	0·2
9	4	36
15·5	2	31
20	4	80
21·5	2	43
20·5	4	82
18·5	2	37
12·5	4	50
1·3	1	1·3
	Sum =	360·5

$$\begin{aligned}
\text{Number of ordinates} &= 9 \\
\therefore \quad ,, \quad ,, \quad \text{spaces} &= 8 \\
\text{Common interval} &= \text{length} \div \text{no. of spaces} \\
&= 180 \div 8 \\
&= 22{\cdot}5 \text{ metres} \\
\text{Water plane area} &= \frac{360{\cdot}5 \times 22{\cdot}5}{3} \\
&= 2704 \text{ m}^2 \quad \text{Ans.}
\end{aligned}$$

For shapes such as the water plane area of a ship which are symmetrical about the longitudinal centre-line, measurements from the centre-line to the hull may be taken and referred to as "half-ordinates". These half-ordinates are put through Simpson's rule, the half-area calculated, then multiplied by two to obtain the full area.

MID-ORDINATE RULE

Another method of finding the area or the mean height of an irregular figure is by the mid-ordinate rule. This method is usually

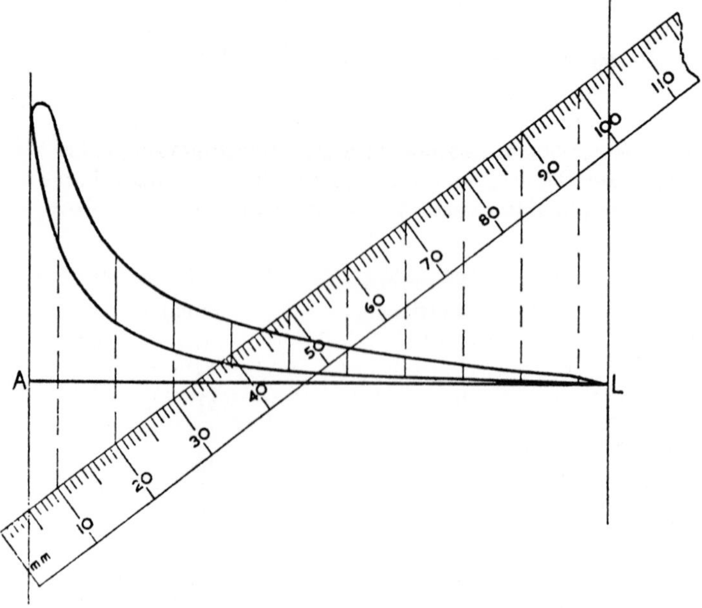

Fig. 100

employed in finding the mean height of a reciprocating engine indicator diagram to obtain the mean effective pressure.

The method is to divide the figure into any number of equally spaced parts, erecting lines midway between these ordinates, these being the mid-ordinates; the mid-ordinates are measured, added together and the sum is divided by the number of mid-ordinates to get their average height.

If the area is required, this can be obtained by multiplying the mean height by the length.

In working out indicator diagrams, ten mid-ordinates are usually erected. The simplest method of dividing the diagram into ten parts and employing the mid-ordinate rule to determine the mean height is explained in the following example.

Referring to Fig. 100 which is a copy of an indicator diagram from the cylinder of an internal combustion engine:

1. Erect a line at each of the two extreme ends of the diagram, perpendicular to the atmospheric line (that is the straight horizontal base line marked AL).
2. Place a rule so that it measures 100 mm between these perpendiculars, if the length of the diagram was exactly 100 mm, the rule would lie parallel with the atmospheric line, but almost all diagrams are less than this therefore the rule must be inclined until it registers 100 mm between the lines.
3. Instead of marking the ten spaces and then the middle of these spaces to indicate the mid-ordinates, it is quicker to slide the rule until it registers 5 mm at the first perpendicular and 105 mm at the other, then mark off every 10 mm point, this gives the position of the mid-ordinates direct.
4. Erect perpendicular lines through these marks across the diagram, these are the mid-ordinates.
5. Measure the mid-ordinates, in this case they are 13·5, 9, 7, 6, 4·5, 3·5, 2·5, 2, 1 and 0·5 mm. Add these measurements together, this gives 49·5 mm. Divide by the number of mid-ordinates, in this case 10, to obtain the mean height of the diagram, thus, $49{\cdot}5 \div 10 = 4{\cdot}95$ mm.

To obtain the mean effective pressure in the engine cylinder from which this diagram was taken, the mean height is multiplied by the pressure scale of the spring used in the indicator. In this diagram, the spring stiffness was such that one millimetre of height represents a pressure of 90 kilonewtons per square metre. Hence, the mean effective pressure is $4{\cdot}95 \times 90 = 445{\cdot}5$ kN/m^2.

TEST EXAMPLES 11

1. In a parallelogram ABCD, the opposite parallel sides AD and BC are each 100 mm long, the other sides are each 60 mm long, and the diagonal AC is 140 mm. Calculate the angles, the short diagonal, the perpendicular height, and the area.

2. The sides of a rhombus are each 32 mm long and the length of the long diagonal is 48 mm. Calculate the angles, the length of the short diagonal, and the area.

3. In a trapezium ABCD, the two parallel sides are AB and CD, and their lengths are 100 and 60 mm respectively. Side BC is perpendicular to the parallel sides and its length is 50 mm.
   (i) Find the area of the trapezium.
   (ii) Find the position of a dividing line EF parallel to AB to divide the trapezium into two equal half areas.

4. The lengths of the sides of a four-sided figure ABCD, are, in metres, AB = 1, BC = 2, CD = 1·5, DA = 3·5, and the angle BCD is 117° 17′. Find the area of the figure.

5. The length of the sides of a regular hexagonal plate is 80 mm. The plate is cut parallel to one of its sides and this reduces the area by 10%, calculate the thickness of the piece cut off.

6. An octagonal plate, the sides of which are each 30 mm long, has a circular hole 50 mm diameter cut out of it. Find the net area of the plate in mm^2.

7. Find the length of the sides and the area of the largest equilateral triangular plate that can be cut out of a circular plate 120 mm diameter.

8. The outer and inner diameters of the collar of a single-collar thrust shaft are 755 and 415 mm respectively, and the effective area of contact with the thrust pads is 0·7 of the face of the collar. Calculate (i) the effective area of contact, in square metres, and (ii) the total force on the collar, in kilonewtons, when the thrust pressure is 2000 kilonewtons on each square metre.

9. Find the area, in cm^2, of the smaller segment of a circle of 200 mm diameter if the length of the chord is 180 mm.

10. The internal dimensions of the liner of an I.C. engine are 762 mm diameter and 1270 mm long. Calculate the surface area in square metres.

11. Find the diameter of a solid hemisphere whose total surface area (including the flat circular base) is 58·9 cm^2.

12. The ball of a Brinell hardness testing machine is 10 mm diameter. Calculate the depth and curved surface area of an indentation in a material under test when the surface diameter of the indentation is 5 mm.

13. It is required to make a hollow cone out of thin flat sheet steel, the base diameter of the cone to be 150 mm and the perpendicular height 125 mm. Find the dimensions of the sector to be cut out of the sheet to make this cone and sketch the pattern.

14. A lampshade has the form of a frustum of a cone, the diameters at the base and top are 320 and 180 mm respectively and the perpendicular height is 170 mm. Calculate the curved surface area.

15. A circular anchor ring made of round bar is 640 mm outside diameter and 440 mm inside diameter. Calculate the surface area to be painted.

16. An equilateral triangular plate has sides 125 mm long, and another similarly shaped plate has sides 175 mm long. By what percentage is the larger plate greater in area than the smaller plate?

17. Regularly spaced semi-ordinates measured transversely across a ship at the load-water-line are as follows: 0·1, 3, 5·85, 7·2, 8·1, 8·4, 8·4, 8·25, 8·1, 7·5, 6·3, 3·75 and 0·5 metres respectively, and the length is 150 metres. Find the area of the water-plane by Simpson's rule.

18. (*a*) Plot the two curves $y = x^2 + 3x + 6$ and $y = 2x^2 - x + 1$ on common axes between the limits $x = 0$ and $x = 4$.

(*b*) By Simpson's rule find the area enclosed between the two curves.

19. An internal combustion engine indicator diagram is divided into ten mid-ordinates and their measurements are: 26, 15, 9·5, 8, 7, 5·5, 4·5, 4, 3 and 0 millimetres respectively. Find (i) the mean height in millimetres by the mid-ordinate rule, (ii) the mean effective pressure if one millimetre represents a pressure of 80 kilonewtons per m^2.

CHAPTER 12

MENSURATION OF VOLUMES AND MASSES

VOLUME is the result of the product of three dimensions measured in similar units, therefore, since the SI unit of length is the metre, then the SI unit of volume is the "cubic metre" (m^3), and the millimetre being a recommended unit of length, then the cubic millimetre (mm^3) is a recommended unit of volume.

However, since one metre is equal to 10^3 millimetres, then one cubic metre is equal to $(10^3)^3$ cubic millimetres, that is 10^9 mm^3. Thus there is a vast difference in size between the cubic metre and the cubic millimetre, consequently intermediate units of volume are accepted. These are the cubic decimetre (dm^3) which is called a *litre* (*l*) and used for fluid measure, and the cubic centimetre (cm^3) which is equal in volume to a *millilitre* (ml). Thus:

$$1 \text{ cm} = 10 \text{ mm}$$
$$\therefore\ 1 \text{ cm}^3 = 1 \text{ ml} = 10^3 \text{ mm}^3$$
$$1 \text{ dm} = 10 \text{ cm} = 10^2 \text{ mm}$$
$$\therefore\ 1 \text{ dm}^3 = 1 \text{ litre} = 10^3 \text{ cm}^3 = 10^3 \text{ ml} = 10^6 \text{ mm}^3$$
$$1 \text{ m} = 10 \text{ dm} = 10^2 \text{ cm} = 10^3 \text{ mm}$$
$$\therefore\ 1 \text{ m}^3 = 10^3 \text{ dm}^3 = 10^3 \text{ litre} = 10^6 \text{ cm}^3 = 10^9 \text{ mm}^3$$

MASS is the quantity of matter possessed by a body and is proportional to the volume and the density of the body. It is a constant quantity, that is, the mass can only be changed by adding more matter or taking matter away.

The abbreviation for mass is m and the SI unit is the kilogramme (kg). For very large or small quantities, multiples or submultiples of the gramme (g) are used. Large masses are common in marine work and these are measured in megagrammes (Mg), one megagramme is equal to 10^3 kilogrammes and called a *tonne* (t).

DENSITY is a measure of the mass per unit volume, the symbol representing density is ρ the Greek letter rho. Since the SI units for mass and volume are kilogramme and cubic metre respectively, then the SI unit of density is kilogramme per cubic metre(kg/m^3). Other units recommended for use are, grammes per cubic centimetre (g/cm^3) for solid materials, grammes per millilitre (g/ml) for liquids, grammes per litre (g/l) for gases. In some cases tonnes per cubic metre (t/m^3) and kilogrammes per litre (kg/l) may be used.

Note that numerically:

$$g/cm^3 = g/ml = kg/l = t/m^3$$

The density of pure water may be taken as 1000 kg/m^3 which is equal to 1 tonne/m^3, 1 kg/litre, and 1 g/ml.

The total mass of a body is therefore the product of the volume and the density. Units must of course be consistent throughout, such as,

$$\text{mass [kg]} = \text{volume [m}^3\text{]} \times \text{density [kg/m}^3\text{]}$$

or

$$\text{mass [g]} = \text{volume [cm}^3\text{]} \times \text{density [g/cm}^3\text{]}$$

and so on.

RELATIVE DENSITY or SPECIFIC GRAVITY of a substance is the ratio of the mass of a volume of the substance to the mass of an equal volume of pure water.

In other words, it is the ratio of the density of the substance to the density of pure water and is therefore numerically equal to g/cm^3 and its equivalents.

PRISMS

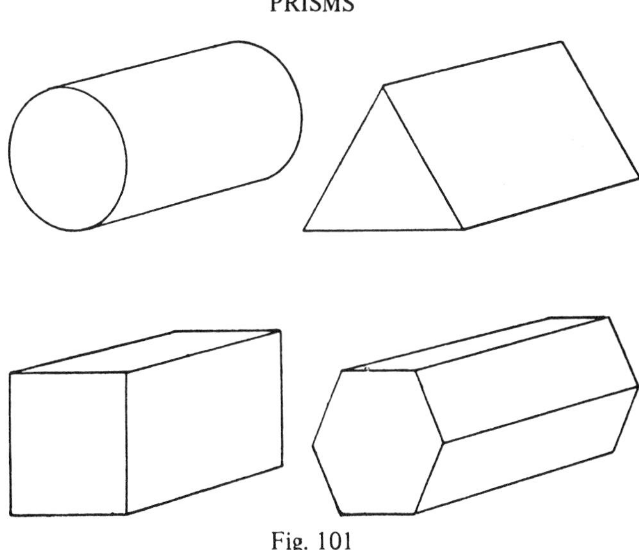

Fig. 101

A regular prism is a bar of regular cross-section, some examples are given in Fig. 101.

In all these cases,

$$\text{volume} = \text{area of cross-section} \times \text{length}$$

Hence, to find the volume of a prism, calculate the area of the end and multiply this by length (or height) of the prism.

Example. A brass bar 250 mm long has a constant hexagonal cross section measuring 90 mm across the face from one corner to the opposite corner. Find (i) the volume of the bar, (ii) the mass in kilogrammes if the density of brass is 8·4 g/cm³.

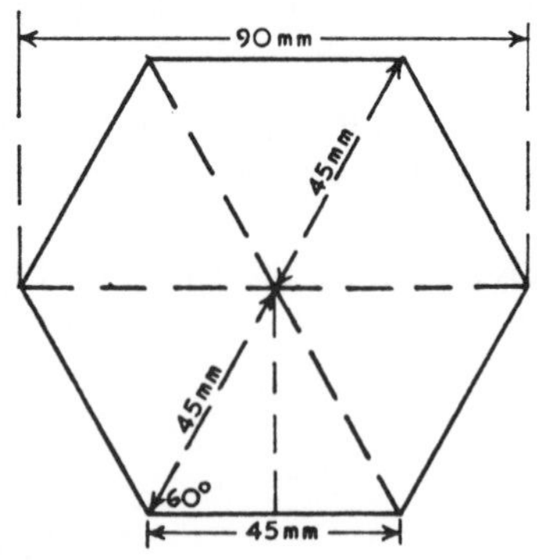

Fig. 102

$$\begin{aligned}
\text{Side of each equilateral triangle} &= 45 \text{ mm} \\
\text{Perpendicular height} &= 45 \times \sin 60 \\
\text{Area of each equilateral triangle} &= \tfrac{1}{2}(\text{base} \times \text{perp. ht.}) \\
&= \tfrac{1}{2} \times 45 \times 45 \times \sin 60 \\
&= 0{\cdot}433 \times 45^2 \\
\text{Area of hexagon} &= 6 \times 0{\cdot}433 \times 45^2 \\
\text{Volume of prism} &= \text{area} \times \text{length} \\
&= 6 \times 0{\cdot}433 \times 45^2 \times 250 \\
&= 1{\cdot}315 \times 10^6 \text{ mm}^3 \\
&\text{or } 1315 \text{ cm}^3 \quad \text{Ans (i)} \\
\text{Mass [g]} &= \text{volume [cm}^3] \times \text{density [g/cm}^3] \\
&= 1315 \times 8{\cdot}4 \\
\text{Mass [kg]} &= 1315 \times 8{\cdot}4 \times 10^{-3} \\
&= 11{\cdot}05 \text{ kg} \quad \text{Ans. (ii)}
\end{aligned}$$

PYRAMIDS

A pyramid is a body standing on a triangular, square or polygonal base, its sides tapering to a point at the apex, some examples are illustrated in Fig. 103. The cone may be considered as a pyramid with a circular base.

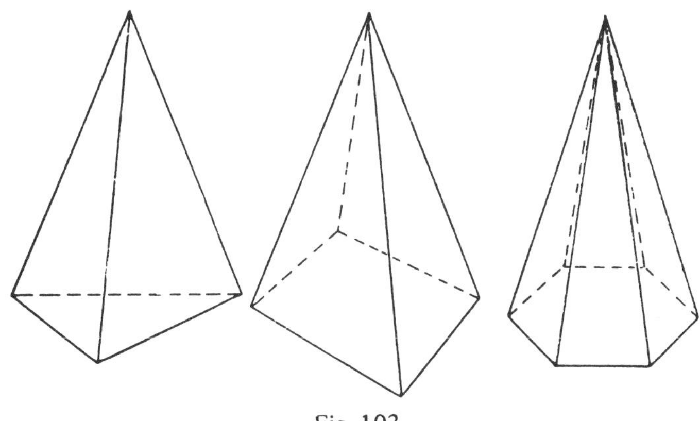

Fig. 103

The volume of a pyramid is one-third of the volume of its circumscribing prism. Thus, the volume of a cone is one-third of the volume of a solid cylinder of the same section as the base of the cone and having the same height; the volume of a square pyramid is one-third the volume of a bar of square section equal to the base of the pyramid and of the same height (or length). In all cases:

$$\text{Volume of pyramid} = \tfrac{1}{3}(\text{area of base} \times \text{perpendicular height})$$

OBLIQUE PRISMS AND PYRAMIDS

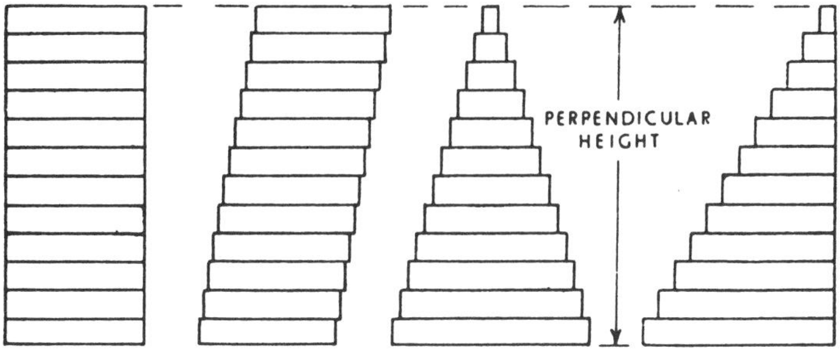

PERPENDICULAR HEIGHT

Fig. 104

If the prism or pyramid be imagined as being made up of a number of discs or laminations and pushed over to one side, it can be seen by reference to Fig. 104 that the same rule for finding the volumes of regular prisms or regular pyramids can be applied provided the *perpendicular* height is used.

FRUSTUMS

A frustum of a pyramid or cone is the bottom piece left, after a portion has been sliced off the top (Fig. 105).

The volume can be found by subtracting the volume of the sliced-off top part from the volume of the complete pyramid.

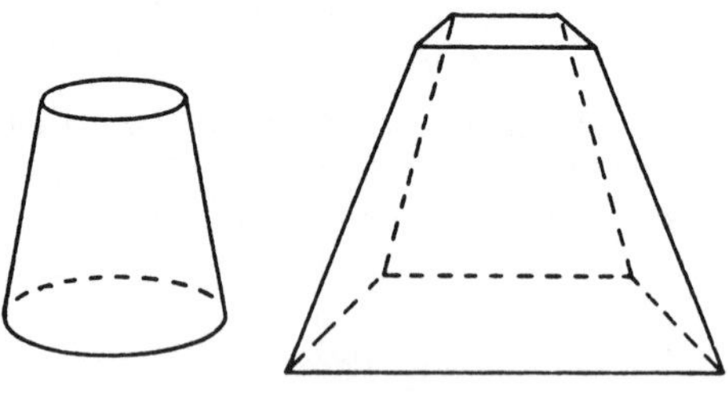

Fig. 105

Example. A frustum of a square pyramid has a height of 4 metres and the lengths of the sides of the square base and top are 5 and 2 metres respectively. Find the volume of the frustum.

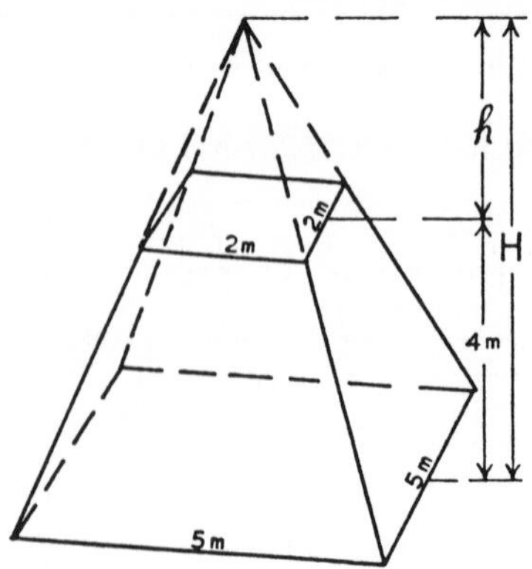

Fig. 106

By similar triangles,

$$\frac{h}{2} = \frac{h+4}{5}$$

$$5h = 2h + 8$$

$$h = 2\tfrac{2}{3}\text{ m}$$

$$H = 2\tfrac{2}{3} + 4 = 6\tfrac{2}{3}\text{ m}$$

Volume of whole pyramid $= \frac{1}{3} \times 5^2 \times 6\frac{2}{3}\text{ m}^3$

Volume of top cut off $= \frac{1}{3} \times 2^2 \times 2\frac{2}{3}\text{ m}^3$

$\therefore$ Volume of frustum $= \frac{1}{3}(5^2 \times 6\frac{2}{3} - 2^2 \times 2\frac{2}{3})$

$= 52\text{ m}^3$ Ans.

SPHERE

The sphere may be considered as being made up of a great number of very small pyramids whose bases lie on the surface of the sphere and their apices all meeting at the centre of the sphere.

Volume of each pyramid

$= \frac{1}{3}$ (area of base $\times$ perpendicular height)

Volume of sphere

= sum of volumes of all pyramids

$= \frac{1}{3} \times$ perpendicular height $\times$ sum of areas of bases

$= \frac{1}{3} \times$ radius of sphere $\times$ curved surface area of sphere

$$= \frac{1}{3} \times \frac{d}{2} \times \pi d^2$$

$$= \frac{\pi}{6} d^3 \text{ or } \frac{4}{3}\pi r^3$$

Volume of hollow sphere

= vol. of outer sphere $-$ vol. of inner spherical space

$$= \frac{\pi}{6} D^3 - \frac{\pi}{6} d^3$$

$$= \frac{\pi}{6}(D^3 - d^3)$$

Example. A solid lead cone, 40 mm diameter at the base and 120 mm perpendicular height is to be melted down and cast into a hollow sphere of 10 mm uniform thickness. Find the inside and outside diameters of the sphere.

Working in centimetres:

$$\text{Let } d = \text{inside diameter,}$$
$$\text{then } (d + 2)\text{ cm} = \text{outside diameter.}$$

$$\text{Volume of hollow sphere} = \frac{\pi}{6}(D^3 - d^3)$$
$$= \frac{\pi}{6}[(d+2)^3 - d^3]$$
$$= \frac{\pi}{6}[d^3 + 6d^2 + 12d + 8 - d^3]$$
$$= \frac{\pi}{6}(6d^2 + 12d + 8)$$

$$\text{Volume of material from cone} = \frac{1}{3} \times \frac{\pi}{4} \times 4^2 \times 12 = \pi \times 16\text{ cm}^3$$

$$\frac{\pi}{6}(6d^2 + 12d + 8) = \pi \times 16$$
$$6d^2 + 12d + 8 = 96$$
$$3d^2 + 6d - 44 = 0$$

Solving this quadratic:

$$\left.\begin{aligned} d &= 2{\cdot}958\text{ cm or }29{\cdot}58\text{ mm} \\ D &= 4{\cdot}958\text{ cm or }49{\cdot}58\text{ mm} \end{aligned}\right\}\text{Ans.}$$

SPHERICAL SEGMENT

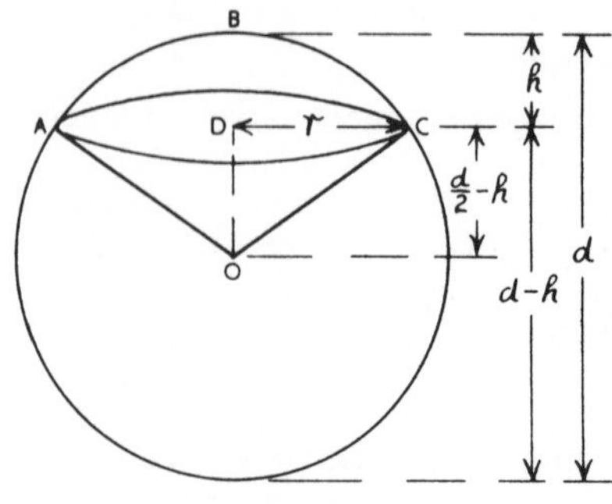

Fig. 107

A segment of a sphere is shown by ABCDA in Fig. 107. Let its radius DC $= r$, thickness BD $= h$, and diameter of sphere $= d$. The volume of the spherical segment can be obtained by subtracting the volume of the cone OADCO from the volume of the spherical sector OABCO.

Volume of sector of sphere. Consider this as being made up of a great number of very small pyramids whose bases lie on the spherical surface of the sector and their apices all meeting at the centre of the sphere.

Volume of pyramid $= \frac{1}{3}$ area of base $\times$ perp. height

Volume of sector $= \frac{1}{3} \times$ perp. height $\times$ sum of areas of bases.

The sum of the areas of the bases is the curved surface area of the slice of the sphere and this has been shown (Chapter 11) to be equal to the curved surface area of the corresponding slice of the circumscribing cylinder, which is πdh.

$$\text{Volume of sector} = \tfrac{1}{3} \times \tfrac{1}{2}d \times \pi dh$$

$$= \frac{\pi}{6} d^2 h$$

It will be seen that the volume of the spherical sector and the volume of the whole sphere are in the ratio of their spherical surface areas, thus,

$$\frac{\text{Volume of sector}}{\text{Volume of sphere}} = \frac{\text{Surface area of sector}}{\text{Surface area of sphere}}$$

$$\text{Volume of sector} = \frac{\pi}{6} d^3 \times \frac{\pi dh}{\pi d^2}$$

$$= \frac{\pi}{6} d^2 h \text{ as above}$$

$$\text{Volume of cone} = \tfrac{1}{3} \text{ area of base } \times \text{ perp. height}$$

$$= \tfrac{1}{3} \pi r^2 \left(\frac{d}{2} - h\right)$$

$$= \frac{\pi}{6} r^2 (d - 2h)$$

Substituting r^2 in terms of d and h, by crossed chords,

$$r^2 = h(d - h)$$

$$\text{Volume of cone} = \frac{\pi}{6} h(d - h)(d - 2h)$$

$$= \frac{\pi}{6} h(d^2 - 3dh + 2h^2)$$

$$\text{Volume of segment} = \text{Vol. of sector} - \text{Vol. of cone}$$

$$= \frac{\pi}{6} d^2 h - \frac{\pi}{6} h(d^2 - 3dh + 2h^2)$$

$$= \frac{\pi}{6} h(d^2 - d^2 + 3dh - 2h^2)$$

$$= \frac{\pi}{6} h^2 (3d - 2h)$$

Example. A segment 2 cm thick is cut off a sphere 10 cm diameter. Find the volumes of the segment cut off, the remaining segment, and the whole sphere.

$$\text{Volume of segment cut off} = \frac{\pi}{6} h^2 (3d - 2h)$$

$$= \frac{\pi}{6} \times 2^2 (3 \times 10 - 2 \times 2)$$

$$= \frac{\pi}{6} \times 4 \times 26 = 54{\cdot}46 \text{ cm}^3$$

Ans. (i)

$$\text{Thickness of remaining segment} = 10 - 2 = 8 \text{ cm}$$

$$\text{Volume} = \frac{\pi}{6} h^2 (3d - 2h)$$

$$= \frac{\pi}{6} \times 8^2 (3 \times 10 - 2 \times 8)$$

$$= \frac{\pi}{6} \times 64 \times 14 = 469{\cdot}2 \text{ cm}^3$$

Ans. (ii)

Volume of sphere $=$ sum of vols. of the two segments

$= 54{\cdot}46 + 469{\cdot}2$

$= 523{\cdot}66 \text{ cm}^3$

Ans. (iii)

As a check:

$$\text{Volume of sphere} = \frac{\pi}{6} d^3 = \frac{\pi}{6} \times 10^3$$

$$= 523{\cdot}66 \text{ cm}^3 \text{ as above}$$

Example. A sphere is sliced into three pieces by two parallel cuts. The top segment is 8 cm thick and 24 cm diameter at its base, the bottom segment is 5 cm thick. Calculate the volume of the zone of sphere between the segments.

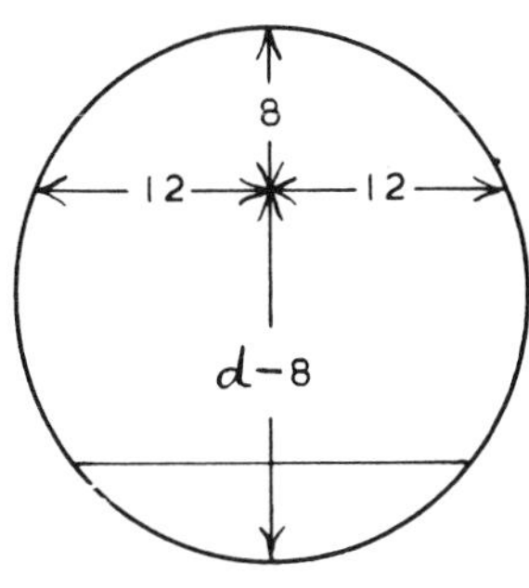

Fig. 108

Let d = diameter of sphere, by crossed chords:

$$8 \times (d - 8) = 12 \times 12$$

$$8d - 64 = 144$$

$$d = 26 \text{ cm}$$

$$\text{Volume of sphere} = \frac{\pi}{6} d^3$$

$$= \frac{\pi}{6} \times 26^3 = 9204 \text{ cm}^3$$

$$\text{Volume of spherical segment} = \frac{\pi}{6} h^2 (3d - 2h)$$

$$\text{Volume of top segment} = \frac{\pi}{6} \times 8^2(3 \times 26 - 2 \times 8)$$

$$= \frac{\pi}{6} \times 64 \times 62 = 2078 \text{ cm}^3$$

$$\text{Volume of bottom segment} = \frac{\pi}{6} \times 5^2(3 \times 26 - 2 \times 5)$$

$$= \frac{\pi}{6} \times 25 \times 68 = 890 \text{ cm}^3$$

$$\text{Volume of zone} = \text{volume of sphere} - \text{top and bottom segments}$$

$$= 9204 - 2078 - 890$$

$$= 6236 \text{ cm}^3 \quad \text{Ans.}$$

THEOREM OF PAPPUS OR GULDINUS APPLIED TO VOLUMES

If an *area,* situated wholly on one side of a fixed axis, be rotated in its own plane about this axis, it will sweep out a *volume* equal to the product of the area and the distance its centre of gravity moves.

In the majority of cases we will deal with areas being swept around one complete revolution and the resultant volumes are referred to as "solids of revolution".

For example, consider a flat circular disc of radius r, its centre being at R from the axis $o\ o$ as in Fig. 109, if this area is swept around the axis through one complete revolution, it will sweep out a solid ring of circular section, the mean radius of the ring being R, and the radius of the cross-section of the material being r.

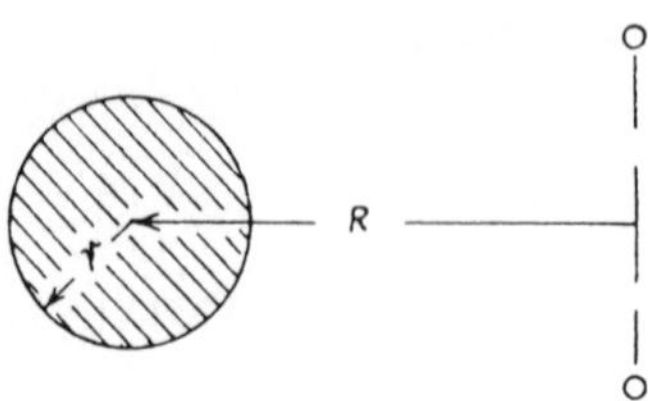

Fig. 109

A practical example of this is the volume of an anchor ring, thus:

$$\text{Volume swept out} = \text{area} \times \text{distance its c.g. moves}$$
$$\text{Volume of anchor ring} = \pi r^2 \times 2\pi R$$
$$= 2\pi^2 Rr^2$$

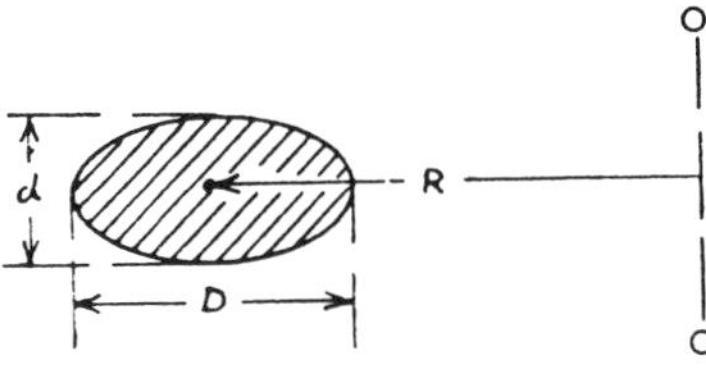

Fig. 110

Now consider an elliptical area of major diameter D, minor diameter d, swept around through one revolution, R being the radius from the axis $o\ o$ to the centre of the ellipse (Fig. 110).

The volume swept out is the shape of a circular lifebuoy of elliptical cross-section.

$$\text{Volume swept out} = \text{area} \times \text{distance its c.g. moves}$$

$$\text{Volume of lifebouy} = \frac{\pi}{4} Dd \times 2\pi R$$

$$= \frac{\pi^2}{2} DdR$$

The formulae produced above for the volumes of an anchor ring and lifebuoy are not intended for the student to memorise, they have been derived as demonstrations on the use of the ring theorem. In such cases as these it is best for students to work from first principles making use of this theorem rather than applying formulae which can so easily be mis-read with regard to which dimension each symbol of the formula represents.

Example. The inside diameter of a solid circular cork lifebuoy is 500 mm and the section is elliptical 160 mm major diameter by 120 mm minor diameter. Find (i) its volume in cubic metres, (ii) its mass if the density of cork is 240 kg/m^3.

Referring to Fig. 110, working in metres:

$$D = 160 \text{ mm} = 0{\cdot}16 \text{ m}$$
$$d = 120 \text{ mm} = 0{\cdot}12 \text{ m}$$
$$R = 250 + 80 = 330 \text{ mm} = 0{\cdot}33 \text{ m}$$

$$\begin{aligned}\text{Volume swept out} &= \text{area} \times \text{distance its c.g. moves}\\ &= 0{\cdot}7854 \times D \times d \times 2\pi R\\ &= 0{\cdot}7854 \times 0{\cdot}16 \times 0{\cdot}12 \times 2\pi \times 0{\cdot}33\\ &= 0{\cdot}03127 \text{ m}^3 \quad \text{Ans. (i)}\end{aligned}$$

$$\begin{aligned}\text{Mass [kg]} &= \text{volume [m}^3] \times \text{density (kg/m}^3)\\ &= 0{\cdot}03127 \times 240\\ &= 7{\cdot}504 \text{ kg} \quad \text{Ans. (ii)}\end{aligned}$$

FORCE, WEIGHT and CENTRE OF GRAVITY

FORCE is that which produces or tends to produce motion in a body. The SI unit of force is the *newton* (abbreviation N) and may be defined as the force required to give unit acceleration (a gain of velocity of one metre per second every second the force is applied) to unit mass (one kilogramme).

The WEIGHT of a body is the gravitational force on the mass of that body, that is, the force of attraction exerted on the body by the earth.

If a body is allowed to fall freely, it will fall with an acceleration of 9·81 metres per (second)^2, this is termed *gravitational acceleration* and is represented by g. Since one newton of force will give one kilogramme of mass an acceleration of one metre per (second)^2, then the force in newtons to give m kg of mass an acceleration of 9·81 m/s^2 is $m \times 9{\cdot}81$. Hence, at the earth's surface, the gravitational force on a mass of m kg is mg newtons, therefore:

$$\text{Weight [N]} = \text{mass [kg]} \times g \text{ [m/s}^2]$$

We see then that weight is a *force* and, in engineering must be expressed in force units, that is, in newtons or in multiples of the newton (kilonewtons etc.). However, outside of design work, such as in the home, in commerce, and in the loading of ships' fuel and cargo, loads will probably be mainly expressed in grammes, kilogrammes and tonnes. When loads are given in mass units, the engineer must convert into force units (1 kilogramme = 9·81 newtons) when performing design calculations.

The CENTRE OF GRAVITY of a body is that point through which the whole weight of the mass can be considered as acting. For instance, if we imagine a body to be compressed in volume into one tiny particle without losing mass, the position of this small heavy particle would be at the centre of gravity of the body to have the same effect. If the body was suspended from this point, or supported on it, it would balance perfectly without tilting.

When dealing with an *area* instead of a solid, an area theoretically has no mass, therefore it is not perfectly correct to use the term centre of gravity, in such cases we use the term *centroid.* However, provided the student understands this there is no harm for the practical man to use centre of gravity almost throughout as he probably has a practical object in mind when dealing with an area, such as a triangular metal plate when considering a triangle, and so on.

Up to the present, some centres of gravity have been introduced only in obvious cases, we will now proceed further.

PARALLELOGRAM

A parallelogram would balance if laid on a knife edge along either one of its two diagonals, therefore its centre of gravity (for a plate) or centroid (for a pure area) is at the intersection of its diagonals as shown in Fig. 111.

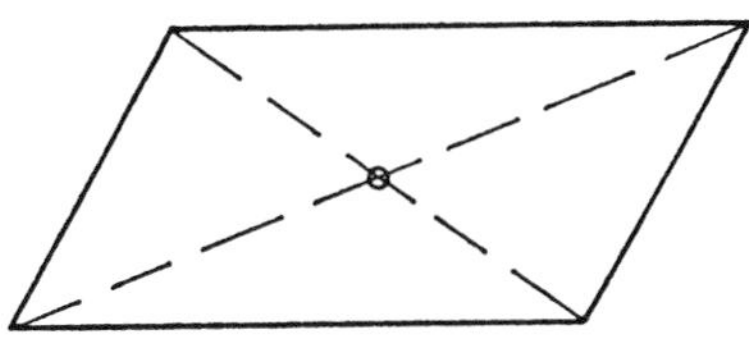

Fig. 111

TRAPEZIUM

The centroid of a trapezium is at the intersection of EF and HG as shown in Fig. 112, found graphically as follows:

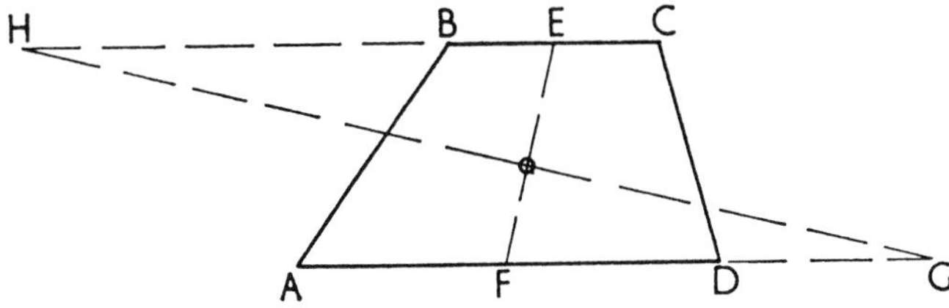

Fig. 112

Bisect BC to get mid-point E

Bisect AD to get mid-point F

Join EF

Produce AD to G, length DG being equal to BC

Produce BC to H, length BH being equal to AD

Join HG

The intersection of EF and HG is the centroid.

TRIANGLE

The centroid of a triangle lies on a line joining a corner of the triangle with the mid-point of its opposite side, at a point at one-third of the height from that side. See Fig. 113.

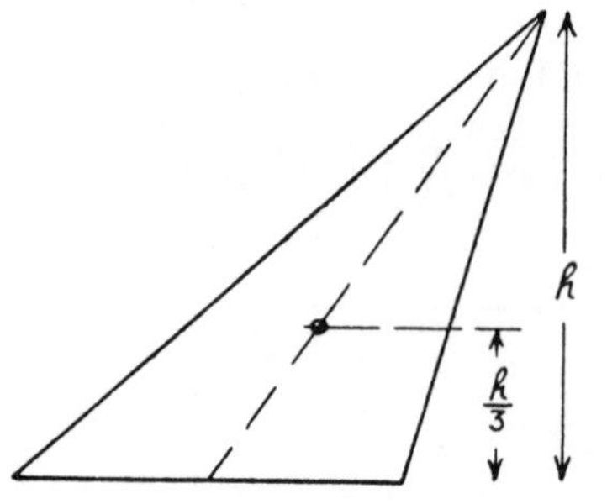

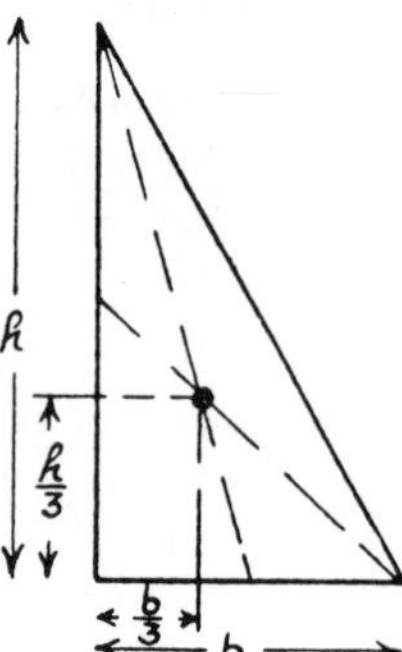

Fig. 113

PYRAMID

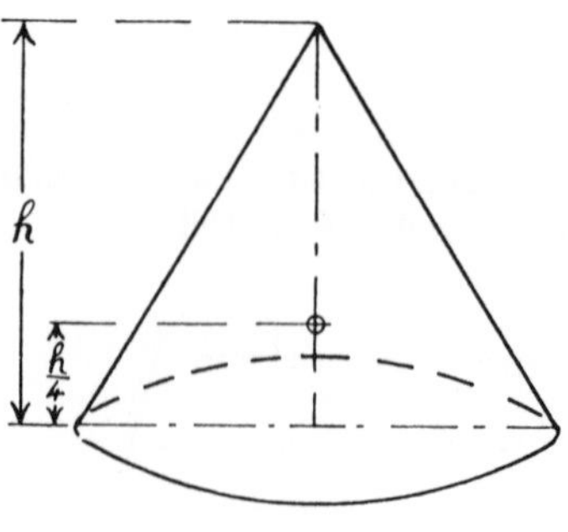

Fig. 114

The centre of gravity of a pyramid or cone is at one-quarter of the height above the base. See Fig. 114.

SEMI-CIRCULAR AREA

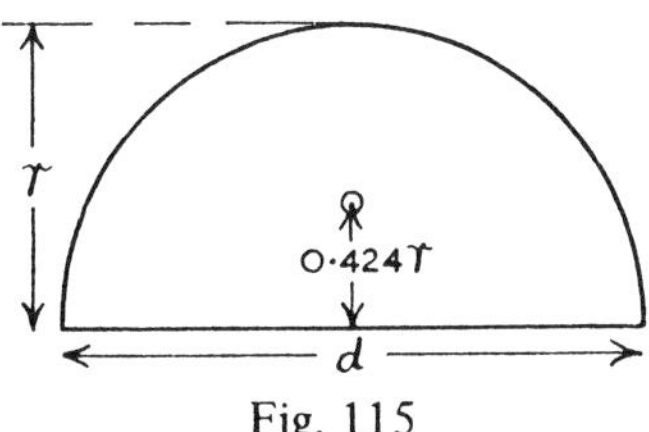

Fig. 115

The centroid of a semi-circular area is at 0·424r from its diameter, as shown in Fig. 115. This can easily be shown by the theorem of Pappus:

If the semi-circular area is swept around through one revolution about an axis on its own diameter, the volume swept out will be that of a sphere. Referring to Fig. 115,

Let the c.g. be at x from the diameter.

Volume swept out = area × distance its c.g. moves

$$\frac{4}{3}\pi r^3 = \tfrac{1}{2}\pi r^2 \times 2\pi x$$

$$x = \frac{4 \times \pi \times r^3 \times 2}{3 \times \pi \times r^2 \times 2\pi}$$

$$= \frac{4r}{3\pi}$$

$$= 0{\cdot}424r$$

HEMISPHERE

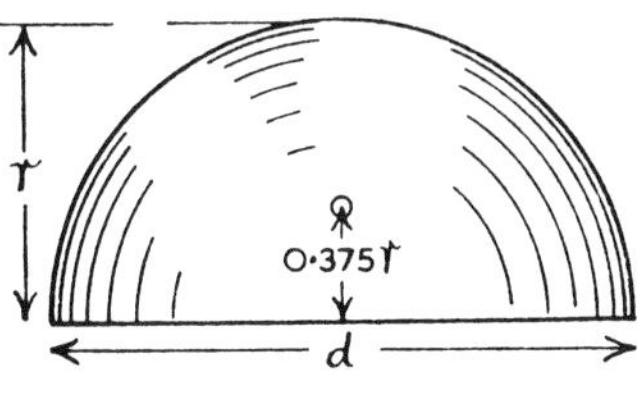

Fig. 116

The centre of gravity of a hemisphere is at three-eighths of the radius above the diameter. See Fig. 116.

VOLUME OF CONE

Now that the position of the centroid of a triangle is known, it can be shown that the volume of a cone can be determined by the ring theorem.

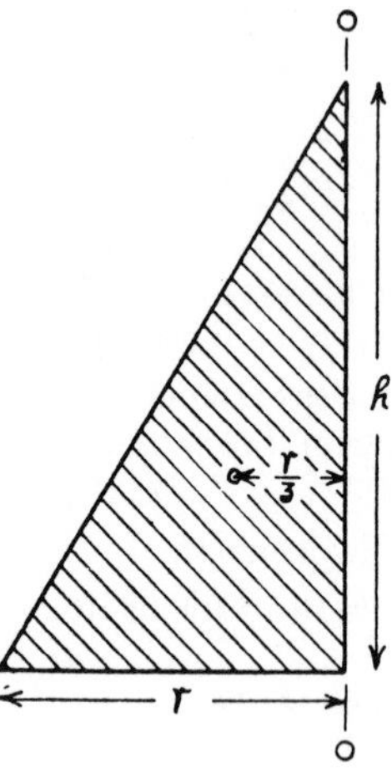

Fig. 117

If a triangular area is swept about an axis *o o* which coincides with its side as shown in Fig. 117, it will, in one revolution, sweep out the volume of a cone. If the radius of the base of the cone is r, the centroid of the triangle is at $\frac{1}{3}r$ from *o o*.

$$\text{Volume swept out} = \text{area of triangle} \times \text{distance its c.g. moves}$$

$$\text{Volume of cone} = \tfrac{1}{2}(\text{base} \times \text{perp. ht.}) \times (2\pi \times \tfrac{1}{3}r)$$

$$= \frac{r \times h \times 2\pi \times r}{2 \times 3}$$

$$= \tfrac{1}{3}\pi r^2 h \text{ (as shown previously)}$$

VOLUME OF FRUSTUM OF A CONE

To derive a formula for the volume of a frustum of a cone, let r = radius at top, R = radius at bottom, and h = perpendicular height.

Consider a trapezium made up of a rectangle and a triangle as shown in Fig. 118. If this area is rotated one revolution about the axis $o\,o$, a frustum of a cone will be swept out.

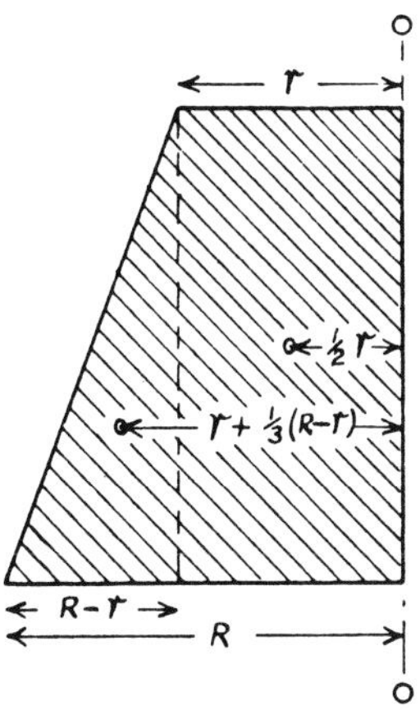

Fig. 118

$$\begin{aligned}
\text{Area of rectangle} &= rh \\
\text{c.g. from } o\,o &= \tfrac{1}{2}r \\
\text{Area of triangle} &= \tfrac{1}{2}(\text{base} \times \text{perp. height}) \\
&= \tfrac{1}{2}(R - r)h \\
\text{c.g. from } o\,o &= r + \tfrac{1}{3}(R - r) \\
&= r + \tfrac{1}{3}R - \tfrac{1}{3}r \\
&= \tfrac{1}{3}R + \tfrac{2}{3}r \\
\text{Volume swept out by rectangle} &= rh \times 2\pi \times \tfrac{1}{2}r \\
&= \pi r^2 h \quad \ldots \quad \ldots \quad \ldots \quad \text{(i)} \\
\text{Volume swept out by triangle} &= \tfrac{1}{2}(R - r)h \times 2\pi(\tfrac{1}{3}R + \tfrac{2}{3}r) \\
&= \pi h(R - r)(\tfrac{1}{3}R + \tfrac{2}{3}r) \\
&= \pi h(\tfrac{1}{3}R^2 + \tfrac{1}{3}Rr - \tfrac{2}{3}r^2) \quad \text{(ii)} \\
\text{Total volume} &= \pi r^2 h + \pi h(\tfrac{1}{3}R^2 + \tfrac{1}{3}Rr - \tfrac{2}{3}r^2) \\
&= \pi h(r^2 + \tfrac{1}{3}R^2 + \tfrac{1}{3}Rr - \tfrac{2}{3}r^2) \\
&= \pi h(\tfrac{1}{3}R^2 + \tfrac{1}{3}Rr + \tfrac{1}{3}r^2) \\
&= \tfrac{1}{3}\pi h(R^2 + Rr + r^2)
\end{aligned}$$

or, since $R = \tfrac{1}{2}D$, and $r = \tfrac{1}{2}d$,

$$\text{Volume} = \tfrac{1}{12}\pi h(D^2 + Dd + d^2)$$

SIMILAR OBJECTS

Volumes of similar objects vary as the cube of their corresponding linear dimensions.

Similar objects mean that they are of the same shape and proportions, but of different size.

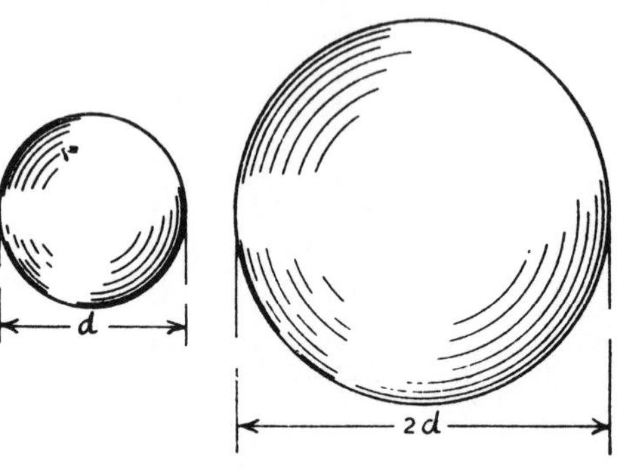

Fig. 119

Consider two solid spheres, one of diameter *d*, the other having a diameter of 2*d*, as in Fig. 119,

$$\text{Volume of small sphere} = \frac{\pi}{6} d^3$$

$$\text{,, ,, large ,,} = \frac{\pi}{6} \times (2d)^3$$

Thus, the ratio of their volumes is:

$$d^3 : (2d)^3$$
$$= d^3 : 2^3 \times d^3$$
$$= 1 : 2^3$$

Hence, the diameter of the larger sphere being twice the diameter of the smaller, all linear dimensions such as the radius and circumference are twice as much, all areas such as the sectional area and curved surface area are $2^2 = 4$ times as much (as explained in last chapter), the volume is $2^3 = 8$ times the volume of the smaller, and consequently, the mass of the larger is 8 times the mass of the smaller if they are made of the same kind of material.

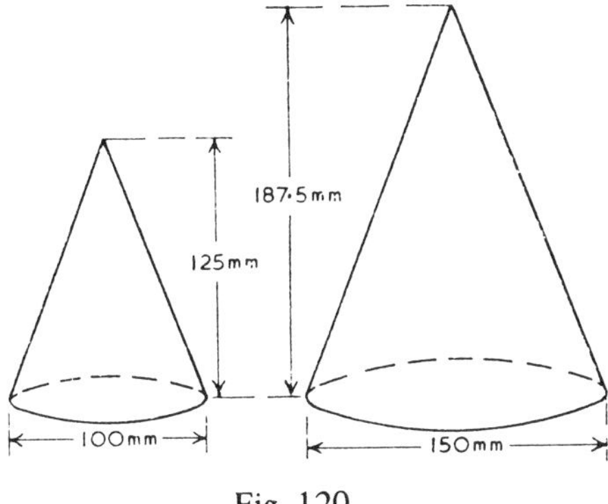

Fig. 120

Fig. 120 shows two similar cones, they are similar because they have the same proportions, for instance, in each case the perpendicular height is 1·25 times the diameter of the base. The dimensions of the larger cone is 1·5 times the corresponding dimensions of the smaller, thus:

Ratio of corresponding dimensions	$= 1$	$: 1{\cdot}5$
,, ,, areas	$= 1^2$	$: 1{\cdot}5^2$
	$= 1$	$: 2{\cdot}25$
,, ,, volumes	$= 1^3$	$: 1{\cdot}5^3$
	$= 1$	$: 3{\cdot}375$

Hence, all linear dimensions (such as diameter of base, circumference of base, perpendicular height, slant height) of the larger cone are 1·5 times those of the smaller cone; all areas (such as area of base, curved surface area, sectional area) of the larger cone are $1{\cdot}5^2 = 2{\cdot}25$ times the corresponding areas of the smaller; and the volume of the larger cone is $1{\cdot}5^3 = 3{\cdot}375$ times the volume of the smaller.

SIMPSON'S RULE APPLIED TO VOLUMES

The procedure of finding the volume of an irregular object by Simpson's rule is the same as for the area of an irregular figure, it is merely a matter of substituting cross-sectional areas for ordinates, Thus:

To the first and last cross-sectional areas, add four times the even cross-sectional areas and twice the odd; multiply this sum by one-third the common interval and the result is the volume of the object.

Example. A casting of light alloy, 750 mm long, has a variable cross-sectional area throughout its length. At regular distances of 125 mm apart, starting from one end, the sectional areas are, 12·2, 17·5, 23·2, 27·9, 21·0, 11·2 and 0 square centimetres respectively. Find the volume and its mass if the density of the material is 3·2 g/cm^3.

Sectional Areas (cm^2)	Simpson's Multipliers	Products
12·2	1	12·2
17·5	4	70·0
23·2	2	46·4
27·9	4	111·6
21·0	2	42·0
11·2	4	44·8
0	1	0
	Sum =	327·0

$$\text{Common interval} = 12{\cdot}5 \text{ cm}$$

$$\begin{aligned} \text{Volume} &= \text{sum of products} \times \tfrac{1}{3} \text{ common interval} \\ &= 327 \times \tfrac{1}{3} \times 12{\cdot}5 \\ &= 1362{\cdot}5 \text{ cm}^3 \quad \text{Ans. (i)} \end{aligned}$$

$$\begin{aligned} \text{Mass [g]} &= \text{volume [cm}^3\text{]} \times \text{density [g/cm}^3\text{]} \\ &= 1362{\cdot}5 \times 3{\cdot}2 \\ &= 4360 \text{ g} = 4{\cdot}36 \text{ kg} \quad \text{Ans. (ii)} \end{aligned}$$

Example. A barrel is 960 mm long; the diameter at each end is 600 mm, at quarter and three-quarter lengths the diameter is 720 mm, and at mid-length the diameter is 800 mm. Find the capacity of the barrel in cubic metres and also in litres.

Working in metres:

Sectional Areas	Multipliers	Products
$0{\cdot}7854 \times 0{\cdot}6^2$	1	$0{\cdot}7854 \times 0{\cdot}36$
$0{\cdot}7854 \times 0{\cdot}72^2$	4	$0{\cdot}7854 \times 2{\cdot}0736$
$0{\cdot}7854 \times 0{\cdot}8^2$	2	$0{\cdot}7854 \times 1{\cdot}28$
$0{\cdot}7854 \times 0{\cdot}72^2$	4	$0{\cdot}7854 \times 2{\cdot}0736$
$0{\cdot}7854 \times 0{\cdot}6^2$	1	$0{\cdot}7854 \times 0{\cdot}36$
	Sum =	$0{\cdot}7854 \times 6{\cdot}1472$

$$\text{Common interval} = 0{\cdot}96 \div 4 = 0{\cdot}24 \text{ m}$$

$$\begin{aligned}\text{Volume} &= 0{\cdot}7854 \times 6{\cdot}1472 \times \tfrac{1}{3} \times 0{\cdot}24 \\ &= 0{\cdot}3863 \text{ m}^3 \quad \text{Ans. (i)} \\ &= 386{\cdot}3 \text{ litres} \quad \text{Ans. (ii)}\end{aligned}$$

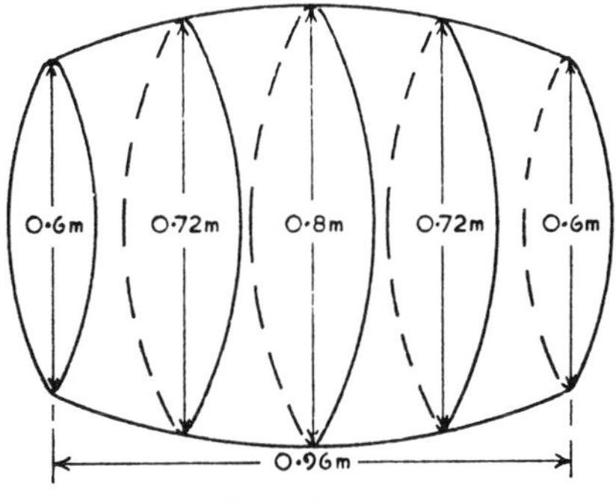

Fig. 121

Note that the common multiplier of 0·7854 is carried through to the final sum of the products to save labour.

This could be expressed in formula fashion thus:

$$\text{Volume} = \frac{h}{3} \times 0{\cdot}7854\,(d_1^2 + 4d_2^2 + 2d_3^2 + \text{etc.})$$

where d_1, d_2, d_3, etc., are the diameters, and h is the common interval between measurements, or

$$\text{Volume} = \frac{h}{3} \times \pi(r_1^2 + 4r_2^2 + 2r_3^2 + \text{etc.})$$

where r_1, r_2, r_3, etc., are the radii.

Example. Plot the graph $xy = 6$ between the limits $x = 2$ and $x = 6$. If the area under this graph is rotated about its x-axis through one complete revolution, calculate, by Simpson's rule, the volume swept out.

$$xy = 6 \quad \therefore y = \frac{6}{x}$$

Plotting points for values of x between the limits $x = 2$ and $x = 6$,

$x =$	2	3	4	5	6
$y = \frac{6}{x}$	3	2	1·5	1·2	1

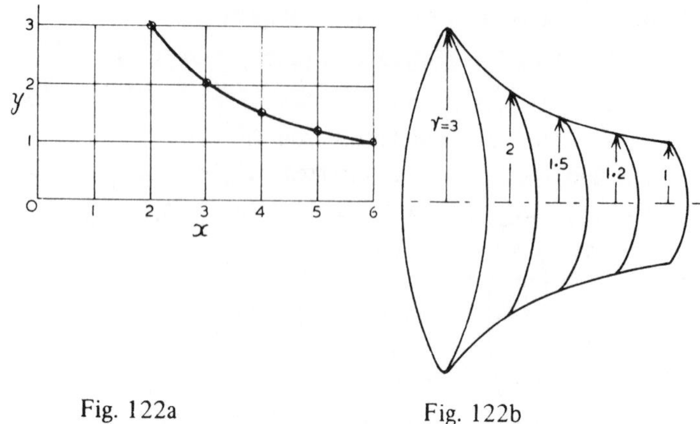

Fig. 122a Fig. 122b

The graph is shown in Fig. 122a. When the area under this graph is rotated about its x-axis through one revolution, the volume swept out appears as shown in Fig. 122b. The y-ordinates of the graph become the radii of the solid at regular intervals along its length. Putting the cross-sectional areas at these regular intervals through Simpson's rule as in the previous example:

Radii r	Cross-sectional Areas $= \pi r^2$	Simpson's Multipliers	Products
3	$\pi \times 9$	1	$\pi \times 9$
2	$\pi \times 4$	4	$\pi \times 16$
1·5	$\pi \times 2{\cdot}25$	2	$\pi \times 4{\cdot}5$
1·2	$\pi \times 1{\cdot}44$	4	$\pi \times 5{\cdot}76$
1	$\pi \times 1$	1	$\pi \times 1$
		Sum =	$\pi \times 36{\cdot}26$

Common interval between ordinates $= 1$

$$\text{Volume} = \pi \times 36{\cdot}26 \times \tfrac{1}{3} \times 1$$
$$= 37{\cdot}98 \text{ units}^3 \quad \text{Ans.}$$

FLOW OF LIQUID THROUGH PIPES, ETC.

VOLUME FLOW is the volume of a fluid flowing past a given point in unit time. The SI unit is cubic metre per second (m^3/s), other commonly used units are cubic metres per minute (m^3/min), cubic metres per hour (m^3/h), and litres per hour (l/h).

The velocity or speed of flow is the "length" of liquid which passes in a given time. For instance, if the velocity of the liquid is 2 metres per second it means that a column of the liquid 2 metres long passes every second, hence, volume flow is the product of the cross-sectional area of the flowing liquid and its velocity. In basic units:

$$\text{Volume flow } [m^3/s] = \text{cross-sect. area } [m^2] \times \text{velocity } [m/s]$$

MASS FLOW is the mass of fluid flowing past a given point in unit time. The SI unit is kilogramme per second (kg/s) and other recommended units are kilogrammes per hour (kg/h) and tonnes per hour (t/h).

Since density is the mass per unit volume, then the mass flow is the product of the volume flow and the density. In SI units:

$$\text{Mass flow } [kg/s] = \text{volume flow } [m^3/s] \times \text{density } [kg/m^3]$$

Example. Oil of density 0·85 g/ml flows full bore through a pipe 50 mm diameter at a velocity of 1·5 m/s. Find the quantity flowing, (i) in cubic metres per hour, (ii) kilogrammes per hour, (iii) tonnes per hour.

$$\text{Velocity} = 1{\cdot}5 \text{ m/s} = 1{\cdot}5 \times 3600 \text{ m/h}$$

$$\begin{aligned}\text{Volume flow } [m^3/h] &= \text{area } [m^2] \times \text{velocity } [m/h]\\ &= 0{\cdot}7854 \times 0{\cdot}05^2 \times 1{\cdot}5 \times 3600\\ &= 10{\cdot}6 \text{ m}^3/\text{h} \quad \text{Ans. (i)}\end{aligned}$$

$$\text{Density} = 0{\cdot}85 \text{ g/ml} = 0{\cdot}85 \times 10^3 \text{ kg/m}^3$$

$$\begin{aligned}\text{Mass flow } [kg/h] &= \text{volume flow } [m^3/h] \times \text{density } [kg/m^3]\\ &= 10{\cdot}6 \times 0{\cdot}85 \times 10^3\\ &= 9012 \text{ kg/h} \quad \text{Ans. (ii)}\\ &= 9{\cdot}012 \text{ tonne/h} \quad \text{Ans. (iii)}\end{aligned}$$

FLOW THROUGH VALVES. When a liquid flows out of the open end of a pipe, the maximum quantity of liquid escaping depends upon the area of the bore of the pipe end. The flow can be restricted by a cover over the pipe end so that the area of escape is less than the area of the pipe bore.

Referring to Fig. 123, the area of escape is the circumferential opening, circumference × lift. The maximum effective lift will be when this circumferential area of escape is equal to the area of the bore, thus,

$$\text{circumference} \times \text{lift} = \text{area of bore}$$

$$\pi \times d \times \text{lift} = \frac{\pi}{4} \times d^2$$

$$\text{lift} = \frac{d}{4}$$

A practical example is a valve over a valve seat, neglecting the area taken up by the wings of the valve, the maximum effective lift is one-quarter of the valve diameter. If the lift is more than this, no more liquid can flow through than that which is allowed by the area of the bore of the seat, but if the lift is less than one-quarter of the diameter, the circumferential area of escape is less than the area of the seat bore, and the quantity of liquid flowing through depends upon the area, circumference × lift.

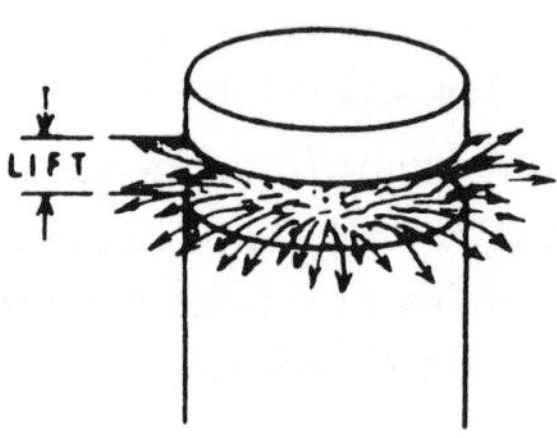

Fig. 123

Example. Calculate the quantity of water flowing, in litres per minute, through a valve 100 mm diameter when the lift is 15 mm and the velocity of the water is 3 metres per second, assuming that the wings of the valve take up one-sixth of the circumference.

One sixth of the area is obstructed by the wings, this leaves five-sixths of the area for the water to flow through. Working in metres:

Circumferential area of escape between valve and seat

$$= \frac{5}{6} \times \pi \times 0{\cdot}1 \times 0{\cdot}015 \text{ m}^2$$

$$\text{Volume flow } [\text{m}^3/\text{s}] = \text{area } [\text{m}^2] \times \text{velocity } [\text{m/s}]$$

$$= \frac{5}{6} \times \pi \times 0{\cdot}1 \times 0{\cdot}015 \times 3 \text{ m}^3/\text{s}$$

$1\text{m}^3 = 10^3$ litres, $\therefore$ litres flowing per minute

$$= \frac{5}{6} \times \pi \times 0{\cdot}1 \times 0{\cdot}015 \times 3 \times 10^3 \times 60$$

$$= 706{\cdot}9 \text{ l/min} \quad \text{Ans.}$$

CENTRES OF GRAVITY BY FIRST MOMENTS

The moment of a force about a given point is the product of the force and the perpendicular distance of its line of action from that point.

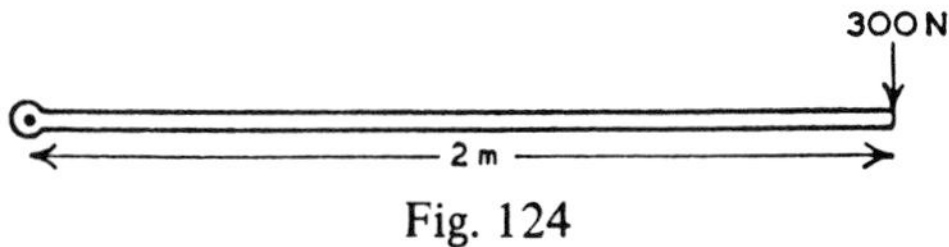

Fig. 124

Fig. 124 illustrates a force of 300 newtons acting on the end of a lever, at a perpendicular distance of 2 metres from o, its moment therefore is 300 × 2 = 600 newton-metres. This is the effect of the force which tends to turn the lever in a clockwise direction around the axis O. A force of 400N acting at 1·5 m, or 600N at 1 m leverage, would have the same turning effect.

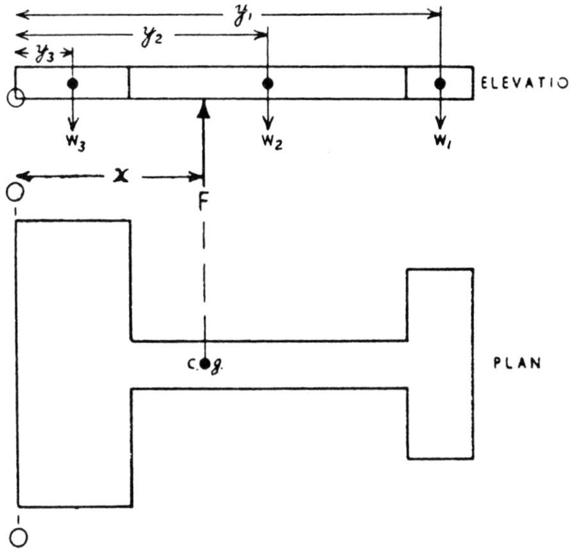

Fig. 125

Consider a piece of plate cut to shape shown in Fig. 125. Imagine this plate supported horizontally on one single support, we call this support the fulcrum. The fulcrum (F) must of course be positioned exactly at the centre of gravity of the plate if the plate is to be perfectly balanced because "the centre of gravity is that position through which the whole weight can be considered as acting".

If we now take moments about the end $o\ o$, this means that we imagine the plate to be temporarily hinged at this end, for perfect equilibrium the moments of all the forces tending to turn the plate clockwise around the hinge must be equal to the moments of the forces tending to turn the plate anticlockwise about the hinge.

Let W_1, W_2 and W_3 represent the weights of the top, centre and bottom parts.

Let y_1, y_2 and y_3 represent the distances of the centres of gravity of these parts from $o\ o$.

Let x represent the position of the fulcrum (F) from $o\ o$, and as previously stated this is the centre of gravity of the whole plate.

Moments about $o\ o$:

$$\text{Clockwise moments} = \text{Anticlockwise moments}$$

$$W_1 \times y_1 + W_2 \times y_2 + W_3 \times y_3 = F \times x$$

Since the total upward force F must be equal to the total downward force, then $F = W_1 + W_2 + W_3$

$$\therefore W_1 \times y_1 + W_2 \times y_2 + W_3 \times y_3 = (W_1 + W_2 + W_3) \times x$$

$$x = \frac{W_1 y_1 + W_2 y_2 + W_3 y_3}{W_1 + W_2 + W_3}$$

In words this is:

$$x = \frac{\text{Summation of all moments of weights}}{\text{Summation of all weights}}$$

The Greek letter Σ (sigma) is the usual symbol to signify "summation of" thus,

$$x = \frac{\Sigma \text{ moments of weights}}{\Sigma \text{ weights}} \quad \ldots \quad \ldots \quad \ldots \quad \ldots \quad \text{(i)}$$

As previously explained, weight is equal to mass $\times$ g, and since g is constant, then weight can be represented by mass:

$$x = \frac{\Sigma \text{ moments of masses}}{\Sigma \text{ masses}} \quad \ldots \quad \ldots \quad \ldots \quad \ldots \quad \text{(ii)}$$

Mass is obtained from, volume × density, so that if the material is the same kind throughout, the same value for the density of the material will be in every term of the above expression and will cancel out, in effect we then have:

$$x = \frac{\Sigma \text{ moments of volumes}}{\Sigma \text{ volumes}} \quad \ldots \quad \ldots \quad \ldots \quad \ldots \quad \text{(iii)}$$

Further, volume = area × thickness, therefore if the thickness of the material is uniform throughout as well as its density, then the thickness will also cancel from every term, leaving us with:

$$x = \frac{\Sigma \text{ moments of areas}}{\Sigma \text{ areas}} \quad \ldots \quad \ldots \quad \ldots \quad \ldots \quad \text{(iv)}$$

Hence, for convenience we often take moments of masses, moments of volumes and moments of areas. The above expressions provide methods of finding the position of the centre of gravity of objects and figures made up from regular shapes.

Moments as used above are usually referred to as "first moments" to distinguish them from "second moments" which are used in some subjects in mechanics and naval architecture. Second moments will be dealt with in the study of those subjects.

Example. To find the position of the centre of gravity of the plate illustrated in Fig. 126 which is of uniform thickness throughout, the dimensions shown being all in millimetres.

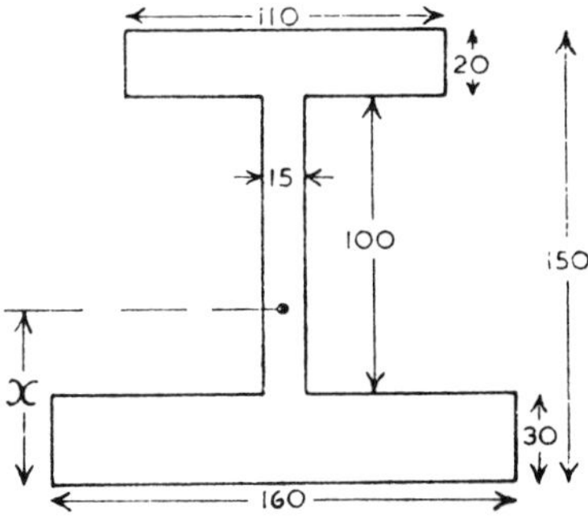

Fig. 126

Working in centimetres:

Area of top flange	=	11×2	= $22\ cm^2$
,, centre web	=	$10 \times 1{\cdot}5$	= 15 ,,
,, bottom flange	=	16×3	= 48 ,,

Distance of c.g. of top flange from base	=	14 cm
,, ,, centre web ,, ,,	=	8 ,,
,, ,, bottom flange ,,	=	1·5 ,,

Taking moments about base,

$$x = \frac{\Sigma \text{ moments of areas}}{\Sigma \text{ areas}}$$

$$= \frac{22 \times 14 + 15 \times 8 + 48 \times 1{\cdot}5}{22 + 15 + 48} = \frac{500}{85}$$

$= 5{\cdot}88$ cm or $58{\cdot}8$ mm above the base. Ans.

Example. A hole 30 mm diameter is bored through a solid disc 90 mm diameter, the centre of the hole being 25 mm from the centre of the disc. Find the position of the centre of gravity of the disc after the hole has been cut out.

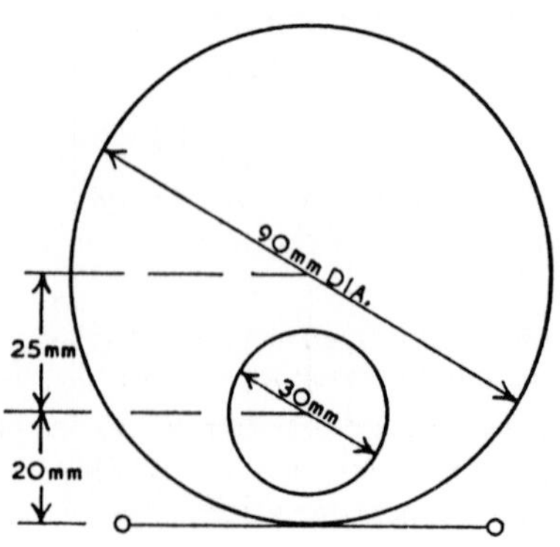

Fig. 127

Moments about base *o o*, working in centimetres:

$$x = \frac{\Sigma \text{ moments of areas}}{\Sigma \text{ areas}}$$

$$x = \frac{0{\cdot}7854 \times 9^2 \times 4{\cdot}5 - 0{\cdot}7854 \times 3^2 \times 2}{0{\cdot}7854 \times 9^2 - 0{\cdot}7854 \times 3^2}$$

Dividing every term by $0{\cdot}7854 \times 3^2$:

$$x = \frac{3^2 \times 4{\cdot}5 - 2}{3^2 - 1} = \frac{38{\cdot}5}{8}$$

$$= 4{\cdot}8125 \text{ cm from bottom,} \quad = 48{\cdot}125 \text{ mm}$$

or, $\quad 48{\cdot}125 - 45$

$$= 3{\cdot}125 \text{ mm from disc centre.} \quad \text{Ans.}$$

(Alternatively, moments can be taken about centre of disc.)

Note, in this example area is lost by boring the hole, therefore the summation of areas is the nett area obtained by subtracting the area of the hole from the area of the disc; also, the summation of moments of areas is the difference between the moments of areas of the disc and hole.

In each of the above two cases, it is obvious that the centre of gravity lies on the vertical centre line because the figures are symmetrical, therefore it is sufficient to calculate the position of the centre of gravity in one direction only. For figures that are not symmetrical, it is necessary to express the position of the centre of gravity in two directions at right angles to each other, say from the base and from one side, this is done by taking moments about these two datum lines separately.

IRREGULAR FIGURES. Simpson's rule can be employed to find the moment of an irregular area in a similar manner to which it is applied in finding the area.

To express the moment about a given point, the perpendicular distance of each ordinate is measured from that point, then:

Add together, the moment of the first ordinate, the moment of the last ordinate, four times the moments of the even ordinates, and twice the moments of the odd ordinates; multiply this sum by one-third of the common interval.

The centroid can then be found by dividing the moment of the area by the area.

As an example, take a right angled triangle of 80 mm base and 48 mm height as illustrated in Fig. 128, to find the area and position of centroid, from base *o o* by Simpson's rule. By taking a regular shape such as this it enables us to compare the results so obtained with those calculated from formulae.

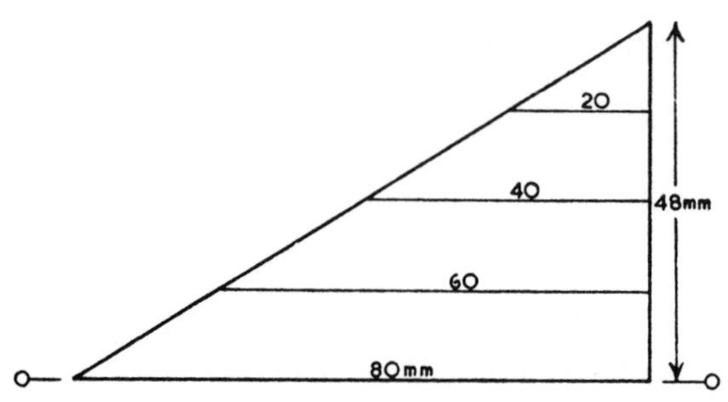

Fig. 128

Measurements from base, in centimetres:

i	ii	iii	iv	v
Ordinates	Simpson's multipliers	Products of i and ii	Distances of ordinates from base	Products of iii and iv
8	1	8	0	0
6	4	24	1·2	28·8
4	2	8	2·4	19·2
2	4	8	3·6	28·8
0	1	0	4·8	0
		Sum = 48		Sum = 76·8

$$\text{Common interval} = 4{\cdot}8 \div 4 = 1{\cdot}2 \text{ cm}$$

$$\text{Area} = 48 \times \tfrac{1}{3} \times 1{\cdot}2 = 19{\cdot}2 \text{ cm}^2 \quad \ldots \quad \text{(i)}$$

Moment of area about base

$$= 76{\cdot}8 \times \tfrac{1}{3} \times 1{\cdot}2 = 30{\cdot}72 \text{ cm}^3$$

$$\text{Centroid from base} = \frac{\text{moment of area}}{\text{area}}$$

$$= \frac{30{\cdot}72}{19{\cdot}2} = 1{\cdot}6 \text{ cm or } 16 \text{ mm} \quad \ldots \quad \text{(ii)}$$

It will be seen that the position of the centroid can be obtained by dividing the sum of column v by the sum of column iii thus,

$$\frac{76{\cdot}8}{48} = 1{\cdot}6 \text{ cm}$$

Further, if this triangle is swept through one complete revolution about its base *o o* it will sweep out the volume of a cone of dimensions 48 mm radius of base and 80 mm perpendicular height.

By theorem of Pappus:

$$\text{Volume swept out} = \text{area} \times \text{distance its centroid moves}$$
$$= 19{\cdot}2 \times 2\pi \times 1{\cdot}6$$
$$= 193{\cdot}1 \text{ cm}^3 \quad \ldots \quad \ldots \quad \ldots \quad \ldots \quad \text{(iii)}$$

The above agrees with results by formulae:

$$\text{Area} = \tfrac{1}{2}(\text{base} \times \text{perpendicular height})$$
$$= \tfrac{1}{2} \times 8 \times 4{\cdot}8 = 19{\cdot}2 \text{ cm}^2$$

$$\text{Centroid} = \tfrac{1}{3} \text{ of perpendicular height}$$
$$= \tfrac{1}{3} \times 4{\cdot}8 = 1{\cdot}6 \text{ cm}$$

$$\text{Volume of cone} = \tfrac{1}{3} \times \text{area of base} \times \text{perpendicular height}$$
$$= \tfrac{1}{3} \times \pi \times 4{\cdot}8^2 \times 8 = 193{\cdot}1 \text{ cm}^3$$

SHIFT OF CENTRE OF GRAVITY DUE TO SHIFT OF LOADS

Consider a system composed of loads which weigh w_1, w_2 and w_3 as shown in Fig. 129, the centre of gravity of each being h_1, h_2 and h_3

respectively from the base *o o*. Let *x* be the distance of the centre of gravity of the whole system from the base, then:

$$x_1 = \frac{\Sigma \text{ moments of weights}}{\Sigma \text{ weights}}$$

$$= \frac{w_1h_1 + w_2h_2 + w_3h_3}{w_1 + w_2 + w_3}$$

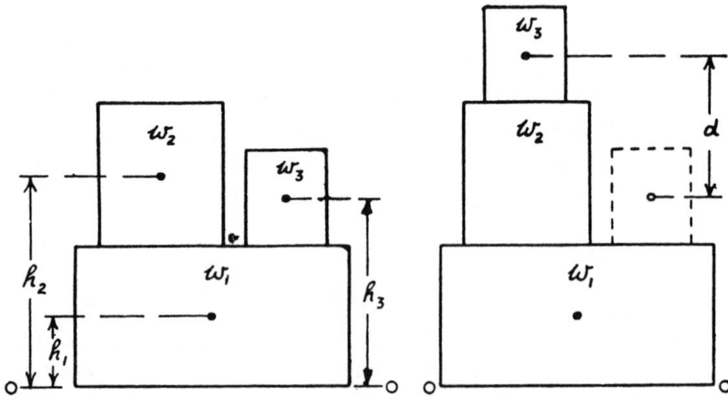

Fig. 129

If w_3 be lifted into the position shown, through a height of *d* the new centre of gravity of the whole system from the base, represented by x_2 is:

$$x_2 = \frac{w_1h_1 + w_2h_2 + w_3(h_3 + d)}{w_1 + w_2 + w_3}$$

$$= \frac{w_1h_1 + w_2h_2 + w_3h_3 + w_3d}{w_1 + w_2 + w_3}$$

The shift of the centre of gravity of the whole system in the direction measured from the base is $x_2 - x_1$ which is:

$$x_2 - x_1 = \frac{w_3d}{w_1 + w_2 + w_3}$$

In words this is:

$$\text{Shift of c.g.} = \frac{\text{weight shifted} \times \text{distance it is moved}}{\textit{total}\ \text{weight}}$$

As shown previously, weight can be represented by mass, therefore we can write:

$$\text{Shift of c.g.} = \frac{\text{mass shifted} \times \text{distance it is moved}}{\textit{total}\ \text{mass}}$$

This simple rule will be found very useful later in the study of Naval Architecture in cases dealing with the effect on the centre of gravity of a ship when cargo, fuel or water, etc., is shifted from one part of the ship to another.

TEST EXAMPLES 12

1. An I-section steel girder of 150 mm overall depth has unequal flanges, the top flange is 100 mm wide by 12 mm thick and the bottom flange is 140 mm wide by 14 mm thick. The centre web is 10 mm thick. Considering the flanges as rectangular in section by neglecting radii and fillets, calculate the mass in kilogrammes per metre run if the density of the material is 7·86 g/cm^3.

2. A hollow steel shaft, 400 mm outside diameter and 200 mm inside diameter, has a coupling 75 mm thick and 760 mm diameter at each end, and the overall length is 6 m. Neglecting fillets and coupling bolt holes, find the mass of the shaft in tonnes taking the density of steel as $7{\cdot}86 \times 10^3$ kg/m^3.

3. A cylinder and sphere and base of a cone are all the same diameter, and the heights of the cylinder and cone are each equal to the diameter of the sphere. Find the ratio of the volumes of the cylinder and sphere relative to the volume of the cone.

4. A piece of flat steel plate, having a mass of 6·5 kg/m^2, is cut to the shape of a sector of a circle of radius 180 mm and subtended angle at the centre 240 degrees, and the sector is rolled into a cone. Find (i) the mass of material used, (ii) the diameter of the base of the cone, (iii) perpendicular height of the cone, (iv) the capacity of the cone in litres.

5. An object is constructed by brazing the base of a solid cone to the flat surface of a solid hemisphere, the diameter of the base of the cone and the diameter of the hemisphere both being 60 mm, and the perpendicular height of the cone 50 mm. Find the mass of the object if the density of the materials is 8·4 g/cm^3.

6. A hollow lead sphere has a uniform thickness of 10 mm and its mass is 3 kg. Taking the specific gravity of lead as 11·4, find its outside diameter.

7. The height of a spherical segment is 4 cm and the diameter of its base is 14 cm. Find its volume.

8. A hole 24 mm diameter is bored centrally through a sphere 51 mm diameter. Calculate the volume of the drilled sphere in cm^3 and its mass if the density of the material is $7{\cdot}86 \times 10^3$ kg/m^3.

9. A tapered hole is bored through a right circular cone, concentric with the axis of the cone. The base diameter of the cone is 64 mm and the perpendicular height is 60 mm. The diameter of the hole at the base of the cone is 28 mm and the diameter where it breaks through the surface of the cone is 16 mm. Calculate the volume and mass of the remaining hollow frustum, taking the density of the material as 8·4 g/cm^3.

10. The lengths of the sides of the base of a regular hexagonal pyramid is 25 mm and the perpendicular height is 60 mm. Find the volume in cm^3. If this pyramid is cut through a plane parallel to its base at half the height, find the volume of the remaining frustum.

11. A vessel in the form of a hollow cone with vertex downwards, is partially filled with water. The volume of the water is 200 cm^3 and the depth of the water is 50 mm. Find the volume of water which must be added to increase the depth to 70 mm.

12. The diameter of the base of a hollow cone is 300 mm and its perpendicular height is 500 mm. It is partly filled with water so that when resting on its base the depth of the water is 250 mm. If the cone is inverted and balanced on its apex, what will then be the depth of the water?

13. The surface area of a solid sphere is $1\frac{1}{2}$ times the surface area of a smaller sphere, and the difference in their volumes is 10 cm^3. Find the volume and diameter of the smaller sphere.

14. The diameters of a barrel are 395 mm at each end, 477 mm at quarter and three-quarter lengths, and 500 mm at mid-length, and the total length is 581 mm. Using Simpson's rule calculate the capacity of the barrel in litres.

15. Plot the graph $y = 5 + 4x - x^2$ between the limits $x = -1$ and $x = +5$. If the area under this curve is rotated about its x-axis, find the volume swept out.

16. A water-trough has a regular isosceles triangular section, the angle at the bottom being 80 degrees. Calculate the volume flow of water along the trough, in cubic metres per hour, when the depth of the water in the trough is 180 mm and it is flowing at a velocity of 0·5 m/s.

17. Find the height of the centre of gravity of a frustum of a cone which is 80 mm diameter at the base, 60 mm diameter at the top, and 40 mm perpendicular height.

18. Find the position of the centre of gravity of the beam knee plate illustrated in Fig. 130 giving the distances from the 375 and 300 mm straight sides as represented by x and y respectively.

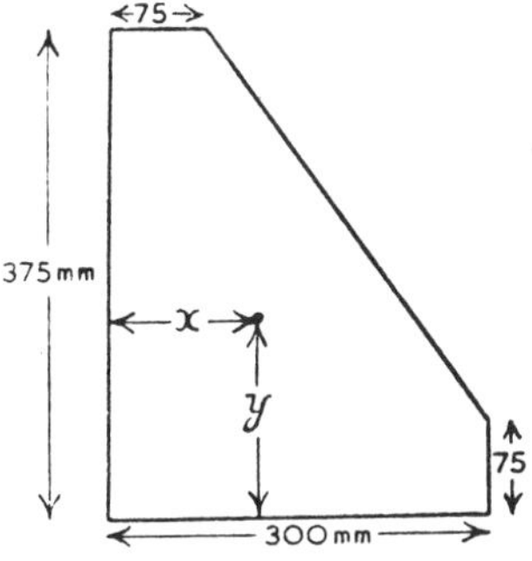

Fig. 130

19. A bolt has a round head and plain round shank, the head is 60 mm diameter and 40 mm deep, the shank is 40 mm diameter and the overall length of the bolt is 270 mm. A ring of the same material as the bolt, and of the same dimensions as the bolt head, is made to slide on the shank to a position of 131 mm from the shank end to the mid-point of the ring. Find the position of the centre of gravity of the whole from end of shank.

CHAPTER 13

INTRODUCTION TO THE DIFFERENTIAL CALCULUS

The development and application of the differential calculus may be explained by the relation between equations and their graphs.

The meaning of the *gradient* at any point on such a graph is of particular importance in this context and it is therefore dealt with first.

GRADIENT OF A LINE

It will be recalled, from previous work on graphs, that the *gradient* (or *slope*) of a line is the ratio $\dfrac{\textit{change in } y}{\textit{change in } x}$ between any two points on the line.

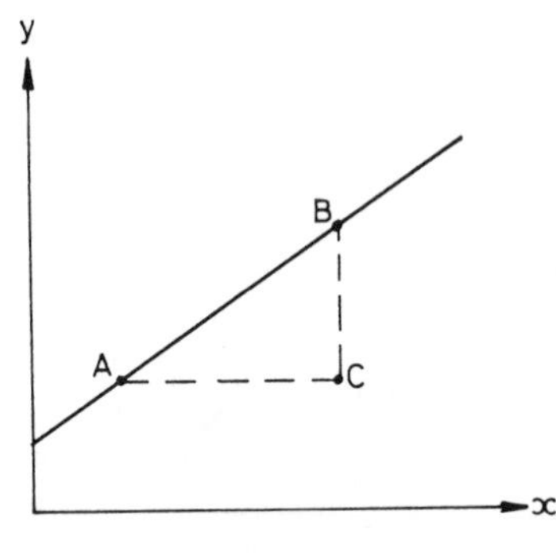

Fig. 131

For the line shown in Fig. 131,

$$\text{gradient} = \frac{\text{change in } y}{\text{change in } x} \text{ between points } A \text{ and } B$$

$$= \frac{BC}{AC}$$

The following examples are given by way of revision, to show how the gradient of a line is calculated:

Example. Plot the graph of the equation $y = 3x$ between the limits $x = 0$ and $x = 4$ and find the gradient of the graph.

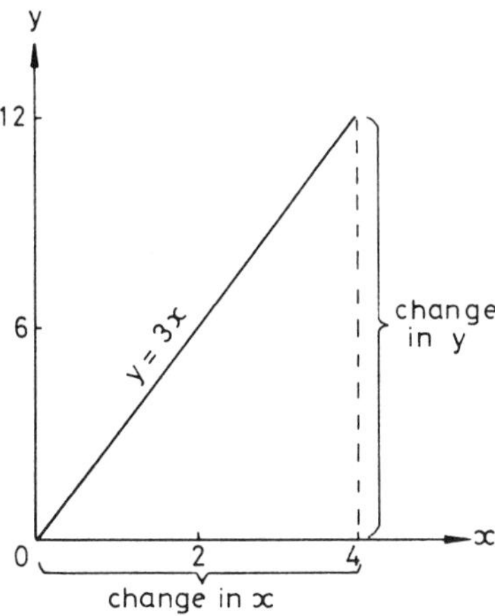

Fig. 132

Referring to Fig. 132:

$$\text{Gradient} = \frac{\text{change in } y}{\text{change in } x}$$

$$= \frac{12}{4}$$

$$= \frac{3}{1}$$

$$= 3 \text{ Answer.}$$

Note: The change in y is three times the change in x.

Thus the *gradient* indicates the *rate of change* of one quantity, *with respect to* another, related, quantity.

Example. Plot the graph of the equation $y = 2 - 0{\cdot}25x$ between the limits $x = 0$ and $x = 8$. Find the gradient of this graph.

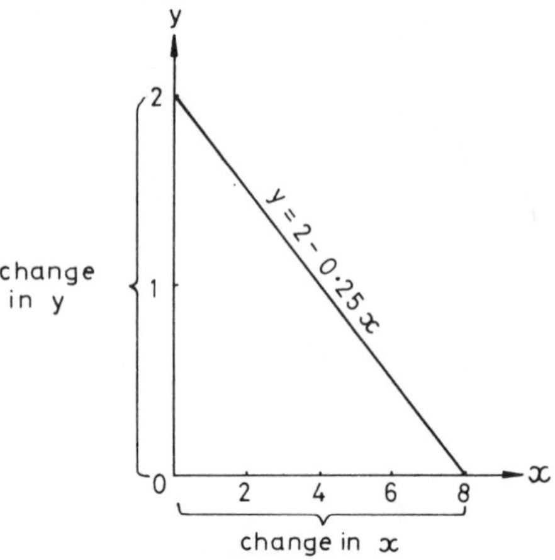

Fig. 133

The graph is shown above, Fig. 133.

$$\text{Gradient} = \frac{\text{change in } y}{\text{change in } x}$$

(Note that the value of y changes from 2 to zero: i.e. the change in y is -2.)

$$\therefore \text{Gradient} = \frac{-2}{8}$$

$$= -\frac{0{\cdot}25}{1}$$

$$= -0{\cdot}25 \quad \text{Answer.}$$

Two important points are noted from the previous examples:

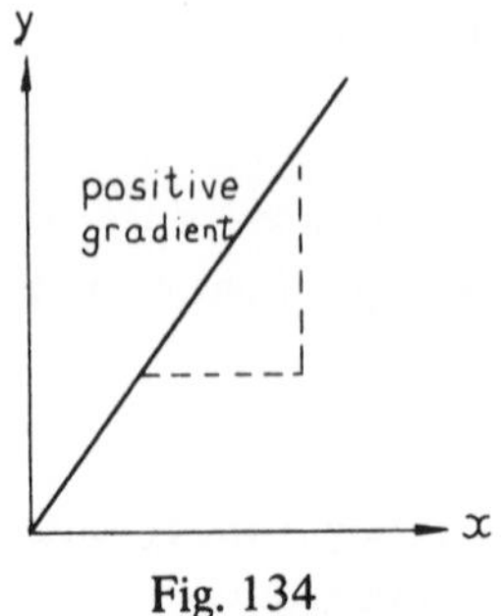

Fig. 134

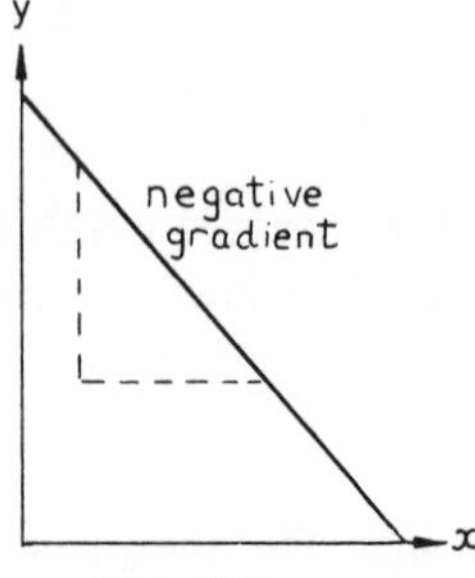

Fig. 135

(i) If y *increases* as x increases (Fig. 134), the gradient is *positive*.

(ii) If y *decreases* as x increases (Fig. 135), the gradient is *negative*.

ZERO GRADIENT

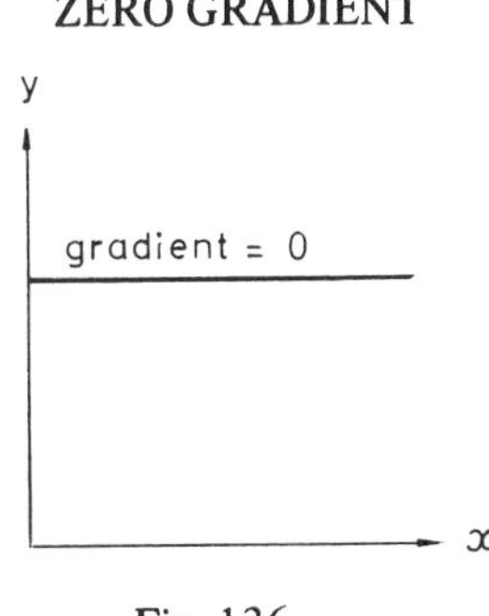

Fig. 136

The gradient of a horizontal straight line is zero, since the change in y is zero for any given change in the value of x.

GRADIENT OF A CURVE

The gradient of a curve is not constant, but is changing from point to point along the length of the curve.

The gradient of a curve at any given point is the gradient of the *tangent* to the curve at that point.

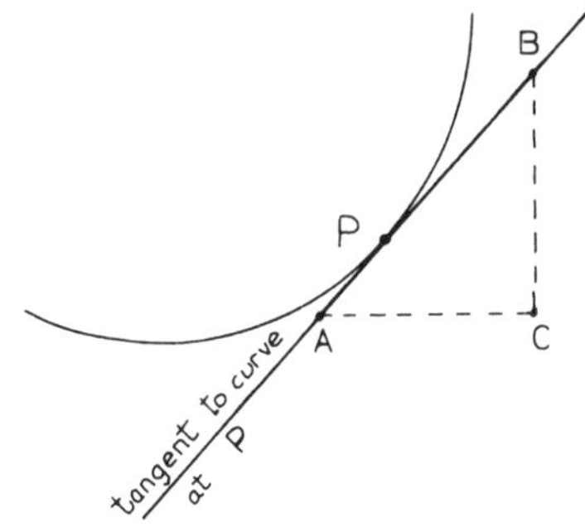

Fig. 137

Referring to Fig. 137:

Gradient of the *curve* at P = gradient of the *tangent* at P

The gradient of a curve may be obtained by drawing the tangent and measuring the gradient, or mathematically, by the *differential calculus*.

DIFFERENTIATION FROM FIRST PRINCIPLES

From Fig. 137 it was seen that the gradient of the tangent at P was required, in order to determine the gradient of the curve at P.

Chord PA is now constructed, as shown (Fig. 138).

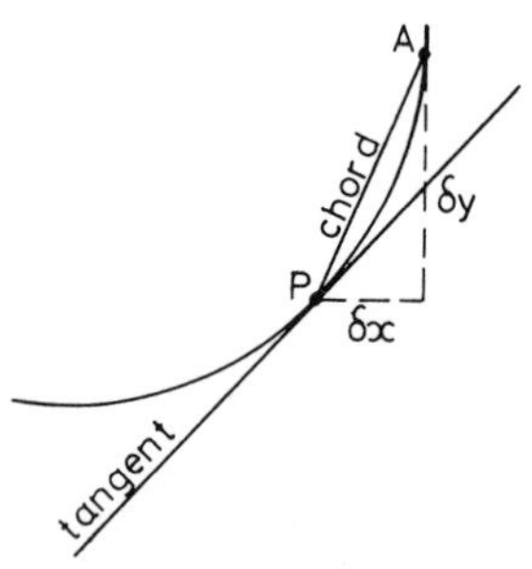

Fig. 138

(The Greek letter δ is used, for convenience, to express mathematically the change in the value of y and x between points P and A.

Thus, δx, pronounced 'delta ex', means a small change in the value of x. It does *not* mean $\delta \times x$.

Similarly, δy means the corresponding change in the value of y.)

From Fig. 138,

$$\text{gradient of chord } PA = \frac{\delta y}{\delta x}$$

The gradient of this chord is obviously quite different to the gradient of the tangent to the curve.

However, if point A is now allowed to move along the curve towards point P, it is seen (Figs. 139a and 139b), that the slope of the *chord* more and more resembles the slope of the *tangent* as A gets closer to P.

In fact, when point A gets very close to point P, the slope of the chord and the tangent are virtually the same.

As A gets very close to P, so length δx is becoming very small, until it finally approaches zero value.

i.e. δx approaches zero value (denoted by $\delta x \rightarrow 0$).

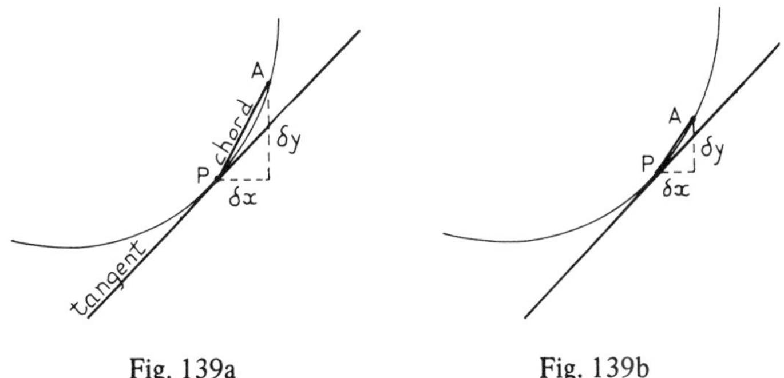

Fig. 139a

Fig. 139b

From the above reasoning, it follows that, when δx has become infinitely small, the ratio $\delta y/\delta x$ has reached a special value, or *limit*, where it represents the gradient of the curve at point P.

In order to indicate that this is a special value of $\delta y/\delta x$, it is identified by changing the notation from $\delta y/\delta x$ to dy/dx. This relationship is expressed mathematically as:–

$$\underset{\delta x \to 0}{\text{limit}} \frac{\delta y}{\delta x} = \frac{dy}{dx}$$

The term dy/dx is called the *differential coefficient* of y with respect to x and the process of finding dy/dx is called *differentiation*.

Example. From first principles, derive an expression for the gradient of the curve $y = x^2$ at any point on the curve.

The graph of $y = x^2$ is shown in Fig. 140. Point P represents any point on the curve.

Chord PA is drawn as shown.

1. The co-ordinates for point P are (x, y).
2. Hence, the co-ordinates for point A are $(x + \delta x), (y + \delta y)$.

3. Now, the equation of this curve is $y = x^2$

$$\therefore \text{At point } P, y = x^2 \quad \dots \quad \dots \quad \dots \quad \dots \quad \text{(i)}$$

$$\text{and, at point } A, (y + \delta y) = (x + \delta x)^2$$

$$\therefore y + \delta y = x^2 + 2x\,.\delta x + (\delta x)^2 \quad \dots \quad \text{(ii)}$$

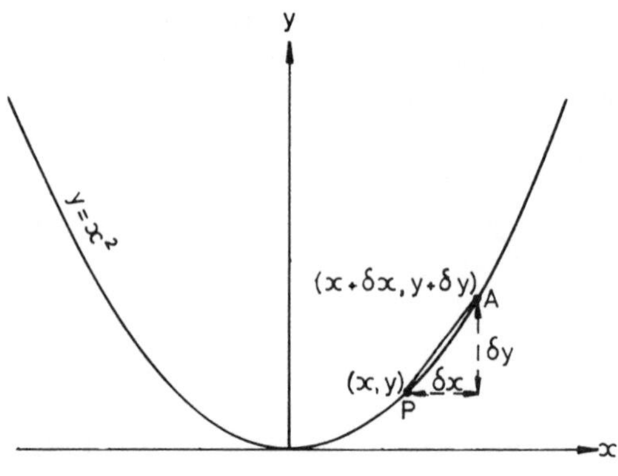

Fig. 140

Subtract equation (i) from equation (ii),

$$y + \delta y = x^2 + 2x \,.\, \delta x + (\delta x)^2$$

$$y = x^2$$

$$\delta y = 2x \,.\, \delta x + (\delta x)^2 \quad \ldots \quad \text{(iii)}$$

Divide equation (iii) by δx, thus obtaining *gradient*,

$$\frac{\delta y}{\delta x} = \frac{2x \,.\, \delta x}{\delta x} + \frac{(\delta x)^2}{\delta x}$$

$$\therefore \frac{\delta y}{\delta x} = 2x + \delta x \quad \ldots \quad \ldots \quad \ldots \quad \text{(iv)}$$

If point *A* now moves very close to point *P*, so length δx will approach zero value.

$$\text{i.e. } \lim_{\delta x \to 0} \frac{\delta y}{\delta x} = 2x$$

$$\therefore \frac{dy}{dx} = 2x$$

Hence, at any point on the curve $y = x^2$

the gradient of the curve $= 2x$ Answer.

Note: Since δx becomes very close to zero value, it can be omitted from the right-hand side of equation (iv) as being negligible compared to the term $2x$. This does not happen on the left side of equation (iv), because, as δx becomes very small, so too does δy, hence the ratio $\delta y/\delta x$ remains a significant quantity.

Example. Calculate the gradient of the curve $y = x^2$ at the points $x = -3$ and $x = +2$.

From the previous example,

$$\text{gradient at any point on the curve} = \frac{dy}{dx} = 2x.$$

$$\begin{aligned}
\therefore \text{At the point } x &= -3 \\
\text{gradient} &= 2 \times (-3) \\
&= -6 \\
\text{At the point } x &= 2, \\
\text{gradient} &= 2 \times (2) \\
&= 4 \qquad \text{Answer.}
\end{aligned}$$

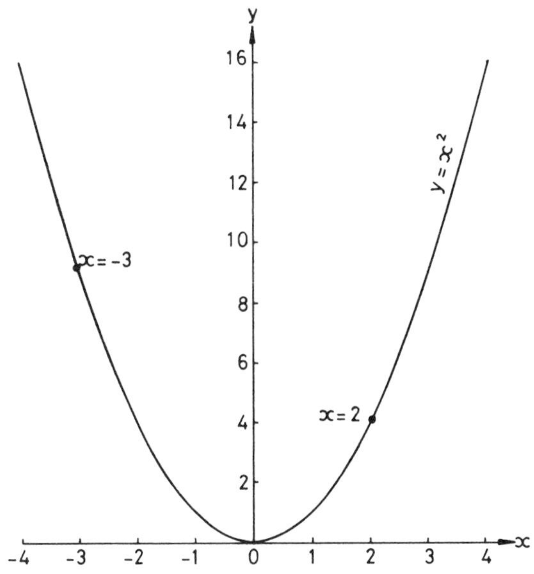

Fig. 141

The student can confirm these results by drawing tangents to the curve $y = x^2$ at $x = -3$ and $x = 2$.

The gradient of each of these tangents, when measured, should be reasonably close to the calculated values. (The tangents can be drawn on Fig. 141.)

Example. From first principles, find the differential coefficient of y, with respect to x, for the equation $y = 3x^3$.

(The differential coefficient or gradient may be calculated without actually drawing a graph, although a sketch of the graph is often useful when solving problems.)

$$y = 3x^3 \quad \dots \quad \dots \quad \dots \quad \dots \qquad \text{(i)}$$

$$\begin{aligned} \therefore (y + \delta y) &= 3(x + \delta x)^3 \\ &= 3\{x^3 + 3x^2 . \delta x + 3x(\delta x)^2 + (\delta x)^3\} \\ &= 3x^3 + 9x^2 . \delta x + 9x(\delta x)^2 + 3(\delta x)^3 \end{aligned} \qquad \text{(ii)}$$

Subtract equation (i) from equation (ii),

$$\begin{aligned} y + \delta y &= 3x^3 + 9x^2 . \delta x + 9x(\delta x)^2 + 3(\delta x)^3 \\ y &= 3x^3 \\ \hline \delta y &= 9x^2 . \delta x + 9x(\delta x)^2 + 3(\delta x)^3 \end{aligned}$$

Divide by δx,

$$\frac{\delta y}{\delta x} = 9x^2 + 9x . \delta x + 3(\delta x)^2$$

$$\lim_{\delta x \to 0} \frac{\delta y}{\delta x} = 9x^2$$

$$\text{i.e.} \frac{dy}{dx} = 9x^2$$

$\therefore$ For the equation $y = 3x^3$,

differential coefficient $= 9x^2$ Answer.

GENERAL RULE FOR DIFFERENTIATION

From the previous section, the following results were obtained:

$$\text{when } y = x^2, \qquad \frac{dy}{dx} = 2x$$

$$\text{and, when } y = 3x^3, \qquad \frac{dy}{dx} = 9x^2$$

From the form of these results (which is confirmed by solving further examples), the following general rule is noted:

$$\text{when } y = ax^n$$

$$\frac{dy}{dx} = nax^{n-1}$$

$$\text{e.g. when } y = 9x^4 \ \ldots \text{(i.e. } a = 9, n = 4)$$

$$\frac{dy}{dx} = (4 \times 9)\,x^{(4-1)}$$

$$\therefore \frac{dy}{dx} = 36\,x^3 \quad \text{Answer}$$

Example. Differentiate the following equations with respect to x:

(a) $y = x^8$ (b) $y = 2x^{-3}$ (c) $y = -7x^{-4}$

(d) $y = x$ (e) $y = \frac{1}{x}$ (f) $y = \frac{x^3}{12}$

(a)

$$y = x^8$$

$$\therefore \frac{dy}{dx} = 8x^7 \quad \text{Answer}$$

(b)

$$y = 2x^{-3} \qquad \text{(note: } -3 - 1 = -4)$$

$$\therefore \frac{dy}{dx} = -6x^{-4}$$

$$= -\frac{6}{x^4} \quad \text{Answer}$$

(c)

$$y = -7x^{-4}$$

$$\therefore \frac{dy}{dx} = 28x^{-5}$$

$$= \frac{28}{x^5} \qquad \text{Answer}$$

(d)

$$y = x$$

$$(\text{i.e. } y = x^1)$$

$$\therefore \frac{dy}{dx} = x^0$$

$$= 1 \qquad \text{Answer}$$

(e)

$$y = \frac{1}{x}$$

$$(\text{i.e. } y = x^{-1})$$

$$\therefore \frac{dy}{dx} = -x^{-2}$$

$$= -\frac{1}{x^2} \qquad \text{Answer}$$

(f)

$$y = \frac{x^3}{12}$$

$$\therefore \frac{dy}{dx} = \frac{3 \times x^2}{12}$$

$$= \frac{x^2}{4} \qquad \text{Answer}$$

DIFFERENTIAL COEFFICIENT OF A CONSTANT

As shown previously, when y is a constant, the resulting graph has zero gradient (see Fig. 136).

$$\text{i.e. if } y = a \text{ (where } a \text{ is any constant)}$$

$$\frac{dy}{dx} = 0$$

$$\text{e.g. if } y = 7$$

$$\frac{dy}{dx} = 0$$

DIFFERENTIAL COEFFICIENT OF A SUM OF TERMS

The differential coefficient of a sum of terms is obtained by differentiating each term separately.

$$\text{e.g. if } y = x^3 - 2x^2 + 4x - 9$$

$$\frac{dy}{dx} = 3x^2 - 4x + 4$$

Example. Differentiate the equation $y = \dfrac{x^3}{3} - \dfrac{3}{x^2} + 3x$ with respect to x.

Rearranging the equation,

$$y = \frac{x^3}{3} - 3x^{-2} + 3x$$

$$\frac{dy}{dx} = x^2 + 6x^{-3} + 3$$

$$= x^2 + \frac{6}{x^3} + 3 \qquad \text{Answer.}$$

Example. Differentiate the equation $s = 3t^2 - 20t + 40$ with respect to t. (N.B. Symbols s and t represent two related quantities, as do x and y.)

$$s = 3t^2 - 20t + 40$$

$$\frac{ds}{dt} = 6t - 20 \qquad \text{Answer.}$$

SECOND DIFFERENTIAL COEFFICIENT

In some cases, having differentiated an expression once, it is necessary to differentiate a second time.

The notation d^2y/dx^2 is used to denote the second differentiation of an expression.

$$\text{e.g. Consider the equation } y = x^4$$

$$\text{differentiating once, } \frac{dy}{dx} = 4x^3$$

$$\text{differentiating twice, } \frac{d^2y}{dx^2} = 12x^2$$

Example. Determine the second differential coefficient of the following equation:

$$y = x^3 - 4x^2 + 3x - 7$$

$$\frac{dy}{dx} = 3x^2 - 8x + 3$$

$$\frac{d^2y}{dx^2} = 6x - 8 \qquad \text{Answer.}$$

DISTANCE, VELOCITY AND ACCELERATION

For a moving body,

$$\text{average velocity} = \frac{\text{distance travelled}}{\text{time taken}}$$

Using symbol v for velocity, s for distance and t for time, this equation may be expressed in calculus notation:

$$\text{average velocity} = \frac{\delta s}{\delta t}$$

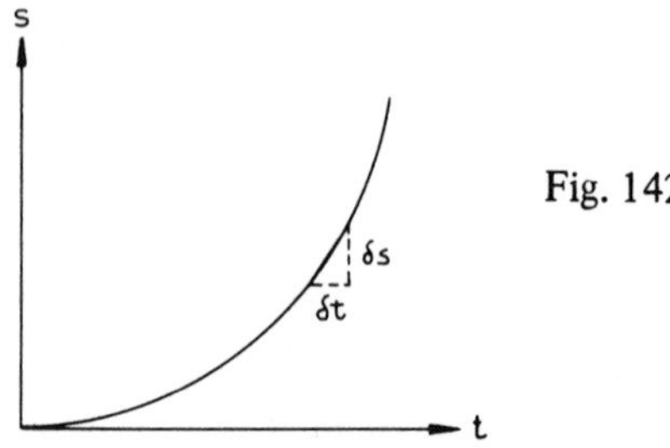

Fig. 142

If the velocity at a given instant (the *instantaneous* velocity) is required, the time interval δt must be very small. i.e. δt must approach zero value (see Fig. 142).

Thus, $$\lim_{\delta t \to 0} \frac{\delta s}{\delta t} = \frac{ds}{dt}$$

Hence, at any given time t, the *instantaneous velocity* of a moving body may be expressed as:

$$v = \frac{ds}{dt}$$

Similarly,

$$\text{average acceleration} = \frac{\text{change of velocity}}{\text{time to change}}$$

By the same reasoning as that above, the *instantaneous acceleration a* is given by the expression:

$$a = \frac{dv}{dt}$$

Example. The distance s moved by a body in time t is given by the formula $s = t^3 - 3t^2$. Find expressions for the velocity and acceleration of the body at any instant.

$$s = t^3 - 3t^2$$

$$\text{velocity, } v = \frac{ds}{dt} = 3t^2 - 6t$$

$$\text{acceleration } a = \frac{dv}{dt} = 6t \qquad \text{Answer.}$$

Note: The expression for the acceleration of the body is seen to be the second differential equation of the original equation relating distance s to time t.

$$\text{i.e. } s = t^3 - 3t^2$$

$$\frac{ds}{dt} = 3t^2 - 6t$$

$$\frac{d^2 s}{dt^2} = 6t$$

Example. A body moves s metres in t seconds according to the relationship $s = t^3 - 7t^2 + 3$.

(a) Derive expressions for the velocity and acceleration of the body at any instant.

(b) Use these expressions to find the velocity and acceleration of the body after 5 seconds.

(a)

$$s = t^3 - 7t^2 + 3$$

$$v = \frac{ds}{dt} = 3t^2 - 14t$$

$$a = \frac{dv}{dt} = 6t - 14$$

∴ At any instant

$$\text{velocity} = 3t^2 - 14t \text{ m/s}$$

$$\text{acceleration} = 6t - 14 \text{ m/s}^2$$ Answer (a)

(b) When

$$t = 5,$$

$$v = 3 \times (5)^2 - 14 \times (5)$$

$$= 5 \text{ m/s}$$

$$a = 6 \times (5) - 14$$

$$= 16 \text{ m/s}^2$$ Answer (b)

Example. The distance s moved by a body in time t is given by the expression

$$s = 40t - 5t^2$$

Calculate (a) The velocity after 2 seconds.
(b) The time taken for the body to come to rest.

(a)

$$s = 40t - 5t^2$$

$$v = \frac{ds}{dt} = 40 - 10t$$

when $t = 2$

$$v = 40 - 20$$

$$= 20 \text{ m/s.}$$

(b) When the body comes to rest, $v = 0$

$$\therefore 40 - 10t = 0$$

$$\therefore t = \frac{40}{10}$$

$$= 4 \text{ seconds} \qquad \text{Answer.}$$

MAXIMA AND MINIMA

An important application of differential calculus involves functions which have maximum or minimum values.

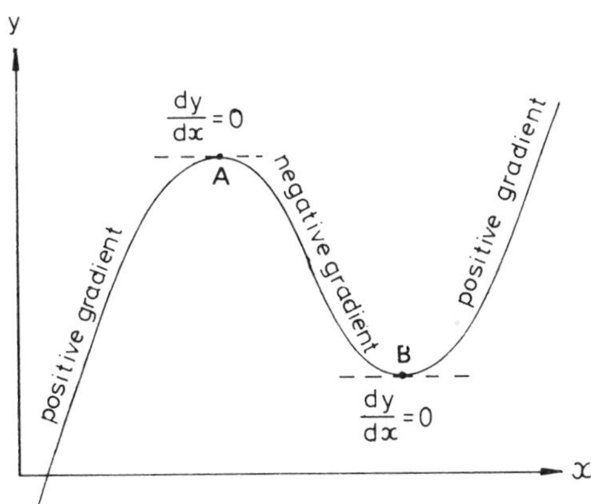

Fig. 143

The graph shown above (Fig. 143) represents a function which has a local *maximum* value of y at point A and a local *minimum* value of y at B.

Such points are called *turning points* and have zero gradient.

$$\text{i.e. at a turning point, } \frac{dy}{dx} = 0.$$

Example. The curve $y = 3x - x^2 + 1$ has a turning point.

(a) Find the value of x at which this occurs.

(b) Determine whether the turning point is a maximum or minimum value.

(c) Find the value of y at the turning point.

(a)
$$y = 3x - x^2 + 1$$
$$\frac{dy}{dx} = 3 - 2x$$
for turning points, $3 - 2x = 0$
$$\therefore 2x = 3$$
$$\therefore x = 1{\cdot}5 \qquad \text{Answer.}$$

(b) The nature of the turning point can be determined by sketching the curve:

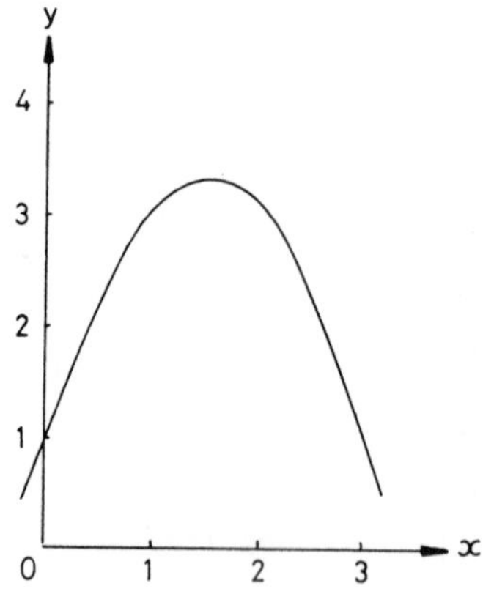

Fig. 144

From the sketch (Fig. 144), at $x = 1{\cdot}5$, a *maximum* value occurs. Answer.

(c) To find the maximum value of y, substitute $x = 1{\cdot}5$ in the original equation:

$$\text{i.e. } y = (3 \times 1{\cdot}5) - (1{\cdot}5)^2 + 1$$
$\therefore$ Maximum value of $y = 3{\cdot}25$ Answer.

Inspection of Fig. 143 shows that the gradient changes sign when passing through a turning point.

This presents an alternative test for maximum or minimum values:–

(i) At a *maximum* value, the gradient dy/dx is *positive* before the turning point and *negative* after it.

Hence, dy/dx is *decreasing* in value, as x increases.
i.e. the *rate of change* of dy/dx is negative

$$\therefore \frac{d^2y}{dx^2} \text{ is negative}$$

(Note: since dy/dx is a gradient, the rate of change of the gradient is d^2y/dx^2).

(ii) At a *minimum* value, the gradient is *negative* before the turning point and *positive* after it.

Hence dy/dx is *increasing* in value and the rate of change of dy/dx is positive.

$$\therefore \frac{d^2y}{dx^2} \text{ is positive.}$$

Summing up:

At a maximum value,

$$\frac{dy}{dx} = 0, \quad \frac{d^2y}{dx^2} \text{ is negative.}$$

At a minimum value,

$$\frac{dy}{dx} = 0, \quad \frac{d^2y}{dx^2} \text{ is positive.}$$

Example.

(a) Determine the values of x which give maximum or minimum values of y in the equation

$$y = x^3 - 6x^2 + 9x$$

(b) Calculate these maximum or minimum values of y.

(a)
$$\frac{dy}{dx} = 3x^2 - 12x + 9$$

$$\frac{d^2y}{dx^2} = 6x - 12.$$

For maximum or minimum values,

$$\frac{dy}{dx} = 0$$

$$\text{i.e. } 3x^2 - 12x + 9 = 0$$

$$\therefore x^2 - 4x + 3 = 0$$

$$\therefore (x-1)(x-3) = 0$$

$$\text{i.e. } (x-1) = 0 \quad \text{or} \quad (x-3) = 0$$

$$\therefore x = 1 \quad \text{or} \quad x = 3$$

∴ Maximum or minimum values of y occur when $x = 1$ or $x = 3$. Answer (a).

(b) When $x = 1$,

$$\frac{d^2y}{dx^2} = 6 - 12 \text{ (i.e. negative)}$$

When $x = 3$,

$$\frac{d^2y}{dx^2} = 18 - 12 \text{ (positive)}$$

$$\therefore \text{When } x = 1,$$

$$y = (1)^3 - 6(1)^2 + 9(1)$$

$$\therefore \text{Maximum value of } y = 4$$

$$\text{When } x = 3$$

$$y = (3)^3 - 6(3)^2 + 9(3)$$

$$\therefore \text{Minimum value of } y = 0$$

Answer (b)

Note: These results are confirmed by the sketch of the graph shown in Fig. 145.

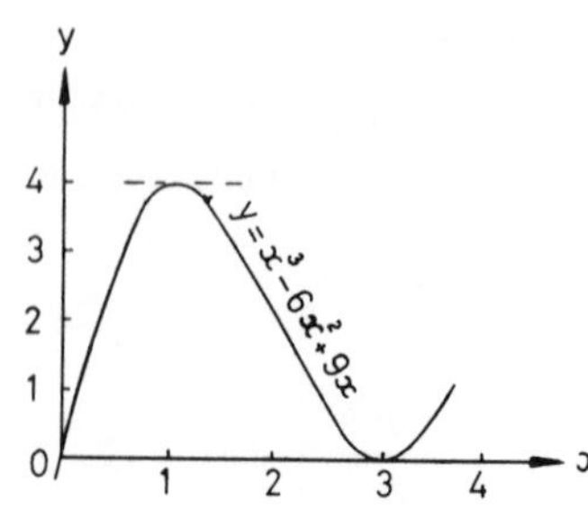

Fig. 145.

Example. A rectangular sheet of steel is 60 cm wide and 28 cm long. Four square portions are to be removed at the corners and the sides turned up to form an open rectangular box.

Prove that, for this box to have maximum volume, its depth should be 6 cm.

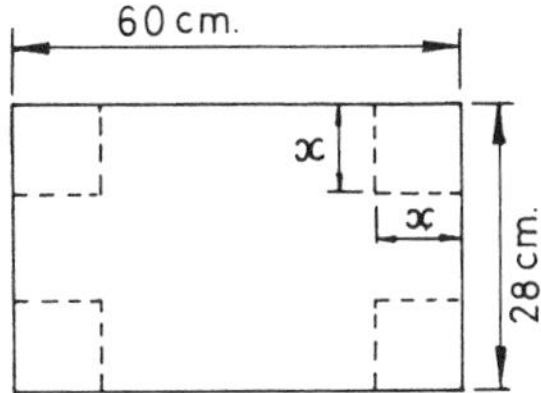

Fig. 146a

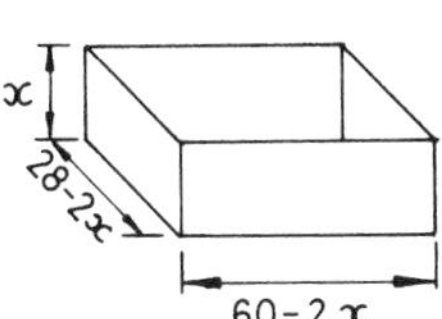

Fig. 146b

Referring to Fig. 146a,

let x = side of square removed (cm)

With the corner portions removed, the box is formed, with the dimensions shown (Fig. 146b)

$$\text{let } V = \text{volume of box } (\text{cm}^3)$$

$$\text{then, } V = x(28-2x)(60-2x)$$

$$\therefore V = 4x^3 - 176x^2 + 1680x$$

$$\therefore \frac{dV}{dx} = 12x^2 - 352x + 1680$$

for maximum (or minimum) value of V,

$$12x^2 - 352x + 1680 = 0$$

solve for x:

$$3x^2 - 88x + 420 = 0$$

$$\therefore (3x-70)(x-6) = 0$$

hence, $3x - 70 = 0$ or, $x - 6 = 0$

$x = 23{\cdot}33$ or $x = 6$

$$\therefore x = 6 \text{ cm (rejecting the alternative solution)}$$

Note: in a problem of this type, it is evident that $x = 6$ cm must give a *maximum* volume, and not a *minimum*.

The result can of course, be confirmed by differentiating again:

$$\frac{d^2y}{dx^2} = 24x - 352$$

When $x = 6$
$$\frac{d^2y}{dx^2} = 144 - 352$$
$$= -208 \text{ (negative)}$$

$\therefore$ Maximum volume is obtained when depth is 6 cm. Answer.

DIFFERENTIATION OF SIN x, COS x

For any angle x,

$$\text{If } y = \sin x, \quad \frac{dy}{dx} = \cos x$$

$$\text{If } y = \cos x, \quad \frac{dy}{dx} = -\sin x$$

These differential coefficients can be demonstrated by reference to the graphs of sin x and cos x (Fig. 147).

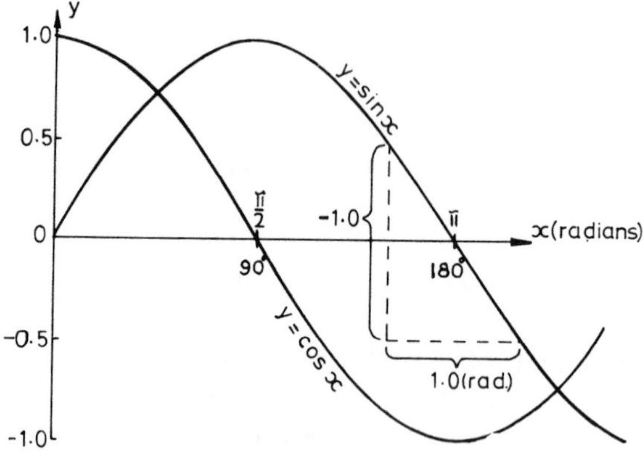

Fig. 147

For example, when $x = 180°$ (i.e. π radians):

from the sin x curve, gradient $= \frac{-1}{1}$

$$\text{i.e. } \frac{dy}{dx} = -1$$

and, from the cos x curve, $\cos x = -1$

i.e. differential coefficient of $\sin x = \cos x$

Similarly, the gradient at any point on the cos x curve is numerically equal to the corresponding value of sin x. In this case, however, the sign must be reversed, because the cos x curve has a *negative* gradient where sin x has *positive* values.

Example. Differentiate the following expressions with respect to x:

(a) $y = -\sin x$

(b) $y = 2 \sin x$

(c) $y = 4 \sin x - 3 \cos x$

(a)
$$y = -\sin x$$
$$\frac{dy}{dx} = -\cos x$$

(b)
$$y = 2 \sin x$$
$$\frac{dy}{dx} = 2 \times \cos x$$
$$= 2 \cos x$$

(c)
$$y = 4 \sin x - 3 \cos x$$
$$\frac{dy}{dx} = 4 \cos x + 3 \sin x \quad \text{Answer.}$$

The *second differential coefficients* of sin x and cos x can be determined quite easily by repeated differentiation:

$$y = \sin x$$
$$\frac{dy}{dx} = \cos x$$
$$\therefore \frac{d^2y}{dx^2} = -\sin x$$

Similarly,
$$y = \cos x$$
$$\frac{dy}{dx} = -\sin x$$
$$\frac{d^2y}{dx^2} = -\cos x$$

Example. Obtain the second differential coefficients of (a) $4 \cos x$, (b) $\cos x - 2 \sin x$.

(a)

$$y = 4 \cos x$$

$$\frac{dy}{dx} = -4 \sin x$$

$$\therefore \frac{d^2y}{dx^2} = -4 \cos x \qquad \text{Answer.}$$

(b)

$$y = \cos x - 2 \sin x$$

$$\frac{dy}{dx} = -\sin x - 2 \cos x$$

$$\frac{d^2y}{dx^2} = -\cos x + 2 \sin x \qquad \text{Answer.}$$

Note: It is seen that, for sines and cosines, the second differential coefficient is equal to the original function, but of opposite sign.

DIFFERENTIAL COEFFICIENT OF Inx and e^x

Napierian logarithms (ln) have special applications, for example, when finding the work done by an expanding gas.

When

$$y = \ln x$$

$$\frac{dy}{dx} = \frac{1}{x}$$

When

$$y = e^x$$

$$\frac{dy}{dx} = e^x$$

The proof of those relationships are beyond the scope of this work.

Example. Obtain the second differential coefficient of the equations $y = \ln x$ and $y = e^x$.

$$y = \ln x$$

$$\frac{dy}{dx} = \frac{1}{x}$$

$$= x^{-1}$$

$$\frac{d^2y}{dx^2} = -x^{-2}$$

$$= -\frac{1}{x^2} \qquad \text{Answer.}$$

$$y = e^x$$

$$\frac{dy}{dx} = e^x$$

$$\frac{d^2y}{dx^2} = e^x \qquad \text{Answer.}$$

FUNCTIONAL NOTATION

In an equation such as $y = 2x$ or $y = 5x^3 + 3x$, the value of y obviously depends upon the value chosen for x. Hence, y is said to be a *function* of x and the general expression for such a relationship is:–

$$y = f(x)$$

Functional notation may also be used to indicate the differentiation process, using the symbols shown below:

$$y = f(x)$$

$$\frac{dy}{dx} = f'(x)$$

$$\frac{d^2y}{dx^2} = f''(x)$$

e.g. If $\qquad y = 5x^3 + 2x$

Then, $\qquad f(x) = 5x^3 + 2x$

$$f'(x) = 15x^2 + 2$$

$$f''(x) = 30x.$$

TEST EXAMPLES 13

1. Differentiate the following equations with respect to x:

   (a) $y = x^3 + 3x^2 - 9x + 4$

   (b) $y = \dfrac{2x^3}{3} - \dfrac{7}{x^2} + x$

   (c) $y = \sqrt[5]{x^3} + 1$

   (d) $y = 5 \cos x - 7 \cos x + 2 \sin x$

2. Calculate the gradient of the curve $x^2 + 3x - 7$ at the points where $x = 3$ and $x = -2$.

3. The displacement s metres of a body from a fixed point is given by the equation $s = 20t - 5t^2 + 4$ where t is the time in seconds. Find (a) The velocity after 2 seconds
   (b) The displacement when the velocity is zero
   (c) The acceleration.

4. Determine the gradient of a tangent to any point in the curve
$$y = 4x + \frac{1}{x}$$
Show that there are two points where the gradient is zero.

5. Find the co-ordinates of the point on the graph of $y = 3x^2 - x + 2$ at which the gradient is equal to -7.

6. The angle θ radians through which a shaft has turned after time t seconds is given by the equation
$$\theta = 2 + 16t - \frac{t^2}{2}.$$
Find the angular velocity after 2 seconds and the time for the shaft to come to rest.
(Note: angular velocity $\omega = \frac{d\theta}{dt}$).

7. (a) Determine the second differential coefficient of the expression $y = x^3 + 3x + \log_\epsilon x$ with respect to x.
   (b) Determine the second differential coefficient of the expression $y = 3\cos\theta - 7\cos\theta + \theta$ with respect to θ.

8. Determine the maximum value of y in the equation $y = 12x + 3x^2 - 2x^3$.

9. For a beam of length l, the bending moment M at any distance x from one end is given by the equation
$$M = \frac{wlx}{2} - \frac{wx^2}{2}$$
where w is the uniform load per unit length. Show that bending moment is a maximum at the centre of the beam.

10. Differentiate the following equations:

(a) $t = 2 \cdot 1 \sqrt[3]{\frac{1}{\theta^2}} - \frac{4}{\sqrt[5]{\theta}}$ (b) $z = \frac{au^n - 1}{c}$

(c) $y = 3x(x^2 - 4)$ (d) $y = 2\cos\theta + 5$

11. Determine the minimum value of $f(x)$ for the equation

$$f(x) = \frac{x^3}{3} - 2x^2 + 3x + 1$$

12. Power (P) and Voltage (V) of a lamp are related by $P = aV^b$, where a and b are constants. Find an expression for the rate of change of power with voltage and power per volt at 100 volts, when $a = 0{\cdot}5 \times 10^{-10}$ and $b = 6$.

13. A line of length l is to be cut up into four parts and put together as a rectangle. Show that the area of the rectangle will be a maximum if each of its sides is equal to one quarter of l (i.e. a square).

14. Determine the second differential coefficient of:
(a) $f(\theta) = \cos\theta - \text{In}\ \theta$
(b) $f(t) = at^2 + 2\ \text{In}\ t$
(c) $f(x) = 5e^x$

15. A body moves so that its displacement x metres, which it travels from a certain point 0, is given by:
$x = 0{\cdot}2\,t^2 + 10{\cdot}4$ where t is the time in seconds.
Find the velocity and acceleration (a) five seconds after the body begins to move and (b) when the displacement is 100 m.

16. Angular displacement (θ radians) from rest of a revolving wheel is given by:
$\theta = 2 \cdot 1 - 3 \cdot 2\,t + 4 \cdot 8\,t^2$ where t is the time in seconds. Find the angular velocity and angular acceleration after $1 \cdot 5$ seconds.

17. Verify that the equation $f(x) = x^5 - 5x$ has a maximum and a minimum value and determine the value of x and $f(x)$ at these points.

CHAPTER 14

INTEGRAL CALCULUS

Integration may be considered as *reversing* the process of differentiation.

That is, given the differential coefficient of a function, we are required to find the original function.

The symbol $\int$ is used to denote the integration process. This is the old-fashioned letter 'S' and the reason for its use becomes evident when integration is used to find the 'sum' of a number of quantities.

CONSTANT OF INTEGRATION

Consider the three equations below:

$$y = x^2, \qquad y = x^2 + 3, \qquad y = x^2 + 7$$

For each of these equations,

$$\frac{dy}{dx} = 2x$$

Obviously, when reversing the differentiation process (i.e. *integrating*), provision must be made for the *possibility* of a *constant* in the original equation.

This is achieved by adding a constant, C, called the *constant of integration.*

$$\text{i.e.} \quad \int 2x\,dx = x^2 + C,$$

which may be interpreted as:
(the integral of) $2x$ (with respect to x) $= x^2 + C$.

GENERAL RULE FOR INTEGRATION

Consider the following differential equations:

$$\text{if } y = \frac{x^2}{2}, \qquad \frac{dy}{dx} = x$$

$$\text{if } y = \frac{x^3}{3}, \qquad \frac{dy}{dx} = x^2$$

$$\text{if } y = \frac{x^4}{4}, \qquad \frac{dy}{dx} = x^3$$

From these equations, the following integrals are obtained:

$$\int x \, . \, dx = \frac{x^2}{2}$$

$$\int x^2 \, . \, dx = \frac{x^3}{3}$$

$$\int x^3 \, . \, dx = \frac{x^4}{4}$$

It is seen that in each case, the procedure is to raise the power of x by 1 and divide by this raised power.

Expressing this mathematically, a general rule for integration is obtained, which includes a constant of integration:

$$\int x^n \, . \, dx = \frac{x^{n+1}}{n+1} + C$$

Note: there is an important exception to this rule, the integral of x^{-1} or $1/x$, which is dealt with later.

Example: Evaluate the following integrals:

(a) $\int x^7 dx$ (b) $\int 3x^5 dx$ (c) $\int \frac{x^{-4}}{3} dx$ (d) $\int 8dx$

(a) $$\int x^7 dx = \frac{x^8}{8} + C \qquad \text{Answer.}$$

(b) $$\int 3x^5 dx = 3 \times \frac{x^6}{6} + C$$

$$= \frac{x^6}{2} + C \qquad \text{Answer.}$$

(c) $$\int \frac{x^{-4}}{3} dx = \frac{x^{-3}}{3 \times -3} + C \qquad (\text{note: } -4 + 1 = -3)$$

$$= -\frac{x^{-3}}{9} + C$$

$$= -\frac{1}{9x^3} + C \qquad \text{Answer.}$$

(d) $$\int 8dx = 8x + C \qquad \text{Answer.}$$

This result may be confirmed by differentiating the answer, or by applying the general rule, as shown:

$\int 8dx$ may be written as $\int (8 \times x^0)dx$, since $x^0 = 1$.

Applying the general rule,

$$\int (8 \times x^0)dx = 8 \times \frac{x^1}{1} + C$$

$$= 8x + C \quad \text{Answer.}$$

INTEGRATION OF A SUM OF TERMS

The integral of a sum of terms is equal to the sum of their separate integrals.

$$\text{e.g.} \int (3x^2 + 7x - 10)dx = \int 3x^2 . dx + \int 7x . dx - \int 10dx$$

$$= x^3 + \frac{7x^2}{2} - 10x + C$$

Note: the constants which should be added to each separate term are combined as a single constant.

Example. Evaluate the integral $\int (x^2 - 3x^3 + 2)dx$

$$\int (x^2 - 3x^3 + 2)dx = \frac{x^3}{3} - \frac{3x^4}{4} + 2x + C \quad \text{Answer.}$$

Example. Integrate the expression $3t^2 + t + 1$ with respect to t.

$$\int (3t^2 + t + 1)dt = t^3 + \frac{t^2}{2} + t + C \quad \text{Answer.}$$

Example. Integrate the expression $3x^2 - 6x + \frac{1}{2}$ with respect to x.

$$\int \left(3x^2 - 6x + \frac{1}{2}\right)dx = x^3 - 3x^2 + \frac{x}{2} + C \quad \text{Answer.}$$

EVALUATING THE CONSTANT OF INTEGRATION

The value of the constant of integration, for a given function, can be calculated, provided a corresponding pair of values of x and y are known.

Example. The gradient of a curve is $4x + 5$. If the curve passes through the point ($x = 0$, $y = -4$), find the equation of the curve.

$$\text{Gradient of curve} = \frac{dy}{dx} = 4x + 5$$

$$\therefore y = \int \left(\frac{dy}{dx}\right) dx$$

$$\therefore y = \int (4x + 5) dx$$

$$\therefore y = 2x^2 + 5x + C \quad \ldots \quad \ldots \quad \text{(i)}$$

To find the constant of integration C, substitute $x = 0$, $y = -4$, in equation (i):

$$\therefore -4 = 2 \times (0)^2 + 5 \times (0) + C$$

$$\therefore C = -4$$

$\therefore$ The equation of the curve is $y = 2x^2 + 5x - 4$ Answer.

Example. The curve of a graph has a gradient of $10x - x^2$. If the curve passes through the point (3, 52), find the equation of the graph.

$$\frac{dy}{dx} = 10x - x^2$$

$$\therefore y = \int (10x - x^2) \, . \, dx$$

$$\therefore y = 5x^2 - \frac{x^3}{3} + C$$

Substituting $x = 3$, $y = 52$,

$$52 = 5(3)^2 - \frac{(3)^3}{3} + C$$

$$\therefore 52 = 45 - 9 + C$$

$$\therefore C = 16$$

$\therefore$ The equation is $y = 5x^2 - \dfrac{x^3}{3} + 16$ Answer.

INTEGRATION OF SIN x, COS x, $1/x$, e^x

The following differential coefficients are known:

$$y = \sin x, \qquad \frac{dy}{dx} = \cos x$$

$$y = \cos x, \qquad \frac{dy}{dx} = -\sin x$$

$$y = \ln x, \qquad \frac{dy}{dx} = \frac{1}{x}$$

$$y = e^x \qquad \frac{dy}{dx} = e^x$$

Reversing the process, the following integrals are obtained:

$$\int \cos x \,.\, dx = \sin x + C$$
$$\int \sin x \,.\, dx = -\cos x + C$$
$$\int \frac{1}{x} \,.\, dx = \ln x + C$$
$$\int e^x \, dx = e^x + C$$

Example. Evaluate the integral $\int (2 \cos x - \sin x)dx$

$$\int (2 \cos x - \sin x)dx = 2 \sin x - (-\cos x) + C$$
$$= 2 \sin x + \cos x + C \quad \text{Answer.}$$

Example. Evaluate the integral $\int (\cos \theta + 5 \sin \theta)$ with respect to θ. (Symbol θ is commonly used to denote an angle which is measured in radians.)

$$\int (\cos \theta + 5 \sin \theta)d\theta = \sin \theta - 5 \cos \theta + C \quad \text{Answer.}$$

Example. Evaluate the integral $\int (5/x)dx$.

$$\int \left(\frac{5}{x}\right)dx = \int \left(5 \times \frac{1}{x}\right)dx$$
$$= 5 \,.\, \ln x + C \qquad \text{Answer.}$$

AREA BY INTEGRATION. DEFINITE INTEGRAL

In Fig. 148, the curve *ac* represents the graph of a function of *x*. The area *abc* may be divided into a large number of strips, or *elements*, one of which is shown.

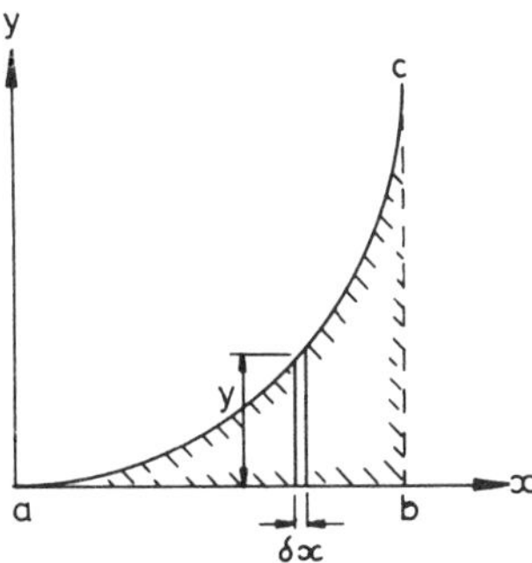

Fig. 148

$$\text{Let } A = \text{area } abc$$

$$\delta A = \text{area of one strip}$$

If δx is made very small, the effect of the curve at the top of the strip becomes negligible. The strip may then be considered as a rectangle of height y and width δx.

$$\text{i.e. } \delta A = y \,.\, dx$$

$$\therefore \frac{\delta A}{\delta x} = y$$

$$\lim_{\delta x \to 0} , \frac{dA}{dx} = y$$

Integrating both sides of this expression with respect to x,

$$\int \left(\frac{dA}{dx}\right) dx = \int (y) dx$$

$$\therefore A = \int y \, dx, \text{ between the limits } x = a \text{ and } x = b$$

This general rule is expressed mathematically as:

$$\text{Area} = \int_a^b y \, dx$$

Example. Find the area between the curve $y = x^2$, the x axis and the ordinates $x = 2$ and $x = 4$ (Fig. 149).

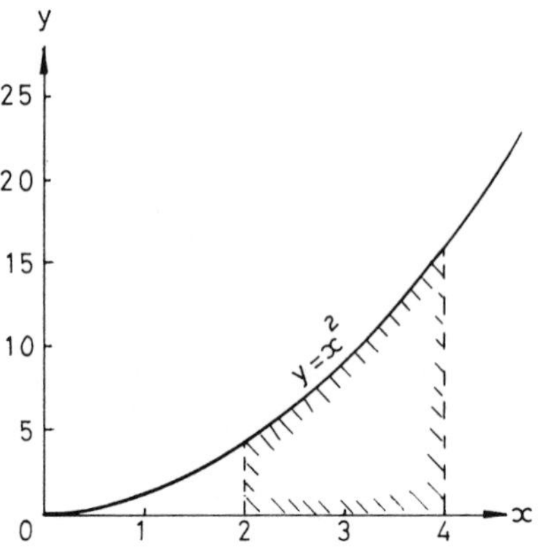

Fig. 149

$$\text{Area} = \int_a^b y \,.\, dx$$

$$= \int_2^4 x^2 \,.\, dx$$

$$= \frac{x^3}{3} + C \text{ between } x = 4 \text{ and } x = 2.$$

when $x = 4$, $\dfrac{x^3}{3} + C = 21\frac{1}{3} + C$

when $x = 2$, $\dfrac{x^3}{3} + C = 2\frac{2}{3} + C$

Subtract:

$$\text{Area} = [21\tfrac{1}{3} + C] - [2\tfrac{2}{3} + C]$$

$$= 21\tfrac{1}{3} + C - 2\tfrac{2}{3} - C$$

$$= 18\tfrac{2}{3} \text{ square units.} \qquad \text{Answer.}$$

Note: The constant of integration C *always* disappears when integrating between limits.

Because of this, when an integral has limits, it is called a *definite integral.*

Example. Find the area enclosed by the curve $y = x^2 + 2x + 1$, the x-axis, and the ordinates $x = 5, x = 2$.

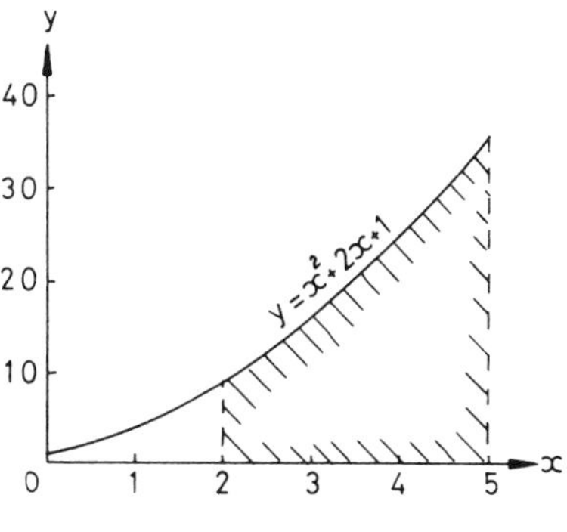

Fig. 150

The graph of the equation is shown above (Fig. 150).

$$\text{Area} = \int_a^b y\,dx$$

$$= \int_2^5 (x^2 + 2x + 1)dx$$

Integrate:

$$\text{Area} = \left[\frac{x^3}{3} + x^2 + x\right]_2^5$$

(The integral is inserted in square brackets, with upper and lower limits of x placed as shown above.)

Solve:

$$\text{Area} = \left[\frac{(5)^3}{3} + (5)^2 + 5\right] - \left[\frac{(2)^3}{3} + (2)^2 + 2\right]$$

$$= \left[41\tfrac{2}{3} + 25 + 5\right] - \left[2\tfrac{2}{3} + 4 + 2\right]$$

$$= 71\tfrac{2}{3} - 8\tfrac{2}{3}$$

$$= 63 \text{ square units. Answer.}$$

Note: When an *area* is to be calculated, a sketch of the graph should always be made.

Other problems involving definite integrals may not require a sketch. In either case, the use of upper and lower limits eliminates the constant of integration.

Example. Evaluate the definite integral $\int_{-1}^{1} (x^4 - 2x^2 + 1)dx$

$$\int_{-1}^{1} (x^4 - 2x^2 + 1)dx = \left[\frac{x^5}{5} - \frac{2x^3}{3} + x\right]_{-1}^{1}$$

$$= \left[\frac{(1)^5}{5} - \frac{2(1)^3}{3} + 1\right] - \left[\frac{(-1)^5}{5} - \frac{2(-1)^3}{3} + (-1)\right]$$

$$= \left[\frac{1}{5} - \frac{2}{3} + 1\right] - \left[-\frac{1}{5} + \frac{2}{3} - 1\right]$$

$$= \left[\frac{8}{15}\right] - \left[-\frac{8}{15}\right]$$

$$= \frac{16}{15} \quad \text{Answer.}$$

The arrangement of these solutions should be carefully noted.

The *lower* value of x must be placed at the *bottom* of the integral sign and the bracket.

The *upper* value of x must be at the *top* of the integral sign and bracket.

Example. Evaluate the definite integral

$$\int_{0}^{\pi} (\sin x)dx$$

Note: For this integral, upper limit = π radians (i.e. 180°)
lower limit = 0 radians (i.e. 0°)

$$\int_{0}^{\pi} (\sin x)dx = [-\cos x]_{0}^{\pi}$$

$$= [-\cos \pi] - [-\cos 0]$$

$$= [-(-1)] - [-1]$$

$$= 1 + 1$$

$$= 2 \quad \text{Answer.}$$

INTEGRATION AS A SUMMATION

In the previous section, it was shown that the area under a curve could be determined by dividing the area into very narrow strips.

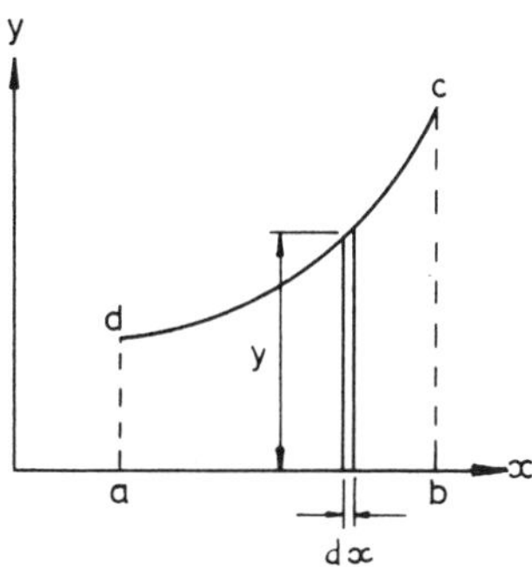

Fig. 151

One such strip is shown in Fig. 151. The width of the strip may be represented by symbol dx if its width approaches zero value.

$$\text{Area of one strip} = y \times dx$$

The total area *abcd* could be obtained by adding together all these small areas, such as the one shown.

i.e. Area *abcd* = the summation of all the areas such as $y \cdot dx$ between the limits $x = a$ and $x = b$

$$\therefore \text{Area} = \int_a^b y\,dx$$

Thus, the symbol ($\int$) may be interpreted as meaning the (summation of) the quantities to which it is applied.

VOLUME OF A SOLID OF REVOLUTION

If a curve is rotated about the x-axis as shown, (Figs. 152a and 152b), the shape generated is called a *solid of revolution.*

The area under the graph is divided into a large number of elemental strips, such as the one shown (Fig. 152a).

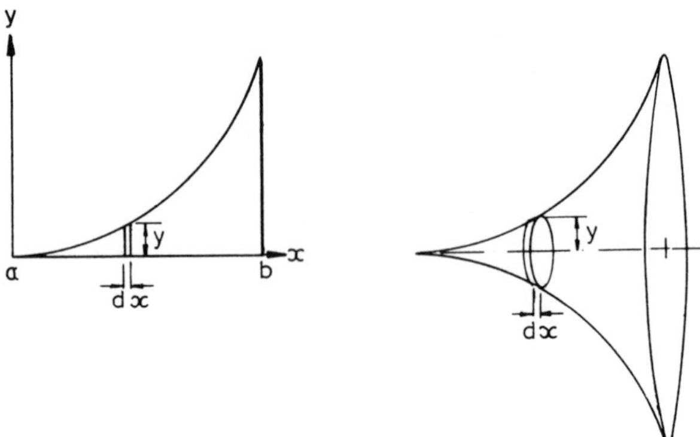

Fig. 152a

Fig. 152b

By rotating this element about the x-axis, a solid disc is generated (Fig. 152b).

Volume of one disc = cross-section area x thickness

$= \pi y^2 . dx$

Total volume = the summation of the volumes of all such discs, between $x = a$ and $x = b$.

$$\text{i.e. Volume} = \int_a^b \pi y^2 \, dx$$

Since the constant factor π is not affected by the integration process, the equation may be written:

$$\text{Volume of a solid of revolution} = \pi \int_a^b y^2 \, dx$$

Example. Calculate the volume generated by rotating the curve $y = x^3$ about the x-axis between $x = 0$ and $x = 2$.

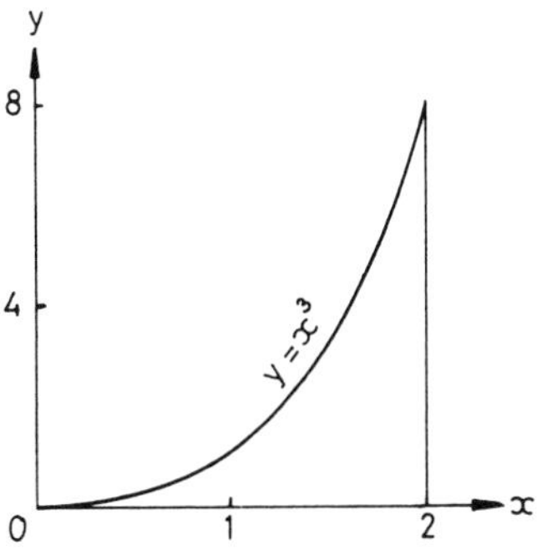

Fig. 153

$$V = \pi \int_a^b y^2 \,.\, dx$$

$$\text{where } y = x^3, \quad a = 0, \quad b = 2.$$

$$\therefore V = \pi \int_0^2 (x^3)^2 \, dx$$

$$= \pi \int_0^2 x^6 \,.\, dx$$

$$= \pi \left[\frac{x^7}{7}\right]_0^2$$

$$= \pi \left[\frac{2^7}{7}\right] - \pi[0]$$

$$= \pi[18{\cdot}28]$$

$$\therefore \text{Volume} = 57{\cdot}45 \text{ cubic units.} \quad \text{Answer.}$$

Example. Use integral calculus to prove that the volume of a sphere is $\frac{4}{3}\pi r^3$ where r is the radius of the sphere.

Note: For convenience, the volume of a *hemi-sphere* is initially calculated.

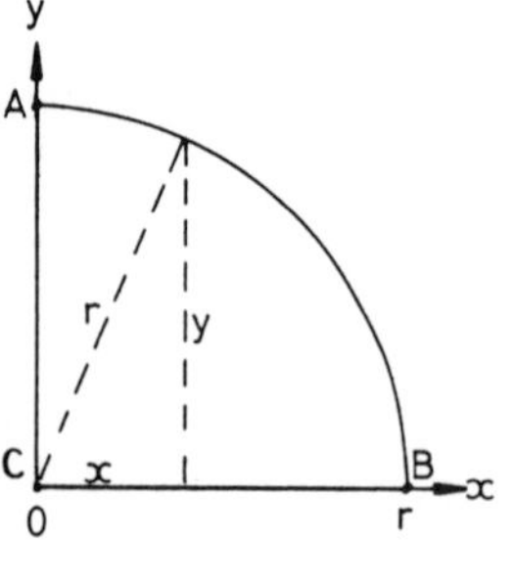

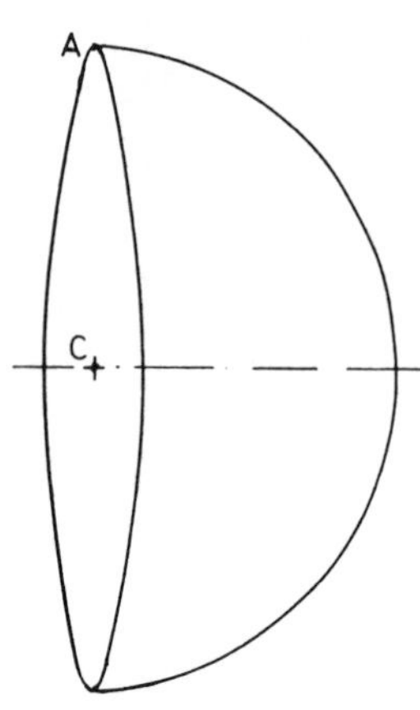

Fig. 154a

Fig. 154b

If the quadrant ABC is rotated about the x-axis, a hemi-sphere is generated. (Lower and upper limits of curve AB are $x = 0, x = r$.)

For any point on the curve AB;

by Pythagoras theorem: $y^2 = r^2 - x^2$

$$\therefore \text{Volume of hemisphere} = \pi \int_a^b y^2 \,.\, dx$$

$$= \pi \int_0^r (r^2 - x^2) \,.\, dx$$

$$= \pi \left[r^2 x - \frac{x^3}{3} \right]_0^r$$

$$= \pi \left[r^3 - \frac{r^3}{3} \right] - 0$$

$$= \frac{2}{3} \pi r^3$$

$$\therefore \text{Volume of sphere} = 2 \times \frac{2}{3} \pi r^3$$

$$= \frac{4}{3} \pi r^3 \quad \text{Answer.}$$

DISTANCE AND VELOCITY BY INTEGRATION

In the previous chapter, it was shown that the relation between distance s, time t and velocity v could be expressed as:

$$\frac{ds}{dt} = v$$

Integrating both sides of this equation with respect to t, the following equation is obtained:

$$s = \int v\,dt \qquad \dots \quad \dots \quad \dots \qquad \text{(i)}$$

Likewise, the relation between acceleration a, velocity and time was defined as:

$$\frac{dv}{dt} = a$$

Integrating both sides of this equation gives the following expression:

$$v = \int a\,dt \qquad \dots \quad \dots \quad \dots \qquad \text{(ii)}$$

Example. The velocity v of a body after time t is given by the expression $v = 10 - 2t$. If displacement $s = 24$ when $t = 2$, obtain an expression for distance s in terms of t.

$$s = \int v\,dt$$

$$= \int (10 - 2t)dt$$

$$\therefore s = 10t - t^2 + C \qquad \dots \quad \dots \qquad \text{(i)}$$

Substituting $s = 24, \quad t = 2$, in equation (i),

$$24 = 20 - 4 + C$$

$$\therefore C = 8 \qquad \dots \quad \dots \quad \dots \quad \dots \qquad \text{(ii)}$$

From equations (i) and (ii),

$$s = 10t - t^2 + 8 \qquad \text{Answer.}$$

Example. The velocity v metres/second after t seconds for a body is given by $v = 4 + 7t$. Find the distance travelled by the body in the interval from $t = 0$ to $t = 5$ seconds.

$$s = \int_0^5 v \,.\, dt \text{ (note: definite integral)}$$

$$\therefore s = \int_0^5 (4 + 7t)\, dt$$

$$= \left[4t + \frac{7t^2}{2}\right]_0^5$$

$$= [107{\cdot}5] - [0]$$

$$\therefore \text{Distance} = 107{\cdot}5 \text{ metres.} \quad \text{Answer.}$$

Example. A body moves such that its acceleration a m/s^2 after time t seconds is given by $a = 18 - 2t$.

(a) Derive an expression for the velocity v m/s of the body, given that $v = 20$ m/s when $t = 0$.

(b) Use the expression to find the velocity of the body after 3 seconds.

(a)

$$v = \int a\,.\,dt$$

$$\therefore v = \int (18 - 2t)dt$$

$$\therefore v = 18t - t^2 + C \quad \ldots \quad \ldots \quad \text{(i)}$$

Substitute $v = 20$, $t = 0$ in equation (i),

$$20 = 0 - 0 + C$$

$$\therefore C = 20 \quad \ldots \quad \ldots \quad \ldots \quad \ldots \quad \text{(ii)}$$

From equations (i) and (ii),

$$v = 18t - t^2 + 20 \quad \text{Answer (a).}$$

(b) When $t = 3$

$$v = 54 - 9 + 20$$

$$\therefore \text{Velocity} = 65 \text{ m/s.} \quad \text{Answer (b).}$$

TEST EXAMPLES 14

1. Evaluate the following integrals:

(a) $\int (x^4 + x^2 - 8x + 5)dx$

(b) $\int \left(\frac{4}{x^2} + \frac{1}{x}\right) dx$

(c) $\int_1^3 (2x - 3x^2 + 1)dx$

2. A curve has a gradient of $x^2 + x - 2$. If the curve passes through the point $x = 2, y = 5$, find its equation.

3. Evaluate the integrals:

   (a) $\int (3 \cos x - 2 \sin x + 4)dx$

   (b) $\int (4 \cos x - \cos x + x)dx$

   (c) $\int_0^{\pi/2} \cos x \, dx$

4. Find the area between the curve $y = x^3 - 4x^2 + 3x$ and the x-axis between the limits $x = 1$ and $x = 3$.

5. The curve shown below (Fig. 155) was plotted during the isothermal expansion of a gas, following the law $pV = c$, between volume V_1 and V_2. Show that the area under the graph $= pV \ln \dfrac{V_2}{V_1}$.

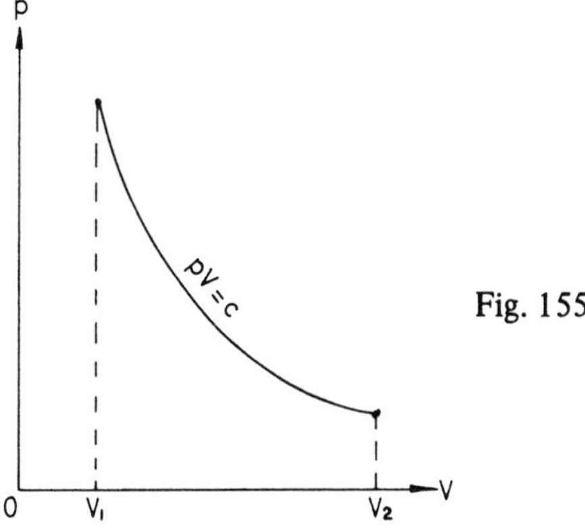

Fig. 155

6. The acceleration a m/s^2 of a body is given by the equation $a = 6t$.

   (a) Use calculus to obtain an expression for velocity v after t seconds given that $v = 0$ when $t = 0$.

   (b) Calculate the *average* velocity during the period $t = 2$ to $t = 3$.

   (c) Calculate the *instantaneous* velocity at $t = 2{\cdot}5$.

7. The velocity v m/s of a body after time t seconds is given by the equation $v = 3t^2 + 8t + 12$. Find the displacement s metres for the body after 10 seconds, given that $s = 10$ when $t = 0$.

8. A cone is generated by rotating the line $y = x/2$ about the x-axis from $x = 0$ to $x = 6$. Calculate the volume of the cone.

9. Find the volume generated by rotating about the x-axis that part of the curve $y = x^2 - x$ which lies between its intersections with the x-axis.

10. Evaluate the following integrals:

(a) $\int (x + 1)(x + 2)\, dx$ (b) $\int \left(\frac{1}{x} - \frac{1}{x^2}\right) dx$

(c) $\int_1^2 (a + 2b)(x + 1)\, dx$ (d) $\int 3e^x\, dx$

11. Evaluate the area enclosed by the curve $f(x) = 0{\cdot}06\, x^2 + 10$, the x axis and the ordinates $x = 6$ and $x = 8$.

12. A gas is compressed from a volume of V_1 to a volume of V_2 according to the law $pV^n = C$, where p is the pressure and C is a constant. Show that the area under the compression curve on a pressure-volume diagram, between the stated volume limits, is given by:

$$\text{Area} = \frac{-C}{n-1}\left(\frac{1}{V_2^{\,n-1}} - \frac{1}{V_1^{\,n-1}}\right)$$

13. Evaluate the following integrals:

(a) $\int \left(\sin\theta - \frac{1}{2}\right)\frac{d\theta}{3}$ (b) $\int_0^{\pi} \sin x\, dx$

(c) $\int_0^{\pi/2} (5\cos x + 3\sin x - x)\, dx$

14. Find the volume of the ellipsoid formed by rotating the ellipse $x^2/a^2 + y^2/b^2 = 1$, of major axis 2a and minor axis 2b, about the major axis.

15. A particle is projected with a horizontal velocity (u) and moves into a resisting medium so that at time (t) its acceleration (dv/dt) equals a constant (k) multiplied by its velocity (v). Show that:

$$\ln\frac{v}{u} = -kt$$

16. Sketch the curve and find an expression for the area between the limits $x = 0$ and $x = 1$, the curve $y = \dfrac{b}{x + a}$ and the x axis. Evaluate this area when $a = 1$ and $b = 2$.

17. Evaluate the following integrals:

(a) $\displaystyle\int \frac{3}{x}\, dx$ (b) $\displaystyle\int \frac{1}{a}\, x^3\, dx$

(c) $\displaystyle\int (4z^3 + 3z^2 + 2z + 1)\, dz$ (d) $\displaystyle\int (2\cos\theta - 5\sin\theta)\, d\theta$

SELECTION OF RULES AND FORMULAE

FROM THE CHAPTERS OF THIS BOOK

LAWS OF INDICES

$$x^m \times x^n = x^{m+n}$$
$$x^m \div x^n = x^{m-n}$$
$$(x^m)^n = x^{m \times n} \text{ written } x^{mn}$$

$$x^{-n} = \frac{1}{x^n}$$

$$x^{1/n} = \sqrt[n]{x}$$
$$x^{m/n} = \sqrt[n]{x^m}$$
$$1^m = 1 \text{ for all values of } m$$
$$x^1 = x$$
$$x^0 = 1 \text{ for all values of } x \text{ except when } x = 0$$

ALGEBRA

$$x^2 - y^2 = (x + y)(x - y)$$
$$x^4 - y^4 = (x^2 + y^2)(x^2 - y^2)$$
$$x^3 + y^3 = (x + y)(x^2 - xy + y^2)$$
$$x^3 - y^3 = (x - y)(x^2 + xy + y^2)$$
$$(x + y)^2 = x^2 + 2xy + y^2$$
$$(x - y)^2 = x^2 - 2xy + y^2$$
$$(x + y)^3 = x^3 + 3x^2y + 3xy^2 + y^3$$
$$(x - y)^3 = x^3 - 3x^2y + 3xy^2 - y^3$$

Linear equation: $y = a + bx$

Quadratic equation: $y = ax^2 + bx + c$

Roots of quadratic equation of the form:

$$ax^2 + bx + c = 0$$

are:

$$x = \frac{-b \pm \sqrt{b^2 - 4ac}}{2a}$$

LOGARITHMS

$$\log_x N = n \text{ means } x^n = N$$
$$\log(x \times y) = \log x + \log y$$
$$\log(x \div y) = \log x - \log y$$
$$\log(x^m) = (\log x) \times m$$
$$\log \sqrt[m]{x} = (\log x) \div m$$
$$\log_b N = \frac{\log_a N}{\log_a b}$$
$$\ln N = 2{\cdot}3026 \times \log_{10} N$$

TRIGONOMETRY

$$\sin\theta = \frac{\text{opposite}}{\text{hypotenuse}}$$
$$\cos\theta = \frac{\text{adjacent}}{\text{hypotenuse}}$$
$$\tan\theta = \frac{\text{opposite}}{\text{adjacent}}$$
$$\operatorname{cosec}\theta = \frac{1}{\sin\theta} = \frac{\text{hypotenuse}}{\text{opposite}}$$
$$\sec\theta = \frac{1}{\cos\theta} = \frac{\text{hypotenuse}}{\text{adjacent}}$$
$$\cot\theta = \frac{1}{\tan\theta} = \frac{\text{adjacent}}{\text{opposite}}$$
$$\text{Hypotenuse}^2 = \text{opposite}^2 + \text{adjacent}^2$$
$$\tan\theta = \frac{\sin\theta}{\cos\theta}$$

$$\sin^2\theta + \cos^2\theta = 1$$
$$\operatorname{cosec}^2\theta - \cot^2\theta = 1$$
$$\sec^2\theta - \tan^2\theta = 1$$

$$\sin(A + B) = \sin A\cos B + \cos A\sin B$$
$$\sin(A - B) = \sin A\cos B - \cos A\sin B$$
$$\cos(A + B) = \cos A\cos B - \sin A\sin B$$
$$\cos(A - B) = \cos A\cos B + \sin A\sin B$$

$$\tan(A + B) = \frac{\tan A + \tan B}{1 - \tan A \tan B}$$

$$\tan(A - B) = \frac{\tan A - \tan B}{1 + \tan A \tan B}$$

$$\sin 2A = 2 \sin A \cos A$$

$$\cos 2A = \cos^2 A - \sin^2 A$$

$$\cos 2A = 1 - 2 \sin^2 A$$

$$\cos 2A = 2 \cos^2 A - 1$$

$$\tan 2A = \frac{2 \tan A}{1 - \tan^2 A}$$

SINE RULE: $$\frac{a}{\sin A} = \frac{b}{\sin B} = \frac{c}{\sin C}$$

COSINE RULE: $$a^2 = b^2 + c^2 - 2bc \cos A$$

$$\cos A = \frac{b^2 + c^2 - a^2}{2bc}$$

NON-PARALLEL CHORDS:

$$ao \times bo = co \times do$$

CYCLIC TRIANGLES:

Angles in the same segment of a circle are equal.

The angle in a semi-circle is a right angle.

The angle at the centre of a circle is double the angle at the circumference for triangles in the same segment and on the same chord.

CIRCULAR MEASURE:

π = 3·142 to nearest four figures.

Approximately, $\pi = \dfrac{22}{7}$

$\dfrac{\pi}{4}$ = 0·7854 to nearest four figures.

Approximately, $\dfrac{\pi}{4} = \dfrac{11}{14}$

Circumference of circle $= \pi d = 2\pi r$

Circumference of ellipse $= \pi \left\{ \dfrac{D + d}{2} \right\}$ approximately

One circle $=$ 360 degrees $= 2\pi$ radians

One radian $=$ 57·3 degrees

Length of arc $= \theta r$

Linear velocity $= \omega r$

PLANE AREAS

Area of circle $= \pi r^2 = \dfrac{\pi}{4} d^2$

,, annulus $= \pi(R^2 - r^2) = \dfrac{\pi}{4}(D^2 - d^2)$

Area of ellipse $= \pi Rr = \dfrac{\pi}{4} Dd$

,, parallelogram $=$ base $\times$ perp. height

,, rhombus $= \frac{1}{2}$ product of diagonals

,, trapezium $= \frac{1}{2}$ sum of parallel sides $\times$ perp distance between

,, sector $= \dfrac{\theta^\circ}{360} \times \pi r^2$

,, segment $= \frac{1}{2} r^2 (\theta - \sin\theta)$

,, triangle $= \frac{1}{2}$(base $\times$ perp. height)

$= \frac{1}{2}(ab \sin C)$

$= \sqrt{s(s - a)(s - b)(s - c)}$

,, equilateral triangle $=$ 0·433 side^2

,, hexagon $=$ six equilateral triangles

,, irregular figure by Simpson's rule: To the first and last ordinates add four times the even ordinates and twice the odd, multiply this sum by one-third the common interval between ordinates. In symbols:

$$= \frac{h}{3}(a + 4b + 2c + 4d + e)$$

SURFACE AREAS

Curved surface of cylinder	=	πdh
,, ,, sphere	=	πd^2
,, ,, cone	=	πrl
,, ,, frustum	=	$\pi l(R + r)$

VOLUMES

Volume of prism	=	area of end × length.
,, cone and pyramid	=	$\frac{1}{3}$(area of base × perp. height)
,, sphere	=	$\frac{\pi}{6}d^3 = \frac{4}{3}\pi r^3$
,, spherical segment	=	$\frac{\pi}{6}h^2(3d - 2h)$
,, frustum of cone	=	$\frac{1}{12}\pi h(D^2 + Dd + d^2)$
	=	$\frac{1}{3}\pi h(R^2 + Rr + r^2)$
,, frustum of pyramid	=	$\frac{1}{3}h(A + \sqrt{Aa} + a)$

CENTRES OF GRAVITY

Triangle	=	$\frac{1}{3}$ height from base
Pyramid	=	$\frac{1}{4}$ height from base
Semi-circular area	=	0·424 r from diameter
Hemisphere	=	0·375 r from diameter

CALCULUS

$\frac{dy}{dx}$	y	$\int y\,dx$
1	x	$\frac{x^2}{2} + C$
0	a	$ax + C$
1	$x + a$	$\frac{x^2}{2} + ax + C$

(contd)

contd.

$\dfrac{dy}{dx}$	y	$\int y\,dx$
a	ax	$\dfrac{ax^2}{2} + C$
nx^{n-1}	x^n	$\dfrac{x^{n+1}}{n+1} + C$
$\cos x$	$\sin x$	$-\cos x + C$
$-\sin x$	$\cos x$	$\sin x + C$
$-x^{-2}$	$\dfrac{1}{x}$ (i.e. x^{-1})	$\ln x + C$
e^x	e^x	$e^x + C$

SOLUTIONS TO TEST EXAMPLES 1

1.
$$\frac{5^3 \times 5 \times 5^5}{5^4 \times 5^2}$$

$$= \frac{5^9}{5^6} = 5^3 = 125 \quad \text{Ans.}$$

2.
$$= \frac{10^{\frac{3}{2}} \times 10^4 \times 10^{\frac{3}{4}}}{10^3 \times 10^{\frac{1}{4}} \times 10^2}$$

$$= \frac{10^{6\frac{1}{4}}}{10^{5\frac{1}{4}}} = 10^1 = 10 \quad \text{Ans.}$$

3.
$$\frac{4^2 \times 4^{\frac{2}{3}} \times 4^{-2}}{4^{\frac{1}{2}} \times 4^{\frac{1}{6}}}$$

$$= \frac{4^{\frac{2}{3}}}{4^{\frac{2}{3}}} = 4^\circ = 1 \quad \text{Ans.}$$

4.
$$\frac{8^{-2}}{8^{-5}}$$

(Note result of subtracting -5 from -2 is $+3$)

$$= 8^3 = 512 \quad \text{Ans.}$$

5.
$$\frac{(2^2)^3 \times (2^{\frac{1}{2}})^4 \times 2^2}{(2^{\frac{2}{3}})^3 \times (2^3)^2 \times 1}$$

$$= \frac{2^6 \times 2^2 \times 2^2}{2^2 \times 2^6 \times 1}$$

$$= \frac{2^{10}}{2^8} = 2^2 = 4 \quad \text{Ans.}$$

6.
$$\frac{(3^{\frac{2}{5}})^{\frac{5}{3}} \times \sqrt{3} \times 3}{3^{-\frac{4}{3}} \times 3^3}$$

$$= \frac{3^{\frac{2}{3}} \times 3^{\frac{1}{2}} \times 3}{3^{-\frac{4}{3}} \times 3^3}$$

$$= \frac{3^{\frac{13}{6}}}{3^{\frac{5}{3}}} = 3^{\frac{1}{2}} = \sqrt{3} = 1{\cdot}732 \quad \text{Ans.}$$

7.
$$(1\tfrac{2}{3})^2 + \sqrt{\tfrac{4}{81}} + \sqrt{200}$$

$$= \left(\frac{5}{3}\right)^2 + \frac{\sqrt{4}}{\sqrt{81}} + \sqrt{2} \times \sqrt{100}$$

$$= \frac{25}{9} + \frac{2}{9} + 1{\cdot}414 \times 10$$

$$= \frac{27}{9} + 14{\cdot}14 = 3 + 14{\cdot}14$$

$$= 17{\cdot}14 \quad \text{Ans.}$$

8.
$$\sqrt[\frac{3}{2}]{27} + \sqrt{6\tfrac{1}{4}}$$

$\sqrt[\frac{3}{2}]{27}$ is the square of the cube root of 27
and may be written $27^{\frac{2}{3}}$
The cube root of 27 is 3, and $3^2 = 9$

$$\sqrt{6\tfrac{1}{4}} = \sqrt{\frac{25}{4}} = \frac{\sqrt{25}}{\sqrt{4}} = \frac{5}{2} = 2\tfrac{1}{2}$$

$$9 + 2\tfrac{1}{2} = 11\tfrac{1}{2} \quad \text{Ans.}$$

9.
$$\frac{(2^2)^3 \times \sqrt{3} \times 27^{\frac{2}{3}} \times \sqrt{12}}{\sqrt[4]{81} \times 4^{\frac{3}{2}}}$$

$$= \frac{2^6 \times \sqrt{3} \times 3^2 \times \sqrt{3} \times \sqrt{4}}{3 \times 2^3}$$

$$= \frac{2^6 \times 3 \times 3^2 \times 2}{3 \times 2^3}$$

$$= 2^4 \times 3^2 = 16 \times 9 = 144 \quad \text{Ans.}$$

10. Let x represent the circumference of the second circle. Ratio:

1st circum. : 2nd circum. : : 1st dia. : 2nd dia.

110 : x : : 35 : 105

$$x \times 35 = 110 \times 105$$

$$x = \frac{110 \times 105}{35} = 330 \text{ mm} \quad \text{Ans.}$$

11. Let x represent volume of second sphere

1st vol. : 2nd vol. : : 1st dia.^{3} : 2nd dia.^{3}

24·25 : x : : 1^3 : 2^3

24·25 : x : : 1 : 8

$$x \times 1 = 24{\cdot}25 \times 8$$

$$x = 194 \text{ cm}^3 \quad \text{Ans.}$$

12. Quantity pumped by 1st pump in 1 hr. = $\frac{1}{12}$ of tank

" " " 2nd " 1 hr. = $\frac{1}{4}$ "

" " " 3rd " 1 hr. = $\frac{1}{9}$ "

When all three pumps are working together,

$$\text{quantity pumped per hour} = \frac{1}{12} + \frac{1}{4} + \frac{1}{9}$$

$$= \frac{3 + 9 + 4}{36} = \frac{16}{36}$$

$$= \frac{4}{9} \text{ of tank}$$

$$\therefore \text{ Time to empty whole tank} = \frac{9}{4} \text{ hours} = 2\tfrac{1}{4} \text{ hrs.} \quad \text{Ans.}$$

13. 6 men to assemble 5 machines = 10 hours

1 man " " 5 machines = 10×6 hours

1 man " " 1 machine = $\frac{10 \times 6}{5}$ "

1 man " " 24 machines = $\frac{10 \times 6 \times 24}{5}$ hours

8 men " " 24 machines = $\frac{10 \times 6 \times 24}{5 \times 8}$ "

= 36 hours Ans.

14. Resistance $\propto$ length

$$\text{also, resistance} \propto \frac{1}{\text{diameter}^2}$$

$$\therefore \frac{\text{resistance} \times \text{diameter}^2}{\text{length}} = \text{constant}$$

$$\therefore \frac{R_1 \times d_1^2}{l_1} = \frac{R_2 \times d_2^2}{l_2}$$

$$\frac{25 \times 1^2}{1000} = \frac{150 \times 0{\cdot}5^2}{l_2}$$

$$l_2 = \frac{1000 \times 150 \times 0{\cdot}25}{25 \times 1}$$

$$\text{length}_2 = 1500 \text{ m} \quad \text{Ans.}$$

15.

$$\text{Mass} \propto \text{length}$$

$$\text{also, mass} \propto \text{diameter}^2$$

$$\therefore \frac{\text{mass}}{\text{length} \times \text{diameter}^2} = \text{constant}$$

$$\frac{m_1}{l_1 \times d_1^2} = \frac{m_2}{l_2 \times d_2^2}$$

$$\frac{280}{4{\cdot}5 \times 100^2} = \frac{m_2}{5 \times 90^2}$$

$$280 \times 5 \times 8100 = m_2 \times 4{\cdot}5 \times 10\,000$$

$$m_2 = \frac{280 \times 5 \times 8100}{4{\cdot}5 \times 10\,000}$$

$$m_2 = 252 \text{ kg} \quad \text{Ans.}$$

16.

$$\text{Strength} \propto \text{breadth} \times \text{depth}^2 \times \frac{1}{\text{length}}$$

$$\therefore \frac{\text{Strength} \times \text{length}}{\text{breadth} \times \text{depth}^2} = \text{constant}$$

$$\frac{S_1 \times l_1}{b_1 \times d_1^{\,2}} = \frac{S_2 \times l_2}{b_2 \times d_2^{\,2}}$$

$S_1 = S_2$ and cancels

$$\frac{5}{40 \times 100^2} = \frac{3}{b_2 \times 80^2}$$

$$b_2 = \frac{40 \times 10\,000 \times 3}{5 \times 6400}$$

$$= 37{\cdot}5 \text{ mm} \quad \text{Ans.}$$

17.

$$\text{Elongation} = 62{\cdot}5 - 50$$
$$= 12{\cdot}5 \text{ mm}$$

$$\%\ \text{elongation} = \frac{\text{elongation}}{\text{original length}} \times 100$$

$$= \frac{12{\cdot}5}{50} \times 100 = 25\% \quad \text{Ans. (i)}$$

$$\text{Reduction in area} = 80 - 48$$
$$= 32 \text{ mm}^2$$

$$\%\ \text{reduction in area} = \frac{\text{reduction}}{\text{original area}} \times 100$$

$$= \frac{32}{80} \times 100 = 40\% \quad \text{Ans. (ii)}$$

18. Increase of strength of solid over hollow shaft

$$= 1 - \tfrac{15}{16} = \tfrac{1}{16}$$

$$\%\ \text{increase} = \frac{\text{increase over hollow}}{\text{strength of hollow}} \times 100$$

$$= \frac{\frac{1}{16}}{\frac{15}{16}} \times 100$$

$$= \frac{1 \times 16}{16 \times 15} \times 100 = 6\tfrac{2}{3}\% \quad \text{Ans. (i)}$$

Decrease of strength of hollow shaft compared with solid shaft

$$= 1 - \tfrac{15}{16} = \tfrac{1}{16}$$

$$\% \text{ decrease} = \frac{\text{decrease from solid}}{\text{strength of solid}}$$

$$= \frac{\tfrac{1}{16}}{1} \times 100$$

$$= \tfrac{1}{16} \times 100 = 6\tfrac{1}{4}\% \quad \text{Ans. (ii)}$$

Increase of mass of solid shaft over hollow shaft

$$= 1 - \tfrac{3}{4} = \tfrac{1}{4}$$

$$\% \text{ increase} = \frac{\text{increase over hollow}}{\text{mass of hollow}}$$

$$= \frac{\tfrac{1}{4}}{\tfrac{3}{4}} \times 100$$

$$\frac{1 \times 4}{4 \times 3} \times 100 = 33\tfrac{1}{3}\% \quad \text{Ans. (iii)}$$

Decrease of mass of hollow shaft compared with solid shaft

$$= 1 - \tfrac{3}{4} = \tfrac{1}{4}$$

$$\% \text{ decrease} = \frac{\text{decrease from solid}}{\text{mass of solid}}$$

$$= \frac{\tfrac{1}{4}}{1} \times 100$$

$$= \tfrac{1}{4} \times 100 = 25\% \quad \text{Ans. (iv)}$$

19. Ratio of powers,

	No. 1	No. 2	No. 3
	115	: 95	: 100
=	1·15	: 0·95	: 1

$$1{\cdot}15 + 0{\cdot}95 + 1 = 3{\cdot}1$$

% of total power developed in No. 1 cylinder

$$= \frac{1{\cdot}15}{3{\cdot}1} \times 100 = 37{\cdot}1\% \quad \text{Ans.}$$

% of total power developed in No. 2 cylinder

$$= \frac{0{\cdot}95}{3{\cdot}1} \times 100 = 30{\cdot}6\% \quad \text{Ans.}$$

% of total power developed in No. 3 cylinder

$$= \frac{1}{3{\cdot}1} \times 100 = 32{\cdot}3\% \quad \text{Ans.}$$

20. Mass of tin = 112 kg

Mass of copper and antimony required

$$= \tfrac{1}{10} \times 112 = 11{\cdot}2 \text{ kg} \quad \text{Ans. (i)}$$

Ratio, tin : copper : antimony

= 10 : 1 : 1

$$\text{Total parts} = 12$$

$$\%\ \text{tin} = \tfrac{10}{12} \times 100 = 83\tfrac{1}{3}\%$$

$$\%\ \text{copper} = \tfrac{1}{12} \times 100 = 8\tfrac{1}{3}\%$$

$$\%\ \text{antimony} = \tfrac{1}{12} \times 100 = 8\tfrac{1}{3}\% \quad \text{Ans. (ii)}$$

21.

$$\%\ \text{zinc content} = 100 - (71 + 1 + 3)$$

$$= 100 - 75 = 25\%$$

$$\text{Total mass of alloy} = 500 \text{ kg}$$

$$\text{Mass of copper} = \frac{71}{100} \times 500 = 355 \text{ kg}$$

$$\text{,, ,, tin} = \frac{1}{100} \times 500 = 5 \text{ kg}$$

$$\text{,, ,, lead} = \frac{3}{100} \times 500 = 15 \text{ kg}$$

$$\text{,, ,, zinc} = \frac{25}{100} \times 500 = 125 \text{ kg}$$

$$\text{Total} = 500 \text{ kg}$$

22.
$$\text{Speed of ship} = \frac{\text{distance}}{\text{time}}$$

$$= \frac{324}{24} = 13{\cdot}5 \text{ knots}$$

$$\text{Slip} = 15 - 13{\cdot}5 = 1{\cdot}5 \text{ knots}$$

$$\%\text{ slip} = \frac{1{\cdot}5}{15} \times 100 = 10\% \quad \text{Ans.}$$

23. Sum of measurements

$= 27 + 39 + 47 + 51 + 48 + 32 + 20 + 11 + 8 + 5$

$= 288$ mm

$$\text{Total number of readings} = 10$$

$$\text{Mean height} = \frac{\text{total of measurements}}{\text{total number of readings}}$$

$$= \frac{288}{10}$$

$$= 28{\cdot}8 \text{ mm} \quad \text{Ans.}$$

24. Total number of revolutions turned

$$= 333\,520 - 312\,460 = 21\,060$$

$$\text{Total time in minutes} = 4 \times 60 - 6 = 240 - 6 = 234$$

$$\text{Mean speed} = \frac{\text{total revolutions}}{\text{total time}}$$

$$= \frac{21\,060}{234} = 90 \text{ rev/min} \quad \text{Ans. (i)}$$

$$\text{Increase of speed} = 10\%$$

$$= \frac{10}{100} \times 90 = 9 \text{ rev/min.}$$

$$\therefore \text{ New speed} = 99 \text{ rev/min.}$$

Total revolutions turned from 12 to 4 p.m.

$$= 99 \times 240 = 23\,760$$

$$\text{Counter reading at 4 p.m.} = 333\,520 + 23\,760$$

$$= 357\,280 \quad \text{Ans. (ii)}$$

25. Cost of 200 tonne @ £60/tonne = £12 000

,, 600 ,, £70/tonne = £42 000

Total cost = £54 000

Total number of tonnes = 200 + 600 = 800

$$\text{Average cost} = \frac{\text{total cost}}{\text{total number of tonnes}}$$

$$= \frac{54\,000}{800}$$

= £67·50 per tonne Ans.

As explained in the text we see again, that it is essential to divide one *total* quantity by another *total* quantity to obtain an average.

Note that the average cost is *not* $\frac{£60 + £70}{2} = £65$

26. 3 minutes 20 seconds = $3\frac{1}{3}$ minutes

$$3\tfrac{1}{3} \div 60 = \frac{10}{3 \times 60} = \frac{1}{18} \text{ hour}$$

$$\text{Speed} = \frac{\text{distance}}{\text{time}}$$

$$= 1 \div \tfrac{1}{18} = \frac{1 \times 18}{1}$$

= 18 knots Ans. (i)

$$3 \text{ minutes} = \frac{3}{60} \text{ hour} = \frac{1}{20} \text{ hour}$$

$$\text{Speed} = 1 \div \tfrac{1}{20} = \frac{1 \times 20}{1}$$

= 20 knots Ans. (ii)

Total distance = 1 + 1 = 2 nautical miles

$$\text{Total time} = \frac{1}{18} + \frac{1}{20} \text{ hour}$$

$$= \frac{10 + 9}{180} = \frac{19}{180} \text{ hour}$$

$$\text{Average speed} = \frac{\text{total distance}}{\text{total time}}$$

$$= 2 \div \frac{19}{180} = \frac{2 \times 180}{19}$$

$$= 18{\cdot}95 \text{ knots} \quad \text{Ans. (iii)}$$

27. Time for the whole journey should be

$$\frac{\text{distance}}{\text{speed}} = \frac{384}{64} = 6 \text{ hours}$$

Time taken over first 192 km

$$= \frac{192}{48} = 4 \text{ hours}$$

Time remaining for second 192 km

$$= 6 - 4 = 2 \text{ hours}$$

∴ Average speed over second half of journey

$$= \frac{192}{2} = 96 \text{ km/h} \quad \text{Ans.}$$

SOLUTIONS TO TEST EXAMPLES 2

1. (i)

$$\begin{array}{rl} & 3x + 4y - 5z \\ \text{add} & -2x - 5y + 4z \\ \hline & x - y - z \quad \text{Ans.} \end{array}$$

(ii)

$$\begin{array}{rl} & 2a^2b - ab + 3ab^2 \\ \text{add} & -a^2b + ab + 5ab^2 \\ \hline & a^2b + 8ab^2 \quad \text{Ans.} \end{array}$$

(iii)

$$\begin{array}{rl} \text{From} & 5x + 3y - 4z \\ \text{subtract} & 2x + 5y - 3z \\ \hline & 3x - 2y - z \quad \text{Ans.} \end{array}$$

(iv)

$$\begin{array}{rl} \text{From} & 3a - 2b + 6c \\ \text{subtract} & -a - 5b - 4c \\ \hline & 4a + 3b + 10c \quad \text{Ans.} \end{array}$$

2. (i)

$$\begin{aligned} & 5x - 3z - 4x - 2y + 4y + 2z - y \\ = \; & 5x - 4x - 2y + 4y - y - 3z + 2z \\ = \; & x + y - z \quad \text{Ans.} \end{aligned}$$

(ii)

$$\begin{aligned} & 2{\cdot}5a + c - 1{\cdot}2a + 2{\cdot}5b - 3c + b + 1{\cdot}7a \\ = \; & 2{\cdot}5a - 1{\cdot}2a + 1{\cdot}7a + 2{\cdot}5b + b + c - 3c \\ = \; & 3a + 3{\cdot}5b - 2c \quad \text{Ans.} \end{aligned}$$

(iii)

$$\begin{aligned} & b^2 - 3ab^2 + 2a^2b - 4a^2 - 2b^2 + 5a^2 - 2ab^2 \\ = \; & -4a^2 + 5a^2 + 2a^2b - 3ab^2 - 2ab^2 + b^2 - 2b^2 \\ = \; & a^2 + 2a^2b - 5ab^2 - b^2 \quad \text{Ans.} \end{aligned}$$

3. (i)

$$\begin{aligned} & \frac{x \times x^3 \times x^5}{x^2 \times x^4} \\ = \; & x^{1+3+5-2-4} \\ = \; & x^3 \quad \text{Ans.} \end{aligned}$$

(ii) $x^3 \div x^5 = x^{3-5}$

$= x^{-2}$ or $\dfrac{1}{x^2}$ Ans.

(iii) $x^5 \times x^{-3} \times x^{-2}$

$= x^{5-3-2} = x^0$

$= 1$ Ans.

(iv) $\sqrt{x} \times x^{\frac{1}{2}} \times x^{-\frac{1}{3}}$

$= x^{\frac{1}{2}+\frac{1}{2}-\frac{1}{3}} = x^{\frac{2}{3}}$

or $\sqrt[3]{x^2}$ Ans.

(v) $\sqrt[2]{(x^4 y^2)}$

$= x^2 y$ Ans.

(vi) $\dfrac{x^3}{x^3 - 0{\cdot}5x^3} = \dfrac{x^3}{0{\cdot}5x^3}$

$= \dfrac{1}{0{\cdot}5} = 2$ Ans.

(vii) $(x^2)^3 \times \sqrt[3]{x^{\frac{1}{2}}} \times x^{-5}$

$= x^6 \times x^{\frac{1}{6}} \times x^{-5} = x^{1\frac{1}{6}}$

or $x^{\frac{7}{6}}$ Ans.

4. (i) $(x + 2y)(2x + y) = 2x^2 + 5xy + 2y^2$ Ans.

(ii) $(2x + y)(3x - 2y) = 6x^2 - xy - 2y^2$ Ans.

(iii) $(3x - 4y)(2x - 3y) = 6x^2 - 17xy + 12y^2$ Ans.

5. (i) $(a + b)^2 = (a + b)(a + b) = a^2 + 2ab + b^2$ Ans.

(ii) $(a - b)^2 = (a - b)(a - b) = a^2 - 2ab + b^2$ Ans.

(iii) $(a + b)^3 = (a + b)^2 \times (a + b) = (a^2 + 2ab + b^2) \times (a + b)$

$$\begin{array}{lllll} a^2 & + 2ab & + b^2 & & \\ & a & + b & & \\ \hline a^3 & + 2a^2b & + ab^2 & & \\ & a^2b & + 2ab^2 & + b^3 & \\ \hline a^3 & + 3a^2b & + 3ab^2 & + b^3 & \text{Ans.} \end{array}$$

(iv) $(a - b)^3 = (a - b)^2 \times (a - b) = (a^2 - 2ab + b^2) \times (a - b)$

$$
\begin{array}{rrrr}
a^2 & - 2ab & + b^2 & \\
 & a & - b & \\
\hline
a^3 & - 2a^2b & + ab^2 & \\
 & - a^2b & + 2ab^2 & - b^3 \\
\hline
a^3 & - 3a^2b & + 3ab^2 & - b^3
\end{array}
\quad \text{Ans.}
$$

6. (i)

$$
\begin{array}{rrrr}
2a - 3b)\,8a^2 & - 8ab & - 6b^2\,(4a + 2b & \text{Ans.} \\
8a^2 & - 12ab & & \\
\hline
 & 4ab & - 6b^2 & \\
 & 4ab & - 6b^2 & \\
\hline
 & \cdot & \cdot &
\end{array}
$$

(ii)

$$
\begin{array}{rrrrr}
3x - 4y)\,9x^3 & - 9x^2y & - 10xy^2 & + 8y^3\,(3x^2 + xy - 2y^2 & \text{Ans.} \\
9x^3 & - 12x^2y & & & \\
\hline
\cdot & 3x^2y & - 10xy^2 & & \\
 & 3x^2y & - 4xy^2 & & \\
\hline
 & \cdot & - 6xy^2 & + 8y^3 & \\
 & & - 6xy^2 & + 8y^3 & \\
\hline
 & & \cdot & \cdot &
\end{array}
$$

(iii)

$$
\begin{array}{rrrrr}
x - y)\,x^3 & & - y^3\,(x^2 + xy + y^2) & & \text{Ans.} \\
x^3 & - x^2y & & & \\
\hline
\cdot & + x^2y & - y^3 & & \\
 & x^2y & - xy^2 & & \\
\hline
 & \cdot & + xy^2 & - y^3 & \\
 & & xy^2 & - y^3 & \\
\hline
 & & \cdot & \cdot &
\end{array}
$$

7. (i) $(a + b) + (c - d) - (a - b) - (c + d) + (a - b)$

$= a + b + c - d - a + b - c - d + a - b$

$= a - a + a + b + b - b + c - c - d - d$

$= a + b - 2d$ Ans.

(ii) $2\{a - 3(a + 2) + 4(2a - 1) + 5\}$
$= 2\{a - 3a - 6 + 8a - 4 + 5\}$
$= 2a - 6a - 12 + 16a - 8 + 10$
$= 12a - 10$ Ans.

(iii) $2x - [2x - \{2x - (2x - 2) - 2\} - 2] - 2$
$= 2x - [2x - \{2x - 2x + 2 - 2\} - 2] - 2$
$= 2x - [2x - 2x + 2x - 2 + 2 - 2] - 2$
$= 2x - 2x + 2x - 2x + 2 - 2 + 2 - 2$
$= 0$ Ans.

8. (i) $3b^2 - 6b + 9$
$= 3(b^2 - 2b + 3)$ Ans.

(ii) $pv + pvx$
$= pv(1 + x)$ Ans.

(iii) $ax^3 - 2bx^2 + 3cx$
$= x(ax^2 - 2bx + 3c)$ Ans.

(iv) $12a^3b^3c^3 - 8a^2b^2c^2 + 4abc$
$= 4abc(3a^2b^2c^2 - 2abc + 1)$ Ans.

9. (i) $D^2 - d^2 = (D + d)(D - d)$ Ans.
(ii) $1 - a^2 = (1 + a)(1 - a)$ Ans.
(iii) $4x^2y^2 - 9z^2 = (2xy + 3z)(2xy - 3z)$ Ans.
(iv) $T_1^4 - T_2^4 = (T_1^2 + T_2^2)(T_1^2 - T_2^2)$
or $(T_1^2 + T_2^2)(T_1 + T_2)(T_1 - T_2)$
Ans.

10. (i) $a^2 + 8a + 16 = (a + 4)(a + 4)$
or $(a + 4)^2$ Ans.

(ii) $d^2 - 10d + 25 = (d - 5)(d - 5)$
or $(d - 5)^2$ Ans.

(iii) $9v^2 + 12v + 4 = (3v + 2)(3v + 2)$
or $(3v + 2)^2$ Ans.

(iv) $4x^2 - 12xy + 9y^2 = (2x - 3y)(2x - 3y)$
or $(2x - 3y)^2$ Ans.

11. (i) $x^2 + 3x + 2 = (x + 2)(x + 1)$ Ans.

(ii) $a^2 - a - 6 = (a - 3)(a + 2)$ Ans.

(iii) $x^2 + xy - 6y^2 = (x + 3y)(x - 2y)$ Ans.

(iv) $6x^2 - 9xy - 6y^2 = (3x - 6y)(2x + y)$

$= 3(x - 2y)(2x + y)$ Ans.

12. (i)

$$\frac{x}{2} + \frac{x}{4} - \frac{2x}{3}$$

$$= \frac{6x + 3x - 8x}{12}$$

$$= \frac{x}{12} \text{ Ans.}$$

(ii)

$$\frac{2a - 1}{b} - \frac{2a + 3}{2b} - \frac{3a - 8}{3b}$$

$$= \frac{6(2a - 1) - 3(2a + 3) - 2(3a - 8)}{6b}$$

$$= \frac{12a - 6 - 6a - 9 - 6a + 16}{6b}$$

$$= \frac{1}{6b} \text{ Ans.}$$

(iii)

$$\frac{4}{x + 2} + \frac{18}{x^2 - 2x - 8} - \frac{3}{x - 4}$$

$$= \frac{4(x - 4) + 18 - 3(x + 2)}{(x + 2)(x - 4)}$$

$$= \frac{4x - 16 + 18 - 3x - 6}{(x + 2)(x - 4)}$$

$$= \frac{x - 4}{(x + 2)(x - 4)}$$

$$= \frac{1}{x + 2} \text{ Ans.}$$

13. $3a^3 - 4a^2 - 2a - 15 \qquad a = 5$

$$= 3(5)^3 - 4(5)^2 - 2(5) - 15$$
$$= 3(125) - 4(25) - 2(5) - 15$$
$$= 375 - 100 - 10 - 15$$
$$= 375 - 125$$
$$= 250 \quad \text{Ans.}$$

14. $2x^3 - x^2y + xy^2 - 3y^3 \quad x = 2, y = -2$

$$= 2(2)^3 - (2)^2(-2) + (2)(-2)^2 - 3(-2)^3$$
$$= 2(8) - (4)(-2) + (2)(4) - 3(-8)$$
$$= 16 - (-8) + 8 + 24$$
$$= 16 + 8 + 8 + 24$$
$$= 56 \quad \text{Ans.}$$

15. $3[4x + 2\{x - 2y - (3x + y)\} - 3x]$

$$= 3[4x + 2\{x - 2y - 3x - y\} - 3x]$$
$$= 3[4x + 2x - 4y - 6x - 2y - 3x]$$
$$= 12x + 6x - 12y - 18x - 6y - 9x$$
$$= -9x - 18y$$

substituting $x = -2, \ y = -3$

$$-9(-2) - 18(-3)$$
$$= +18 + 54$$
$$= 72 \quad \text{Ans.}$$

16. Remainder theorem:

$R = f(a)$ when $f(x)$ is divided by $(x - a)$

$$\text{Let } x - a = x - 3$$
$$\text{then } a = +3$$
$$f(x) = 2x^3 - 4x^2 + 5x - 6$$
$$R = f(a) = 2(3)^3 - 4(3)^2 + 5(3) - 6$$
$$= 54 - 36 + 15 - 6$$
$$= 27 \quad \text{Ans.}$$

17. Remainder theorem:

$R = f(a)$ when $f(x)$ is divided by $(x - a)$

$$\text{Let } x - a = x + 5$$
$$\text{then } a = -5$$
$$f(x) = x^3 + 6x^2 + bx + 9$$

$$R = f(a) = (-5)^3 + 6(-5)^2 + b(-5) + 9$$
$$4 = -125 + 150 - 5b + 9$$
$$5b = 150 + 9 - 125 - 4 = 30$$
$$b = 6 \quad \text{Ans.}$$

18. Factor theorem: If $(x - a)$ is a factor of $f(x)$, the remainder is zero, and $f(a) = 0$

$$\text{Let } x - a = x + 4$$
$$\text{then } a = -4$$
$$x + 4 \text{ is a factor, } \therefore f(-4) = 0$$
$$f(x) = x^3 + x^2 - 10x + c$$
$$f(-4) = (-4)^3 + (-4)^2 - 10(-4) + c$$
$$0 = -64 + 16 + 40 + c$$
$$c = 64 - 16 - 40$$
$$= 8 \quad \text{Ans.}$$

19. Factor theorem: If $(x - a)$ is a factor of $f(x)$, the remainder is zero, and $f(a) = 0$

$$\text{Let } x - a = x + 1$$
$$\text{then } a = -1$$
$$x + 1 \text{ is a factor, } \therefore f(-1) = 0$$
$$f(x) = x^3 + 3x^2 - bx - c$$
$$f(-1) = (-1)^3 + 3(-1)^2 - b(-1) - c$$
$$0 = -1 + 3 + b - c$$
$$c = b + 2 \quad \ldots \quad \ldots \quad \ldots \quad \ldots \quad \ldots \quad \text{(i)}$$

Remainder theorem:

$R = f(a)$ when $f(x)$ is divided by $x - a$

$$\text{Let } x - a = x + 2$$
$$\text{then } a = -2$$
$$f(x) = x^3 + 3x^2 - bx - c$$
$$f(-2) = (-2)^3 + 3(-2)^2 - b(-2) - c$$
$$15 = -8 + 12 + 2b - c$$
$$c = 2b - 11 \quad \ldots \quad \ldots \quad \ldots \quad \ldots \quad \ldots \quad \text{(ii)}$$

From (i) and (ii),

$$2b - 11 = b + 2$$
$$b = 13$$

Substituting for b into (i)

$$c = 13 + 2 = 15$$

Constants are 13 and 15 Ans.

SOLUTIONS TO TEST EXAMPLES 3

	NUMBER	LOGARITHM	
1. 562·2 × 0·4323	562·2	2·7499	
Ans. = 243·1	0·4323	$\bar{1}$·6358	Add
	243·1	2·3857	
2. 1·957 ÷ 0·486	1·957	0·2915	
Ans. = 4·026	0·486	$\bar{1}$·6866	Subtract
	4·026	0·6049	
3. $\dfrac{11{\cdot}24 \times 0{\cdot}06836}{2254 \times 0{\cdot}00753}$	11·24	1·0508	
	0·06836	$\bar{2}$·8348	Add
Ans. = 0·04527		$\bar{1}$·8856	
	2254	3·3530	
	0·00753	$\bar{3}$·8768	Add
		1·2298	
		$\bar{1}$·8856	
		1·2298	Subtract
	0·04527	$\bar{2}$·6558	
4. $(2{\cdot}54)^3$	2·54	0·4048	
Ans. = 16·39		3	Multiply
	16·39	1·2144	
5. $(0{\cdot}3937)^2$	0·3937	$\bar{1}$·5952	
Ans. = 0·155		2	Multiply
	0·1550	$\bar{1}$·1904	
6. $(0{\cdot}0807)^{2{\cdot}8}$	0·0807	$\bar{2}$·9069	
Ans. = 0·0008696		= –1·0931	
		2·8	Multiply
		87448	
		21862	
		–3·06068	
	0·0008696	= $\bar{4}$·93932	

		NUMBER	LOGARITHM	
7.	$\sqrt[3]{238 \cdot 3}$	238·3	$3)\overline{2 \cdot 3772}$	Divide by 3
	Ans. = 6·2	6·200	0·7924	
8.	$0 \cdot 0188^{\frac{1}{2}}$	0·0188	$2)\overline{\bar{2} \cdot 2742}$	Divide by 2
	$= \sqrt{0 \cdot 0188}$	0·1371	$\bar{1} \cdot 1371$	
	Ans. = 0·1371			
9.	$\sqrt{0 \cdot 009025}$	0·009025	$2)\overline{\bar{3} \cdot 9554}$	Divide by 2
	Ans. = 0·09499	0·09499	$\bar{2} \cdot 9777$	
10.	$\sqrt[1 \cdot 2]{2 \cdot 514}$	2·514	0·4004	Divide by 1·2
	Ans. = 2·156		$12)\overline{4 \cdot 004}$	
		2·156	0·3337	
11.	$0 \cdot 06626^{\frac{1}{1 \cdot 4}}$	0·06626	$\bar{2} \cdot 8213$	Divide by 1·4
	$= \sqrt[1 \cdot 4]{0 \cdot 06626}$		$= -1 \cdot 1787$	
			$-\ 0 \cdot 8419$ $14)\overline{-11 \cdot 787}$ 112 58 56 27 14 130 126	
	$\left(\frac{1}{1 \cdot 4} = \frac{10}{14} = \frac{5}{7}\right.$ ∴ the log may be multiplied by 5 and divided by 7 instead of dividing by 1·4 as shown.)			
			$-0 \cdot 8419$	
	Ans. = 0·1439	0·1439	$= \bar{1} \cdot 1581$	
12.	$712 \cdot 5^{\frac{2}{3}}$	712·5	2·8528	
			2	Multiply
	Ans. = 79·79		$3)\overline{5 \cdot 7056}$	Divide by 3
		79·79	1·9019	

13. $V = \dfrac{mR(\theta + 273)}{p}$

Putting in values:

$= \dfrac{0{\cdot}05 \times 0{\cdot}287 \times (17+273)}{0{\cdot}97 \times 10^2}$

$= \dfrac{0{\cdot}05 \times 0{\cdot}287 \times 290}{97}$

$= 0{\cdot}0429$ Ans.

NUMBER	LOGARITHM	
0·05	$\bar{2}{\cdot}6990$	
0·287	$\bar{1}{\cdot}4579$	
290	2·4624	Add
	0·6193	
97	1·9868	Subtract
0·0429	$\bar{2}{\cdot}6325$	

14. $15{\cdot}52^2 \times 1{\cdot}75(0{\cdot}766+0{\cdot}0434)$

$= 15{\cdot}52^2 \times 1{\cdot}75 \times 0{\cdot}8094$

$= 341{\cdot}1$ Ans.

NUMBER	LOGARITHM	
15·52	1·1909	
15·52	1·1909	
1·75	0·2430	
0·8094	$\bar{1}{\cdot}9081$	Add
341·1	2·5329	

15. $E_1 - E_2 = 0{\cdot}5mk^2(\omega_1{}^2 - \omega_2{}^2)$

Simplify quantity inside brackets first, note that it is the difference between two squares.

$\omega_1{}^2 - \omega_2{}^2 = (\omega_1 + \omega_2)(\omega_1 - \omega_2)$

$= (20{\cdot}94 + 18{\cdot}85)(20{\cdot}94 - 18{\cdot}85)$

$= 39{\cdot}79 \times 2{\cdot}09$

Putting in values:

$E_1 - E_2$

$= 0{\cdot}5 \times 1220 \times 0{\cdot}58^2 \times 39{\cdot}79 \times 2{\cdot}09$

$= 17\,060$ Ans.

NUMBER	LOGARITHM	
0·5	$\bar{1}{\cdot}6990$	
1220	3·0864	
0·58	$\bar{1}{\cdot}7634$	
0·58	$\bar{1}{\cdot}7634$	
39·79	$\bar{1}{\cdot}5998$	
2·09	0·3201	Add
17060	4·2321	

16. $\sqrt[3]{\dfrac{225 \times 27 \times 72^2}{1110(2 + 2{\cdot}65^2)}}$	2·65	0·4232	
		2	Multiply
	7·021	0·8464	
$= \sqrt[3]{\dfrac{225 \times 27 \times 72^2}{1110 \times 9{\cdot}021}}$	225	2·3522	
	27	1·4314	
	72	1·8573	
	72	1·8573	Add
= 14·65 Ans.		7·4982	
	1110	3·0453	
	9·021	0·9552	Add
		4·0005	
		7·4982	
		4·0005	Subtract
		3)3·4977	Divide by 3
		1·1659	
	14·65		

17. 12·5 x 0·45(1+ln 2·55)	12·5	1·0969	
= 12·5 x 0·45(1+0·9361)	0·45	$\bar{1}$·6532	
= 12·5 x 0·45 x 1·9361	1·936	0·2869	Add
= 10·89 Ans.	10·89	1·0370	

18. $\ln\dfrac{850}{492} + \dfrac{840}{850} + 0{\cdot}6\ln\dfrac{1160}{850}$	850	2·9294	
	492	2·6920	Subtract
	1·728	0·2374	
= ln 1·728 + 0·9883 + 0·61 ln 1·365			
= 0·5469 + 0·9883 + 0·6 x	840	2·9243	
0·3112	850	2·9294	Subtract
= 0·5469 + 0·9883 + 0·18672	0·9883	$\bar{1}$·9949	
= 1·72192 Ans.			
	1160	3·0645	
	850	2·9294	Subtract
	1·365	0·1351	

19.
$$\log_b N = \frac{\log_a N}{\log_a b}$$

$$\log_2 N = \frac{\log_{10} N}{\log_{10} 2}$$

$$= \frac{\log_{10} N}{0{\cdot}3010} = \log_{10} N \times \frac{1}{0{\cdot}3010}$$

$$= \log_{10} N \times 3{\cdot}322$$

∴ Converting multiplier is 3·322 Ans. (i)

$$\log_2 1{\cdot}5 = \log_{10} 1{\cdot}5 \times 3{\cdot}322$$
$$= 0{\cdot}1761 \times 3{\cdot}322$$
$$= 0{\cdot}5851 \quad \text{Ans. (ii)}$$

$$\log_2 2{\cdot}5 = \log_{10} 2{\cdot}5 \times 3{\cdot}322$$
$$= 0{\cdot}3979 \times 3{\cdot}322$$
$$= 1{\cdot}322 \quad \text{Ans. (iii)}$$

$$\log_2 3{\cdot}5 = \log_{10} 3{\cdot}5 \times 3{\cdot}322$$
$$= 0{\cdot}5441 \times 3{\cdot}322$$
$$= 1{\cdot}807 \quad \text{Ans. (iv)}$$

20. NOTE: full working is shown in the following but the values of $\ln 10^n$ and $\ln 10^{-n}$ could be read directly from the supplementary tables as explained in the text.

$$16 = 1{\cdot}6 \times 10$$
$$\ln 16 = \ln 1{\cdot}6 + \ln 10$$
$$= 0{\cdot}4700 + 2{\cdot}3026$$
$$= 2{\cdot}7726 \quad \text{Ans. (i)}$$

$$987{\cdot}6 = 9{\cdot}876 \times 10^2$$
$$\ln 987{\cdot}6 = \ln 9{\cdot}876 + \ln 10^2$$
$$= 2{\cdot}2901 + 2 \times 2{\cdot}3026$$
$$= 2{\cdot}2901 + 4{\cdot}6052$$
$$= 6{\cdot}8953 \quad \text{Ans. (ii)}$$

$$0{\cdot}5 = 5 \times 10^{-1} \text{ or } 5 \div 10$$
$$\ln 0{\cdot}5 = \ln 5 - \ln 10$$
$$= 1{\cdot}6094 - 2{\cdot}3026$$
$$= 1{\cdot}6094 + \bar{3}{\cdot}6974$$
$$= \bar{1}{\cdot}3068 \quad \text{Ans. (iii)}$$

$$0{\cdot}008667 = 8{\cdot}667 \times 10^{-3} \text{ or } 8{\cdot}667 \div 10^3$$

$$\ln 0{\cdot}008667 = \ln 8{\cdot}667 - \ln 10^3$$
$$= 2{\cdot}1595 - 3 \times 2{\cdot}3026$$
$$= 2{\cdot}1595 - 6{\cdot}9078$$
$$= 2{\cdot}1595 + \bar{7}{\cdot}0922$$
$$= \bar{5}{\cdot}2517 \quad \text{Ans. (iv)}$$

21. Antilog of 0·9925:

direct from main table $= 2{\cdot}698$ Ans. (i)

Antilog of 2·1453:

direct from main table $= 8{\cdot}544$ Ans. (ii)

Antilog of 6·8739:

subtract $\ln 10^2$ which is $2 \times 2{\cdot}3026$

$$6{\cdot}8739 - 4{\cdot}6052 = 2{\cdot}2687$$

antilog of 2·2687 from main table $= 9{\cdot}667$

antilog of 6·8739 $= 9{\cdot}667 \times 10^2$

$= 966{\cdot}7$ Ans. (iii)

Antilog of $\bar{9}{\cdot}9836$:

subtract $\ln 10^{-4}$ which is $-4 \times 2{\cdot}3026$

$= -9{\cdot}2104 = \overline{10}{\cdot}7896$

(or $\overline{10}{\cdot}7897$ from supplementary table)

$$\bar{9}{\cdot}9836 - \overline{10}{\cdot}7896 = 1{\cdot}1940$$

antilog of 1·1940 from main table $= 3{\cdot}3$

antilog of $\bar{9}{\cdot}9836 = 3{\cdot}3 \times 10^{-4}$

$= 0{\cdot}00033$ Ans. (iv)

22. By common logs:

$$\log_{10} 0{\cdot}06326^{-0{\cdot}25} = \bar{2}{\cdot}8011 \times (-0{\cdot}25)$$
$$= -1{\cdot}1989 \times (-0{\cdot}25)$$
$$= 0{\cdot}2997$$

Antilog $= 1{\cdot}994$ Ans. (*a*)

By Napierian logs:

$$\ln 0{\cdot}06326^{-0{\cdot}25} = (\ln 6{\cdot}326 + \ln 10^{-2}) \times (-0{\cdot}25)$$
$$= \{1{\cdot}8446 + 2{\cdot}3026 \times (-2)\} \times (-0{\cdot}25)$$
$$= \{1{\cdot}8446 - 4{\cdot}6052\} \times (-0{\cdot}25)$$
$$= -2{\cdot}7606 \times (-0{\cdot}25)$$
$$= 0{\cdot}6901$$

Antilog $= 1{\cdot}994$ Ans. (*b*)

23.

$$
\begin{aligned}
y &= 29{\cdot}5^2(\ln 2{\cdot}411 + \ln 6{\cdot}234) \\
&= 29{\cdot}5^2(0{\cdot}88 + 1{\cdot}83) \\
&= 29{\cdot}5^2 \times 2{\cdot}71 \\
\ln y &= 2(\ln 2{\cdot}95 + \ln 10) + \ln 2{\cdot}71 \\
&= 2(1{\cdot}0818 + 2{\cdot}3026) + 0{\cdot}9969 \\
&= 2 \times 3{\cdot}3844 + 0{\cdot}9969 \\
&= 6{\cdot}7688 + 0{\cdot}9969 \\
&= 7{\cdot}7657
\end{aligned}
$$

To find antilog,

subtract $\ln 10^3$ which is $3 \times 2{\cdot}3026$,

$$
\begin{aligned}
7{\cdot}7657 - 6{\cdot}9078 &= 0{\cdot}8579 \\
\text{Antilog of } 0{\cdot}8579 &= 2{\cdot}358 \\
\text{Antilog of } 7{\cdot}7657 &= 2{\cdot}358 \times 10^3 \\
y &= 2358 \quad \text{Ans.}
\end{aligned}
$$

24. Finding the logs:

$$
\begin{aligned}
\ln 2{\cdot}55 &= 0{\cdot}9361 \\
\ln 0{\cdot}08234 &= \ln 8{\cdot}234 + \ln 10^{-2} \\
&= 2{\cdot}1083 + 2{\cdot}3026 \times (-2) \\
&= 2{\cdot}1083 + (-4{\cdot}6052) \\
&= 2{\cdot}1083 + \bar{5}{\cdot}3948 \\
&= \bar{3}{\cdot}5031 \\
\ln 0{\cdot}36 &= \ln 3{\cdot}6 + \ln 10^{-1} \\
&= 1{\cdot}2809 + (-2{\cdot}3026) \\
&= 1{\cdot}2809 + \bar{3}{\cdot}6974 \\
&= \bar{2}{\cdot}9783 \\
\ln 10{\cdot}2 &= \ln 1{\cdot}02 + \ln 10 \\
&= 0{\cdot}0198 + 2{\cdot}3026 \\
&= 2{\cdot}3224
\end{aligned}
$$

$$
k = 2{\cdot}55 \times \sqrt{\frac{0{\cdot}08234}{0{\cdot}36 \times 10{\cdot}2}}
$$

$$
\begin{aligned}
\ln k &= 0{\cdot}9361 + \{\bar{3}{\cdot}5031 - (\bar{2}{\cdot}9783 + 2{\cdot}3224)\} \div 2 \\
&= 0{\cdot}9361 + \{\bar{3}{\cdot}5031 - 1{\cdot}3007\} \div 2 \\
&= 0{\cdot}9361 + \bar{4}{\cdot}2024 \div 2 \\
&= 0{\cdot}9361 + \bar{2}{\cdot}1012 \\
&= \bar{1}{\cdot}0373
\end{aligned}
$$

To find antilog:

subtract $\ln 10^{-1}$ which is $-2{\cdot}3026 = \bar{3}{\cdot}6974$

$$\bar{1}{\cdot}0373 - \bar{3}{\cdot}6974 = 1{\cdot}3399$$
$$\text{Antilog of } 1{\cdot}3399 = 3{\cdot}819$$
$$\text{Antilog of } \bar{1}{\cdot}0373 = 3{\cdot}819 \times 10^{-1}$$
$$k = 0{\cdot}3819 \quad \text{Ans.}$$

25.

$$\frac{233{\cdot}6^{-\frac{1}{2}}}{0{\cdot}07541^{-0{\cdot}4}}$$
$$= \frac{0{\cdot}07541^{0{\cdot}4}}{233{\cdot}6^{0{\cdot}5}}$$

(a) By common logs:

$$0{\cdot}4(\log 0{\cdot}07541) - 0{\cdot}5(\log 233{\cdot}6)$$
$$= 0{\cdot}4 \times \bar{2}{\cdot}8775 - 0{\cdot}5 \times 2{\cdot}3685$$
$$= 0{\cdot}4 \times (-1{\cdot}1225) - 1{\cdot}1843$$
$$= --0{\cdot}449 - 1{\cdot}1843$$
$$= -1{\cdot}6333$$
$$= \bar{2}{\cdot}3667$$
$$\text{Antilog} = 0{\cdot}02327 \quad \text{Ans. (a)}$$

(b) By Napierian logs:

$$0{\cdot}4(\ln 0{\cdot}07541) - 0{\cdot}5(\ln 233{\cdot}6)$$
$$= 0{\cdot}4(\ln 7{\cdot}541 + \ln 10^{-2}) - 0{\cdot}5(\ln 2{\cdot}336 + \ln 10^{2})$$
$$= 0{\cdot}4(2{\cdot}0203 - 2 \times 2{\cdot}3026) - 0{\cdot}5(0{\cdot}8485 + 2 \times 2{\cdot}3026)$$
$$= 0{\cdot}4(2{\cdot}0203 - 4{\cdot}6052) - 0{\cdot}5(0{\cdot}8485 + 4{\cdot}6052)$$
$$= 0{\cdot}4 \times (-2{\cdot}5849) - 0{\cdot}5 \times 5{\cdot}4537$$
$$= -1{\cdot}03396 - 2{\cdot}72685$$
$$= -3{\cdot}76081$$
$$= \bar{4}{\cdot}2392$$

To find antilog,

subtract $\ln 10^{-2}$ which is $-2 \times 2{\cdot}3026$

$$= -4{\cdot}6052 = \bar{5}{\cdot}3948$$
$$\bar{4}{\cdot}2392 - \bar{5}{\cdot}3948 = 0{\cdot}8444$$
$$\text{antilog of } 0{\cdot}8444 = 2{\cdot}327$$
$$\text{antilog of } \bar{4}{\cdot}2392 = 2{\cdot}327 \times 10^{-2}$$
$$= 0{\cdot}02327 \quad \text{Ans. (b)}$$

26. $x = 0{\cdot}003436^{-2\cdot2} \times 0{\cdot}02938^{1\cdot4}$

(a) By common logs:

$\log x = -2{\cdot}2 \times \bar{3}{\cdot}5361 + 1{\cdot}4 \times \bar{2}{\cdot}4681$

$= -2{\cdot}2 \times (-2{\cdot}4639) + 1{\cdot}4 \times (-1{\cdot}5319)$

$= 5{\cdot}4206 - 2{\cdot}1447$

$= 3{\cdot}2759$

(Antilog) $x = 1888$ Ans. (a)

(b) By Napierian logs:

$\ln x = -2{\cdot}2 \times \ln 0{\cdot}003436 + 1{\cdot}4 \times \ln 0{\cdot}02938$

$= -2{\cdot}2(\ln 3{\cdot}436 + \ln 10^{-3}) + 1{\cdot}4(\ln 2{\cdot}938 + \ln 10^{-2})$

$= -2{\cdot}2(1{\cdot}2343 - 3 \times 2{\cdot}3026) + 1{\cdot}4(1{\cdot}0777 - 2 \times 2{\cdot}3026)$

$= -2{\cdot}2(1{\cdot}2343 - 6{\cdot}9078) + 1{\cdot}4(1{\cdot}0777 - 4{\cdot}6052)$

$= -2{\cdot}2(-5{\cdot}6735) + 1{\cdot}4(-3{\cdot}5275)$

$= 12{\cdot}4817 - 4{\cdot}9385$

$= 7{\cdot}5432$

Finding antilog,

subtract $\ln 10^3$ which is $3 \times 2{\cdot}3026 = 6{\cdot}9078$

$7{\cdot}5432 - 6{\cdot}9078 = 0{\cdot}6354$

Antilog of $0{\cdot}6354 = 1{\cdot}888$

Antilog of $7{\cdot}5432 = 1{\cdot}888 \times 10^3$

$= 1888$ Ans. (b)

27. $V = 0{\cdot}02377^{-1\cdot35} \div 0{\cdot}2455^{2\cdot35}$

or $0{\cdot}02377^{-1\cdot35} \times 0{\cdot}2455^{-2\cdot35}$

(a) By common logs:

$\log V = -1{\cdot}35(\log 0{\cdot}02377) - 2{\cdot}35(\log 0{\cdot}2455)$

$= -1{\cdot}35(\bar{2}{\cdot}3760) - 2{\cdot}35(\bar{1}{\cdot}3901)$

$= -1{\cdot}35 \times (-1{\cdot}624) - 2{\cdot}35 \times (-0{\cdot}6099)$

$= 2{\cdot}1924 + 1{\cdot}4333$

$= 3{\cdot}6257$

(Antilog) $V = 4224$ Ans. (a)

(b) By Napierian logs:

$$
\begin{aligned}
\ln V &= -1{\cdot}35(\text{in } 0{\cdot}02377) - 2{\cdot}35(\ln 0{\cdot}2455) \\
&= -1{\cdot}35(\ln 2{\cdot}377 + \ln 10^{-2}) - 2{\cdot}35(\ln 2{\cdot}455 + \ln 10^{-1}) \\
&= -1{\cdot}35(0{\cdot}8659 - 2 \times 2{\cdot}3026) - 2{\cdot}35(0{\cdot}8981 - 1 \times 2{\cdot}3026) \\
&= -1{\cdot}35(0{\cdot}8659 - 4{\cdot}6052) - 2{\cdot}35(0{\cdot}8981 - 2{\cdot}3026) \\
&= -1{\cdot}35 \times (-3{\cdot}7393) - 2{\cdot}35 \times (-1{\cdot}4045) \\
&= 5{\cdot}0480 + 3{\cdot}3006 \\
&= 8{\cdot}3486
\end{aligned}
$$

Finding antilog,

subtract $\ln 10^3$ which is $3 \times 2{\cdot}3026 = 6{\cdot}9078$

$$
\begin{aligned}
8{\cdot}3486 - 6{\cdot}9078 &= 1{\cdot}4408 \\
\text{Antilog of } 1{\cdot}4408 &= 4{\cdot}224 \\
\text{Antilog of } 8{\cdot}3486 &= 4{\cdot}224 \times 10^3 \\
&= 4224 \quad \text{Ans. (b)}
\end{aligned}
$$

SOLUTIONS TO TEST EXAMPLES 4

1.
$$\begin{aligned} 8 + 5x - 7 &= 3x + 9 \\ 5x - 3x &= 9 + 7 - 8 \\ 2x &= 8 \\ x &= 4 \quad \text{Ans.} \end{aligned}$$

2.
$$\begin{aligned} 2(a + 3) + 3(2a - 4) &= 4(11 - 3a) \\ 2a + 6 + 6a - 12 &= 44 - 12a \\ 2a + 6a + 12a &= 44 + 12 - 6 \\ 20a &= 50 \\ a &= 2\tfrac{1}{2} \quad \text{Ans.} \end{aligned}$$

3.
$$\begin{aligned} (a + 5b) - 2(a - 4) &= 3(2a + b) - 7(a - 3) \\ a + 5b - 2a + 8 &= 6a + 3b - 7a + 21 \\ a - 2a - 6a + 7a + 5b - 3b &= 21 - 8 \end{aligned}$$

(note that we have $+8a - 8a$ which cancels)

$$\begin{aligned} 2b &= 13 \\ b &= 6\tfrac{1}{2} \quad \text{Ans.} \end{aligned}$$

4.
$$\begin{aligned} 3[3 - \{x + 2(1 - x)\} - 4x] &= 2[x - 3(2 + x) - 4] \\ 3[3 - \{x + 2 - 2x\} - 4x] &= 2[x - 6 - 3x - 4] \\ 3[3 - x - 2 + 2x - 4x] &= 2[x - 6 - 3x - 4] \\ 9 - 3x - 6 + 6x - 12x &= 2x - 12 - 6x - 8 \\ -3x + 6x - 12x - 2x + 6x &= -12 - 8 - 9 + 6 \\ -5x &= -23 \\ x &= 4{\cdot}6 \quad \text{Ans.} \end{aligned}$$

5.
$$\frac{2x}{3} + \frac{x}{4} - \frac{5}{6} = \frac{4x}{5} + \frac{1}{3}$$

multiplying throughout by 60,

$$\begin{aligned} 40x + 15x - 50 &= 48x + 20 \\ 40x + 15x - 48x &= 20 + 50 \\ 7x &= 70 \\ x &= 10 \quad \text{Ans.} \end{aligned}$$

6.
$$\frac{3}{x} + \frac{1}{6} = \frac{6}{x} - \frac{4}{3x}$$

multiplying throughout by $6x$,

$$18 + x = 36 - 8$$
$$x = 36 - 8 - 18$$
$$x = 10 \quad \text{Ans.}$$

7.
$$\frac{(1 - 2a^2)}{6a} - \frac{1}{4} + \frac{a}{3} = \frac{1}{6} - \frac{5}{2a}$$

multiplying throughout by $12a$,

$$2(1 - 2a^2) - 3a + 4a^2 = 2a - 30$$
$$2 - 4a^2 - 3a + 4a^2 = 2a - 30$$

$-4a^2$ and $+4a^2$ cancels,

$$-3a - 2a = -30 - 2$$
$$-5a = -32$$
$$a = 6{\cdot}4 \quad \text{Ans.}$$

8.
$$\frac{5}{x + 6} = \frac{3}{x - 4}$$

multiplying throughout by $(x + 6)(x - 4)$,

$$5(x - 4) = 3(x + 6)$$
$$5x - 20 = 3x + 18$$
$$5x - 3x = 18 + 20$$
$$2x = 38$$
$$x = 19 \quad \text{Ans.}$$

9.
$$\frac{2}{x - 3} + \frac{3}{x + 2} = \frac{4}{x^2 - x - 6}$$

Note that $(x - 3)(x + 2)$ are the factors of $x^2 - x - 6$, therefore the L.C.M. of the denominators is $x^2 - x - 6$, or more conveniently $(x - 3)(x + 2)$ for the purposes of multiplying throughout this equation.

Multiplying by $(x - 3)(x + 2)$,

$$
\begin{aligned}
2(x + 2) + 3(x - 3) &= 4 \\
2x + 4 + 3x - 9 &= 4 \\
2x + 3x &= 4 + 9 - 4 \\
5x &= 9 \\
x &= 1{\cdot}8 \quad \text{Ans.}
\end{aligned}
$$

10. Let m = mass of cargo transferred

Hold A will then contain $(250 + m)$ tonne

,, B ,, ,, $(620 - m)$,,

And A is five times as much as B,

$$
\begin{aligned}
\therefore \quad 250 + m &= 5(620 - m) \\
250 + m &= 3100 - 5m \\
m + 5m &= 3100 - 250 \\
6m &= 2850 \\
m &= 475 \text{ tonne} \quad \text{Ans.}
\end{aligned}
$$

11. The equality of the ships is that the *distance* from port to meeting place is the same, therefore the equation is: Distance travelled by fast ship = Distance travelled by slow ship, and for distance we can write speed × time.

Let x = time taken by fast ship, in hours

then $x + 4\frac{1}{2}$ = ,, ,, slow ,, ,,

$$
\begin{aligned}
\text{speed} \times \text{time (of fast ship)} &= \text{speed} \times \text{time (of slow ship)} \\
17{\cdot}5 \times x &= 16 \times (x + 4\tfrac{1}{2}) \\
17{\cdot}5x &= 16x + 72 \\
17{\cdot}5x - 16x &= 72 \\
1{\cdot}5x &= 72 \\
x &= 48
\end{aligned}
$$

$\therefore$ time for fast ship to overtake = 48 hours Ans. (i)

$$
\begin{aligned}
\text{distance} &= \text{speed} \times \text{time} \\
&= 17{\cdot}5 \times 48 \\
&= 840 \text{ naut. miles}
\end{aligned}
$$

$\therefore$ Distance from port = 840 naut. miles Ans. (ii)

12. Let x = distance between points, in nautical miles.

Time to go up river (hours) + Time to go down river (hours) = Total time (hours).

and for time we can write $\dfrac{\text{distance}}{\text{speed}}$

$$\frac{\text{distance up river}}{\text{speed up river}} + \frac{\text{distance down river}}{\text{speed down river}} = \text{total time}$$

$$\frac{x}{6} + \frac{x}{9} = 2{\cdot}5$$

multiplying throughout by 18,

$$3x + 2x = 45$$
$$5x = 45$$
$$x = 9$$

∴ Distance between points = 9 naut. miles. Ans.

13. Let x = distance between air-ports, in kilometers

Time taken by slow plane — Time taken by fast plane = Difference in times.

$$\frac{\text{distance travelled by slow plane}}{\text{speed of slow plane}} - \frac{\text{distance by fast plane}}{\text{speed of fast plane}} = \frac{1}{2}\text{ hour}$$

$$\frac{x}{672} - \frac{x}{960} = \frac{1}{2}$$

multiplying throughout by 6720,

$$10x - 7x = 3360$$
$$3x = 3360$$
$$x = 1120$$

∴ Distance between ports = 1120 kilometres Ans.

14. Let b = breadth of plate, in metres, then
$4b$ = length ,, ,, ,, ,,

$$\text{length} \times \text{breadth} = \text{area}$$
$$4b \times b = 1$$
$$4b^2 = 1$$
$$b^2 = 0{\cdot}25$$
$$b = \sqrt{0{\cdot}25} = 0{\cdot}5$$

∴ breadth is 0·5 m
length = 4 × 0·5 = 2 m } Ans.

15. Let x = speed of current, in knots

Speed of ship against current = $(18 - x)$ knots

" " " with " = $(18 + x)$ "

time to go against current + time to go with current = total time

$$\frac{\text{distance against current}}{\text{speed against current}} + \frac{\text{distance with current}}{\text{speed with current}} = \frac{70}{60}\text{ hours}$$

$$\frac{10}{18 - x} + \frac{10}{18 + x} = \frac{7}{6}$$

multiplying throughout by $6 \times (18 - x)(18 + x)$,

$$60(18 + x) + 60(18 - x) = 7(18 - x)(18 + x)$$

$$1080 + 60x + 1080 - 60x = 7(324 - x^2)$$

$+60x$ and $-60x$ cancels,

$$1080 + 1080 = 2268 - 7x^2$$

$$7x^2 = 2268 - 2160$$

$$7x^2 = 108$$

$$x^2 = 15{\cdot}43$$

$$x = \sqrt{15{\cdot}43}$$

$$x = 3{\cdot}928$$

∴ speed of current = 3·928 knots. Ans.

16. Note that R_0 cancels,

$$\therefore \frac{R_2}{R_1} = \frac{1 + \alpha\theta_2}{1 + \alpha\theta_1}$$

$$R_1(1 + \alpha\theta_2) = R_2(1 + \alpha\theta_1)$$

$$1 + \alpha\theta_2 = \frac{R_2}{R_1}(1 + \alpha\theta_1)$$

$$\alpha\theta_2 = \frac{R_2}{R_1}(1 + \alpha\theta_1) - 1$$

$$\theta_2 = \frac{1}{\alpha}\left[\frac{R_2}{R_1}(1 + \alpha\theta_1) - 1\right] \quad \text{Ans. (i)}$$

Inserting values,

$$\theta_2 = \frac{1}{0{\cdot}0042}\left[\frac{240}{200}(1 + 0{\cdot}0042 \times 15) - 1\right]$$

$$= \frac{1}{0{\cdot}0042}\left[1{\cdot}2 \times 1{\cdot}063 - 1\right]$$

$$= \frac{1}{0{\cdot}0042} \times 0{\cdot}2756$$

$$= 65{\cdot}6 \quad \text{Ans. (ii)}$$

17.
$$p_1 \times V_1^n = p_2 \times V_2^n$$

Dividing both sides by V_2^n:

$$\frac{p_1 \times V_1^n}{V_2^n} = p_2$$

$$\therefore p_2 = p_1 \times \left(\frac{V_1}{V_2}\right)^n \quad \text{Ans. (i)}$$

Substituting values,

$$p_2 = 1{\cdot}015 \times \left(\frac{2{\cdot}5}{0{\cdot}2}\right)^{1{\cdot}4}$$

$$= 1{\cdot}015 \times 12{\cdot}5^{1{\cdot}4}$$

$$= 34{\cdot}84 \quad \text{Ans. (ii)}$$

18.
$$p_1 \times V_1^n = p_2 \times V_2^n$$

Dividing both sides by p_2

$$\frac{p_1 \times V_1^n}{p_2} = V_2^n$$

Taking nth root of both sides,

$$\sqrt[n]{\frac{p_1}{p_2}} \times V_1 = V_2$$

$$\text{or } V_2 = V_1 \times \sqrt[n]{\frac{p_1}{p_2}} \quad \text{Ans. (i)}$$

Inserting values:

$$V_2 = 6{\cdot}5 \times \sqrt[1\cdot38]{\frac{0{\cdot}966}{34{\cdot}5}}$$

$$= 6{\cdot}5 \times \sqrt[1\cdot38]{0{\cdot}028}$$

Taking a root of a number less than unity, whose log has a negative characteristic, has been shown and the following serves as a reminder.

$$\log 0{\cdot}028 = \bar{2}{\cdot}4472$$

Change this to an all-minus quantity,

$$\bar{2}{\cdot}4472 = -1{\cdot}5528$$

$-1{\cdot}5528$ is to be divided by $1{\cdot}38$, make the divisor a whole number, 138, and shift the decimal point of the dividend an equal number of places, $-155{\cdot}28$.

$$-155{\cdot}28 \div 138 = -1{\cdot}1252$$

Change $-1{\cdot}1252$ into logarithmic form,

$$-1{\cdot}1252 = \bar{2}{\cdot}8748$$

$$\log 6{\cdot}5 = 0{\cdot}8129 \text{ add}$$

$$\bar{1}{\cdot}6877$$

$$\text{Antilog of } \bar{1}{\cdot}6877 = 0{\cdot}4872$$

$$\therefore V_2 = 0{\cdot}4872 \quad \text{Ans. (ii)}$$

19.

$$d = \sqrt{\frac{D^3}{3{\cdot}5 \times n \times R}}$$

Square both sides:

$$d^2 = \frac{D^3}{3{\cdot}5 \times n \times R}$$

Multiply both sides by R and divide by d^2,

$$R = \frac{D^3}{3{\cdot}5\,nd^2} \quad \text{Ans. (i)}$$

Substituting values,

$$R = \frac{381^3}{3{\cdot}5 \times 8 \times 82{\cdot}5^2}$$

$$= 290 \text{ mm} \quad \text{Ans. (ii)}$$

20.

$$
\begin{aligned}
3{\cdot}4^n &= 5{\cdot}547 \\
(\log \text{ of } 3{\cdot}4) \times n &= \log \text{ of } 5{\cdot}547 \\
0{\cdot}5315 \times n &= 0{\cdot}7440 \\
n &= \frac{0{\cdot}7440}{0{\cdot}5315} = 1{\cdot}4 \quad \text{Ans.}
\end{aligned}
$$

21.

$$
\begin{aligned}
4^{x+3} &= 4096 \\
\log 4 \times (x + 3) &= \log 4096 \\
0{\cdot}6021(x + 3) &= 3{\cdot}6123 \\
0{\cdot}6021x + 1{\cdot}8063 &= 3{\cdot}6123 \\
0{\cdot}6021x &= 3{\cdot}6123 - 1{\cdot}8063 = 1{\cdot}806 \\
x &= \frac{1{\cdot}806}{0{\cdot}6021} = 3 \quad \text{Ans.}
\end{aligned}
$$

22.

$$p_1 \times V_1{}^n = p_2 \times V_2{}^n$$

Inserting values,

$$1{\cdot}07 \times 2{\cdot}15^n = 12{\cdot}43 \times 0{\cdot}35^n$$

Collect all terms to the nth power on one side, and pure numbers on the other, this is done by dividing both sides by 1·07 and $0{\cdot}35^n$.

$$
\begin{aligned}
\frac{2{\cdot}15^n}{0{\cdot}35^n} &= \frac{12{\cdot}43}{1{\cdot}07} \\
\left(\frac{2{\cdot}15}{0{\cdot}35}\right)^n &= \frac{12{\cdot}43}{1{\cdot}07} \\
6{\cdot}142^n &= 11{\cdot}61 \\
(\log \text{ of } 6{\cdot}142) \times n &= \log \text{ of } 11{\cdot}61 \\
0{\cdot}7883 \times n &= 1{\cdot}0651 \\
n &= \frac{1{\cdot}0651}{0{\cdot}7883} = 1{\cdot}351 \quad \text{Ans.}
\end{aligned}
$$

23.

$$
\begin{aligned}
\frac{T_1}{T_2} &= \left\{\frac{p_1}{p_2}\right\}^{\frac{n-1}{n}} \\
\frac{670}{324} &= \left\{\frac{21}{1{\cdot}25}\right\}^{\frac{n-1}{n}} \\
2{\cdot}068 &= 16{\cdot}8^{\frac{n-1}{n}}
\end{aligned}
$$

$$\text{log of } 2{\cdot}068 = (\text{log of } 16{\cdot}8) \times \left(\frac{n - 1}{n}\right)$$

$$0{\cdot}3156 = 1{\cdot}2253 \times \left(\frac{n - 1}{n}\right)$$

Multiply both sides by n,

$$0{\cdot}3156n = 1{\cdot}2253(n - 1)$$

Taking brackets away,

$$0{\cdot}3156n = 1{\cdot}2253n - 1{\cdot}2253$$
$$1{\cdot}2253 = 1{\cdot}2253n - 0{\cdot}3156n$$
$$1{\cdot}2253 = 0{\cdot}9097n$$

$$n = \frac{1{\cdot}2253}{0{\cdot}9097} = 1{\cdot}347 \quad \text{Ans.}$$

24.

$$0{\cdot}75^n = 0{\cdot}4873$$
$$(\text{log of } 0{\cdot}75) \times n = \text{log of } 0{\cdot}4873$$
$$\bar{1}{\cdot}8751 \times n = \bar{1}{\cdot}6878$$

$$n = \frac{\bar{1}{\cdot}6878}{\bar{1}{\cdot}8751} = \frac{-0{\cdot}3122}{-0{\cdot}1249}$$

$$n = \frac{0{\cdot}3122}{0{\cdot}1249} = 2{\cdot}5 \quad \text{Ans.}$$

25.

$$1 - (0{\cdot}125)^{\gamma - 1} = 0{\cdot}5647$$
$$1 - 0{\cdot}5647 = (0{\cdot}125)^{\gamma - 1}$$
$$0{\cdot}4353 = (0{\cdot}125)^{\gamma - 1}$$
$$\text{log of } 0{\cdot}4353 = (\text{log of } 0{\cdot}125) \times (\gamma - 1)$$

$$\gamma - 1 = \frac{\text{log of } 0{\cdot}4353}{\text{log of } 0{\cdot}125}$$

$$\gamma - 1 = \frac{\bar{1}{\cdot}6388}{\bar{1}{\cdot}0969}$$

$$\gamma - 1 = \frac{-0{\cdot}3612}{-0{\cdot}9031} = 0{\cdot}4$$

$$\gamma = 0{\cdot}4 + 1 = 1{\cdot}4 \quad \text{Ans.}$$

26.

$$4^{0.59x} = 56.36$$
$$\log 4 \times 0.59x = \log 56.36$$
$$0.6021 \times 0.59x = 1.751$$
$$x = \frac{1.751}{0.6021 \times 0.59}$$
$$= 4.929 \quad \text{Ans. (a)}$$

$$x^{1.95} = 12.4x^{0.53}$$
$$\log x \times 1.95 = \log 12.4 + \log x \times 0.53$$
$$1.95 \log x - 0.53 \log x = \log 12.4$$
$$1.42 \log x = 1.0934$$
$$\log x = \frac{1.0934}{1.42}$$
$$\log x = 0.7700$$
$$(\text{antilog})x = 5.888 \quad \text{Ans. (b)}$$

$$\epsilon^{5x} = 46.38^{2.6}$$
$$2.718^{5x} = 46.38^{2.6}$$
$$\log 2.718 \times 5x = \log 46.38 \times 2.6$$
$$0.4343 \times 5x = 1.6663 \times 2.6$$
$$x = \frac{1.6663 \times 2.6}{0.4343 \times 5}$$
$$= 1.995 \quad \text{Ans. (c)}$$

$$\sqrt[3]{x} = \epsilon^{2.5}$$
$$\sqrt[3]{x} = 2.718^{2.5}$$
$$\log x \div 3 = \log 2.718 \times 2.5$$
$$\log x = 0.4343 \times 2.5 \times 3$$
$$\log x = 3.2573$$
$$(\text{antilog})x = 1808 \quad \text{Ans. (d)}$$

SOLUTIONS TO TEST EXAMPLES 5

1. Let x = one number, double this number = $2x$

   Let y = other number, treble this number = $3y$

   sum: $2x + 3y = 16$ ... ... (i)

   difference $2x - 3y = 4$ ... ... (ii)

   subtract (ii) from (i), $6y = 12$

   $y = 2$

   substitute $y = 2$ into (i),

   $2x + 3 \times 2 = 16$

   $2x = 16 - 6$

   $2x = 10$

   $x = 5$

   The numbers are 5 and 2 Ans.

2. Let x = first number, and y = second number.

   From the first statement we have,

   $2\frac{1}{2}x + 3\frac{1}{2}y = 19$ ... ... (i)

   From the second statement we have,

   $2\frac{1}{2}y - 3\frac{1}{2}x = 3$ ... ... (ii)

   Re-arranging terms in (i) in the same order as (ii),

   $3\frac{1}{2}y + 2\frac{1}{2}x = 19$ ... ... (iii)

   $2\frac{1}{2}y - 3\frac{1}{2}x = 3$ ... ... (iv)

   Multiplying (iii) by 5 and multiplying (iv) by 7,

   $17\frac{1}{2}y + 12\frac{1}{2}x = 95$ ... ... (v)

   $17\frac{1}{2}y - 24\frac{1}{2}x = 21$ ... ... (vi)

   Subtracting (vi from (v) $37x = 74$

   $x = 2$

Substituting $x = 2$ into (i)

$$2\tfrac{1}{2} \times 2 + 3\tfrac{1}{2}y = 19$$
$$3\tfrac{1}{2}y = 19 - 5$$
$$3\tfrac{1}{2}y = 14$$
$$y = 4$$

The numbers are 2 and 4 Ans.

3.
$$\frac{2x}{3} - \frac{3y}{5} = \frac{3}{4} \quad \dots \quad \dots \quad \dots \quad \dots \quad \text{(i)}$$
$$\frac{x}{2} - \frac{y}{4} = \frac{13}{16} \quad \dots \quad \dots \quad \dots \quad \dots \quad \text{(ii)}$$

Multiplying (i) by 60, and multiplying (ii) by 16,

$$40x - 36y = 45 \quad \dots \quad \dots \quad \dots \quad \dots \quad \text{(iii)}$$
$$8x - 4y = 13 \quad \dots \quad \dots \quad \dots \quad \dots \quad \text{(iv)}$$

Multiplying (iv) by 5 and subtracting (iii) from the result,

$$40x - 20y = 65 \quad \dots \quad \dots \quad \dots \quad \dots \quad \text{(v)}$$
$$40x - 36y = 45$$
$$16y = 20$$
$$y = 1\tfrac{1}{4}$$

Substituting $y = 1\tfrac{1}{4}$ into (iv),

$$8x - 4 \times 1\tfrac{1}{4} = 13$$
$$8x = 13 + 5$$
$$8x = 18$$
$$x = 2\tfrac{1}{4}$$

The values of x and y are $2\tfrac{1}{4}$ and $1\tfrac{1}{4}$ respectively. Ans.

4.
$$a(1 + 2b) = 3$$
$$a(1 - 3b) = 0{\cdot}5$$

Removing brackets,

$$a + 2ab = 3 \quad \dots \quad \dots \quad \dots \quad \dots \quad \text{(i)}$$
$$a - 3ab = 0{\cdot}5 \quad \dots \quad \dots \quad \dots \quad \dots \quad \text{(ii)}$$

If we subtracted (ii) from (i) to eliminate a, we would still be left with a term containing two unknowns, therefore proceed to eliminate the terms containing ab by multiplying (i) by 3, and (ii) by 2.

$$\begin{aligned} 3a + 6ab &= 9 \\ 2a - 6ab &= 1 \\ \hline \text{add } 5a &= 10 \\ a &= 2 \end{aligned}$$

Substitute $a = 2$ into (i),

$$\begin{aligned} 2 + 2 \times 2 \times b &= 3 \\ 4b &= 3 - 2 \\ 4b &= 1 \\ b &= 0{\cdot}25 \end{aligned}$$

The values of a and b are 2 and 0·25 respectively. Ans.

Alternatively we could divide the original equations thus,

$$\frac{a(1 + 2b)}{a(1 - 3b)} = \frac{3}{0{\cdot}5}$$

a cancels,

$$\begin{aligned} \frac{1 + 2b}{1 - 3b} &= 6 \\ 1 + 2b &= 6(1 - 3b) \\ 1 + 2b &= 6 - 18b \\ 20b &= 5 \\ b &= 0{\cdot}25 \end{aligned}$$

Substituting $b = 0{\cdot}25$ into the first equation,

$$\begin{aligned} a(1 + 2 \times 0{\cdot}25) &= 3 \\ 1{\cdot}5a &= 3 \\ a &= 2 \end{aligned}$$

$$\therefore a = 2 \text{ and } b = 0{\cdot}25 \text{ as before.}$$

5. Let x = age of grandson
and y = age of grand-daughter

$$4x + 3y = 72 \quad \ldots \quad \ldots \quad \ldots \quad \ldots \quad \text{(i)}$$
$$3x + 4y = 68 \quad \ldots \quad \ldots \quad \ldots \quad \ldots \quad \text{(ii)}$$

Multiplying (i) by 4 and (ii) by 3 and subtracting,

$$\begin{aligned} 16x + 12y &= 288 \\ 9x + 12y &= 204 \\ \hline 7x &= 84 \\ x &= 12 \end{aligned}$$

Substituting $x = 12$ into (ii),

$$\begin{aligned} 3 \times 12 + 4y &= 68 \\ 4y &= 68 - 36 \\ 4y &= 32 \\ y &= 8 \end{aligned}$$

Their ages are 12 and 8 years respectively. Ans.

6. Let x = one number,
and y = other number,

$$x - y = 2 \quad \ldots \quad \ldots \quad \ldots \quad \ldots \quad \text{(i)}$$

$$x^2 - y^2 = 6 \quad \ldots \quad \ldots \quad \ldots \quad \ldots \quad \text{(ii)}$$

Factorise (ii) and divide by (i),

$$\frac{(x + y)(x - y)}{(x - y)} = \frac{6}{2}$$

$x - y$ cancels,

$$x + y = 3 \quad \ldots \quad \ldots \quad \ldots \quad \ldots \quad \text{(iii)}$$

Add (iii) to (i),

$$\begin{aligned} x - y &= 2 \\ x + y &= 3 \\ \hline 2x &= 5 \\ x &= 2\tfrac{1}{2} \end{aligned}$$

Substitute $x = 2\frac{1}{2}$ into (iii),

$$\begin{aligned} 2\tfrac{1}{2} + y &= 3 \\ y &= \tfrac{1}{2} \end{aligned}$$

The numbers are $2\frac{1}{2}$ and $\frac{1}{2}$ Ans.

7. Inserting the two sets of values of F and m,

$$55 = a + b \times 70 \quad \ldots \quad \ldots \quad \text{(i)}$$

$$\text{subtract } 35 = a + b \times 30 \quad \ldots \quad \ldots \quad \text{(ii)}$$

$$20 = 40b$$

$$b = 0{\cdot}5$$

Substituting $b = 0{\cdot}5$ into (i),

$$55 = a + 0{\cdot}5 \times 70$$

$$55 = a + 35$$

$$a = 20$$

Constants are, $a = 20$ and $b = 0{\cdot}5$ Ans. (*a*)

Linear law is, $F = 20 + 0{\cdot}5\,m$ Ans. (*b*)

When $m = 60$,

$$F = 20 + 0{\cdot}5 \times 60$$

$$= 20 + 30$$

$$= 50 \text{ N Ans. } (c)$$

8. Let x = speed of one ship, in knots,
and y = speed of other ship, „

If travelling towards each other,

$$\text{speed of approach} = (x + y) \text{ knots}$$

and this is equal to,

$$\frac{\text{distance}}{\text{time}} = \frac{99}{3} = 33 \text{ knots}$$

$$\therefore x + y = 33 \quad \ldots \quad \ldots \quad \ldots \quad \ldots \quad \text{(i)}$$

If travelling in same direction in line,

$$\text{speed of approach} = (x - y) \text{ knots}$$

and this is equal to, $\dfrac{99}{49\frac{1}{2}} = 2$ knots

$$\therefore x - y = 2 \quad \ldots \quad \ldots \quad \ldots \quad \ldots \quad \text{(ii)}$$

Adding (i) and (ii),

$$x + y = 33$$

$$x - y = 2$$

$$2x = 35$$

$$x = 17\tfrac{1}{2}$$

Substituting $x = 17\frac{1}{2}$ into (i),

$$17\frac{1}{2} + y = 33$$
$$y = 33 - 17\frac{1}{2}$$
$$y = 15\frac{1}{2}$$

The ships' speeds are $17\frac{1}{2}$ and $15\frac{1}{2}$ knots respectively. Ans.

9. Let x = original speed of A
and y = original speed of B

Time for A to be overtaken = 9 hours
Time for B to overtake A = 8 hours
Distance from port to point of overtaking is the same for each ship, and distance = speed × time.

$$\therefore x \times 9 = y \times 8$$

$$y = \frac{9x}{8} \quad \ldots \quad \ldots \quad \ldots \quad \ldots \quad \text{(i)}$$

If speed had been 4 knots slower,

Speed of A would be $(x - 4)$ knots
" B " $(y - 4)$ knots

Time for A to be overtaken = 7 hours
Time for B to overtake A = 6 hours

$$\therefore (x - 4) \times 7 = (y - 4) \times 6$$
$$7x - 28 = 6y - 24$$
$$7x - 6y = 4 \quad \ldots \quad \ldots \quad \ldots \quad \ldots \quad \text{(ii)}$$

Substituting value of y from (i) into (ii),

$$7x - \frac{6 \times 9x}{8} = 4$$

$$7x - 6\frac{3}{4}x = 4$$
$$\frac{1}{4}x = 4$$
$$x = 16$$

Substituting $x = 16$ into (i),

$$y = \frac{9 \times 16}{8}$$

$$y = 18$$

∴ Original speeds of ships are 16 and 18 knots respectively. Ans.

10. Substituting the given pairs of values of m and P into the general equation $m = a + bP$ and subtracting (ii) from (i),

$$2025 = a + b \times 250 \quad \ldots \quad \ldots \quad \text{(i)}$$
$$1515 = a + b \times 175 \quad \ldots \quad \ldots \quad \text{(ii)}$$

$$510 = 75b$$
$$b = 6{\cdot}8$$

Substituting $b = 6{\cdot}8$ into (ii),

$$1515 = a + 6{\cdot}8 \times 175$$
$$a = 1515 - 1190$$
$$a = 325$$

$\therefore$ Willans' law is: $m = 325 + 6{\cdot}8P$

When $P = 200$,

$$m = 325 + 6{\cdot}8 \times 200$$
$$= 1685 \text{ kilogrammes/hour. Ans.}$$

11. Let a represent $\dfrac{x}{2 - y}$ and let b represent $\dfrac{1}{x}$:

then the two equations may be written,

$$a + 6b = 4 \quad \ldots \quad \ldots \quad \ldots \quad \ldots \quad \text{(i)}$$
$$2a - 9b = 1 \quad \ldots \quad \ldots \quad \ldots \quad \ldots \quad \text{(ii)}$$

Multiplying (i) by 2 and subtracting (ii) from the result,

$$2a + 12b = 8$$
$$2a - 9b = 1$$

$$21b = 7$$
$$b = \tfrac{1}{3}$$

Substituting $b = \frac{1}{3}$ into (i),

$$a + 6 \times \tfrac{1}{3} = 4$$
$$a = 4 - 2$$
$$a = 2$$

$$a = 2, \ \therefore \ \frac{x}{2 - y} = 2 \text{ Ans. (a)}$$

$$b = \frac{1}{3}, \ \therefore \ \frac{1}{x} = \frac{1}{3} \text{ Ans. (b)}$$

$$\frac{1}{x} = \frac{1}{3}, \qquad \therefore x = 3 \text{ Ans. (c)}$$

$$\frac{x}{2-y} = 2, \therefore \quad \frac{3}{2-y} = 2$$

$$3 = 4 - 2y$$
$$2y = 4 - 3$$
$$2y = 1$$
$$y = \tfrac{1}{2} \text{ Ans. (d)}$$

12. From the first equation,

$$2^x = 4^y$$
$$2^x = 2^{2y}$$
$$\therefore x = 2y \quad \dots \quad \dots \quad \dots \quad \dots \quad \text{(i)}$$

From the second equation,

$$4^{x-1} = 2^{y+1}$$
$$2^{2x-2} = 2^{y+1}$$
$$\therefore 2x - 2 = y + 1$$
$$2x - y = 3 \quad \dots \quad \dots \quad \dots \quad \dots \quad \text{(ii)}$$

Substituting $x = 2y$ from (i) into (ii)

$$2 \times 2y - y = 3$$
$$3y = 3$$
$$y = 1$$

Substituting $y = 1$ into (i),

$$x = 2$$

The values of x and y are 2 and 1 respectively. Ans.

13.
$$1{\cdot}259^{x+1} \times 1{\cdot}175^{y-1} = 2{\cdot}323$$
$$3{\cdot}162^x \times 1{\cdot}778^y = 25{\cdot}12$$

Expressing the equations in log form,

$$(\log 1{\cdot}259)(x + 1) + (\log 1{\cdot}175)(y - 1) = \log 2{\cdot}323$$
$$(\log 3{\cdot}162) \times x + (\log 1{\cdot}778) \times y = \log 25{\cdot}12$$

Inserting log values,

$$0{\cdot}1000(x + 1) + 0{\cdot}0700(y - 1) = 0{\cdot}3660$$
$$0{\cdot}5000 \times x + 0{\cdot}2500 \times y = 1{\cdot}4000$$

Simplifying the equations,

$$0{\cdot}1x + 0{\cdot}07y = 0{\cdot}336 \quad \ldots \quad \ldots \quad \ldots \quad \text{(i)}$$
$$0{\cdot}5x + 0{\cdot}25y = 1{\cdot}4 \quad \ldots \quad \ldots \quad \ldots \quad \ldots \quad \text{(ii)}$$

Multiplying (i) by 5, subtracting (ii) from the result, and solving y,

$$0{\cdot}5x + 0{\cdot}35y = 1{\cdot}68$$
$$0{\cdot}5x + 0{\cdot}25y = 1{\cdot}4$$
$$0{\cdot}1y = 0{\cdot}28$$
$$y = 2{\cdot}8$$

Substituting $y = 2{\cdot}8$ into (i) and solving x,

$$0{\cdot}1x + 0{\cdot}07 \times 2{\cdot}8 = 0{\cdot}336$$
$$0{\cdot}1x = 0{\cdot}336 - 0{\cdot}196$$
$$0{\cdot}1x = 0{\cdot}14$$
$$x = 1{\cdot}4$$

The values of x and y are 1·4 and 2·8 respectively. Ans.

14. Taking square root throughout,

$$\frac{d}{x} = \frac{x}{y} = \frac{y}{D}$$

From $\frac{d}{x} = \frac{y}{D}$, $\quad y = \frac{dD}{x} \quad \ldots \quad \ldots \quad \ldots \quad \ldots \quad$ (i)

From $\frac{d}{x} = \frac{x}{y}$, $\quad x^2 = dy \quad \ldots \quad \ldots \quad \ldots \quad \ldots \quad$ (ii)

Substituting the value of y in (i) into (ii),

$$x^2 = \frac{d \times dD}{x}$$
$$\therefore x^3 = d^2D$$
$$x = \sqrt[3]{d^2D} \text{ Ans. (a)}$$

Inserting values of d and D,

$$x = \sqrt[3]{25^2 \times 75} = 36{\cdot}06$$

Inserting values of d, D and x into (i),

$$y = \frac{25 \times 75}{36{\cdot}06} = 52$$

The values of x and y are 36·06 and 52 respectively. Ans.

15.

$$3a + 6b - 2c = 7{\cdot}25 \quad \ldots \quad \text{(i)}$$
$$2a + 3b + 4c = 26 \quad \ldots \quad \text{(ii)}$$
$$4a - 2b + c = 10{\cdot}25 \quad \ldots \quad \text{(iii)}$$

Multiplying (i) by 2 and adding (ii) to the result,

$$6a + 12b - 4c = 14{\cdot}5$$
$$2a + 3b + 4c = 26$$

$$8a + 15b = 40{\cdot}5 \quad \ldots \quad \text{(iv)}$$

Multiplying (iii) by 4 and subtracting (ii) from the result,

$$16a - 8b + 4c = 41$$
$$2a + 3b + 4c = 26$$

$$14a - 11b = 15 \quad \ldots \quad \text{(v)}$$

Multiplying (iv) by 7, multiplying (v) by 4, and subtracting,

$$56a + 105b = 283{\cdot}5$$
$$56a - 44b = 60$$

$$149b = 223{\cdot}5$$
$$b = 1{\cdot}5$$

Substituting $b = 1{\cdot}5$ into (iv),

$$8a + 15 \times 1{\cdot}5 = 40{\cdot}5$$
$$8a = 40{\cdot}5 - 22{\cdot}5$$
$$8a = 18$$
$$a = 2{\cdot}25$$

Substituting $a = 2{\cdot}25$ and $b = 1{\cdot}5$ into (iii),

$$4 \times 2{\cdot}25 - 2 \times 1{\cdot}5 + c = 10{\cdot}25$$
$$c = 10{\cdot}25 - 9 + 3$$
$$c = 4{\cdot}25$$

Values of a, b and c are 2·25, 1·5 and 4·25 respectively. Ans.

SOLUTIONS TO TEST EXAMPLES 6

1\. (i)

$$(2x + 8)(3x - 5) = 0$$

either $2x + 8 = 0$		or $3x - 5 = 0$
then $2x = -8$		then $3x = 5$
$x = -4$		$x = 1\frac{2}{3}$

$$x = -4 \text{ or } 1\tfrac{2}{3} \text{ Ans.}$$

(ii)

$$(0{\cdot}5x - 10)(0{\cdot}25x + 5) = 0$$

either $0{\cdot}5x - 10 = 0$		or $0{\cdot}25x + 5 = 0$
then $0{\cdot}5x = 10$		then $0{\cdot}25x = -5$
$x = 20$		$x = -20$

$$x = \pm 20 \text{ Ans.}$$

(iii)

$$(5x + 0{\cdot}5)(4x + 0{\cdot}8) = 0$$

either $5x + 0{\cdot}5 = 0$		or $4x + 0{\cdot}8 = 0$
then $5x = -0{\cdot}5$		then $4x = -0{\cdot}8$
$x = -0{\cdot}1$		$x = -0{\cdot}2$

$$x = -0{\cdot}1 \text{ or } -0{\cdot}2 \text{ Ans.}$$

2\. (i)

$$x^2 + 5x + 6 = 0$$
$$(x + 2)(x + 3) = 0$$

either $x + 2 = 0$		or $x + 3 = 0$
then $x = -2$		then $x = -3$

$$x = -2 \text{ or } -3 \text{ Ans.}$$

(ii)

$$x^2 - 10x + 24 = 0$$
$$(x - 4)(x - 6) = 0$$

either $x - 4 = 0$		or $x - 6 = 0$
then $x = 4$		then $x = 6$

$$x = 4 \text{ or } 6 \text{ Ans.}$$

(iii)

$$x^2 + 2x - 15 = 0$$
$$(x - 3)(x + 5) = 0$$

either $x - 3 = 0$		or $x + 5 = 0$
then $x = 3$		then $x = -5$

$$x = 3 \text{ or } -5 \text{ Ans.}$$

3. (i)

$$3x^2 + 2x - 33 = 0$$
$$(x - 3)(3x + 11) = 0$$

either $x - 3 = 0$, then $x = 3$ | or $3x + 11 = 0$, then $3x = -11$, $x = -3\frac{2}{3}$

$$x = 3 \text{ or } -3\tfrac{2}{3} \text{ Ans.}$$

(ii)

$$4x^2 - 17x + 4 = 0$$
$$(x - 4)(4x - 1) = 0$$

either $x - 4 = 0$, then $x = 4$ | or $4x - 1 = 0$, then $4x = 1$, $x = \frac{1}{4}$

$$x = 4 \text{ or } \tfrac{1}{4} \text{ Ans.}$$

(iii)

$$12x^2 + 10x - 12 = 0$$
$$2(6x^2 + 5x - 6) = 0$$
$$2(3x - 2)(2x + 3) = 0$$

either $3x - 2 = 0$, then $3x = 2$, $x = \frac{2}{3}$ | or $2x + 3 = 0$, then $2x = -3$, $x = -1\frac{1}{2}$

$$x = \tfrac{2}{3} \text{ or } -1\tfrac{1}{2} \text{ Ans.}$$

4. (i)

$$x^2 - x - 3\tfrac{3}{4} = 0$$
$$x^2 - x + (\tfrac{1}{2})^2 = 3\tfrac{3}{4} + (\tfrac{1}{2})^2$$
$$x - \tfrac{1}{2} = \pm\sqrt{4}$$
$$x = \pm 2 + \tfrac{1}{2}$$
$$x = 2\tfrac{1}{2} \text{ or } -1\tfrac{1}{2} \text{ Ans.}$$

(ii)

$$3x^2 + 2x - 1 = 0$$
$$x^2 + \tfrac{2}{3}x - \tfrac{1}{3} = 0$$
$$x^2 + \tfrac{2}{3}x + (\tfrac{1}{3})^2 = \tfrac{1}{3} + (\tfrac{1}{3})^2$$
$$x + \tfrac{1}{3} = \pm\sqrt{\tfrac{4}{9}}$$
$$x = \pm\tfrac{2}{3} - \tfrac{1}{3}$$
$$x = \tfrac{1}{3} \text{ or } -1 \text{ Ans.}$$

(iii)

$$4x^2 - 9x + 2 = 0$$
$$x^2 - \frac{9}{4}x + \frac{1}{2} = 0$$

$$x^2 - \frac{9x}{4} + \left(\frac{9}{8}\right)^2 = \left(\frac{9}{8}\right)^2 - \frac{1}{2}$$

$$x - \frac{9}{8} = \pm\sqrt{\frac{81 - 32}{64}}$$

$$x - \frac{9}{8} = \pm\sqrt{\frac{49}{64}}$$

$$x = \pm\frac{7}{8} + \frac{9}{8}$$

$$x = 2 \text{ or } \tfrac{1}{4} \text{ Ans.}$$

5. (i) $3x^2 - 2x + 0{\cdot}25 = 0$

$a = 3,\ b = -2,\ c = 0{\cdot}25$

$$x = \frac{2 \pm \sqrt{4 - 3}}{2 \times 3}$$

$$x = \frac{2 \pm 1}{6}$$

$$x = \frac{1}{2} \text{ or } \frac{1}{6} \text{ Ans.}$$

(ii) $5x^2 + 4x - 5{\cdot}52 = 0$

$a = 5,\ b = 4,\ c = -5{\cdot}52$

$$x = \frac{-4 \pm \sqrt{16 + 110{\cdot}4}}{2 \times 5}$$

$$x = \frac{-4 \pm 11{\cdot}24}{10}$$

$$x = \frac{7{\cdot}24}{10} \text{ or } \frac{-15{\cdot}24}{10}$$

$$x = 0{\cdot}724 \text{ or } -1{\cdot}524 \text{ Ans.}$$

(iii) $10x^2 - x - 0{\cdot}2 = 0$

$a = 10,\ b = -1,\ c = -0{\cdot}2$

$$x = \frac{1 \pm \sqrt{1 + 8}}{2 \times 10}$$

$$x = \frac{1 \pm 3}{20}$$

$$x = \frac{4}{20} \text{ or } \frac{-2}{20}$$

$$x = 0{\cdot}2 \text{ or } -0{\cdot}1 \text{ Ans.}$$

6.
$$\log 0{\cdot}5x = 2 \times \log(x - 6)$$
$$0{\cdot}5x = (x - 6)^2$$
$$0{\cdot}5x = x^2 - 12x + 36$$
$$x^2 - 12{\cdot}5x + 36 = 0$$

Solving this quadratic,

$$x = 8 \text{ or } 4{\cdot}5 \text{ Ans.}$$

7.
$$\log(x^2 + 2) - \log(x - 1) = 2$$
$$\log\left\{\frac{x^2 + 2}{x - 1}\right\} = 2$$
$$\frac{x^2 + 2}{x - 1} = \text{antilog of } 2 = 100$$
$$x^2 + 2 = 100(x - 1)$$
$$x^2 + 2 = 100x - 100$$
$$x^2 - 100x + 102 = 0$$

Solving this quadratic,

$$x = 98{\cdot}97 \text{ or } 1{\cdot}03 \text{ Ans.}$$

8.
$$6b^4 - 2{\cdot}46b^2 + 0{\cdot}24 = 0$$

Dividing throughout by 6,

$$b^4 - 0{\cdot}41b^2 + 0{\cdot}04 = 0$$

Let x represent b^2,

$$x^2 - 0{\cdot}41x + 0{\cdot}04 = 0$$

The solution of this quadratic gives,

$$x = 0{\cdot}25 \text{ or } 0{\cdot}16$$
$$x = b^2, \therefore b = \sqrt{x}$$
$$b = \pm\sqrt{0{\cdot}25} \text{ or } \pm\sqrt{0{\cdot}16}$$
$$b = \pm 0{\cdot}5 \text{ or } \pm 0{\cdot}4 \text{ Ans.}$$

9. $V^{2\cdot 8} - 5\cdot 1 V^{1\cdot 4} + 5\cdot 6 = 0$

Let x represent $V^{1\cdot 4}$

$$x^2 - 5\cdot 1x + 5\cdot 6 = 0$$

The solution of this quadratic gives,

$$x = 3\cdot 5 \text{ or } 1\cdot 6$$

$$x = V^{1\cdot 4}, \therefore V = \sqrt[1\cdot 4]{x}$$

$$V = \sqrt[1\cdot 4]{3\cdot 5} \text{ or } \sqrt[1\cdot 4]{1\cdot 6}$$

$$V \doteq 2\cdot 446 \text{ or } 1\cdot 399 \text{ Ans.}$$

10. $2g - 5\sqrt{g} - 3\cdot 96 = 0$

Let x represent $\sqrt{g}$, then x^2 represents g,

$$2x^2 - 5x - 3\cdot 96 = 0$$

The solution of this quadratic gives,

$$x = 3\cdot 132 \text{ or } -0\cdot 632$$

$g = x^2$,

$$g = 3\cdot 132^2 \text{ or } -0\cdot 632^2$$

$$g = 9\cdot 81 \text{ or } 0\cdot 3994 \text{ Ans.}$$

11. $x^2 - xy + 2y^2 = 16 \quad \ldots \quad \ldots \quad \ldots \quad \ldots \quad$ (i)

$x + 2y = 8 \quad \ldots \quad \ldots \quad \ldots \quad \ldots \quad$ (ii)

From (ii)

$$2y = 8 - x$$

$y = 4 - 0\cdot 5x \quad \ldots \quad \ldots \quad \ldots \quad$ (iii)

Substitute this value of y into equation (i),

$$x^2 - x(4 - 0\cdot 5x) + 2(4 - 0\cdot 5x)^2 = 16$$

$$x^2 - 4x + 0\cdot 5x^2 + 2(16 - 4x + 0\cdot 25x^2) = 16$$

$$x^2 - 4x + 0\cdot 5x^2 + 32 - 8x + 0\cdot 5x^2 = 16$$

$$x^2 + 0\cdot 5x^2 + 0\cdot 5x^2 - 4x - 8x + 32 - 16 = 0$$

$$2x^2 - 12x + 16 = 0$$

$$x^2 - 6x + 8 = 0$$

By factors,

$$(x - 2)(x - 4) = 0$$

either $x - 2 = 0$, then $x = 2$

or $x - 4 = 0$, then $x = 4$

From (iii)

if $x = 2,\ y = 4 - 0\cdot 5 \times 2 = 3$

if $x = 4,\ y = 4 - 0\cdot 5 \times 4 = 2$

Values are:

$$\left.\begin{array}{ll} & x = 2 \text{ and } y = 3 \\ \text{or} & x = 4 \text{ and } y = 2 \end{array}\right\} \text{Ans.}$$

12. Let x = actual speed of ship, in knots.

$$\text{Time} = \frac{\text{distance}}{\text{speed}} = \frac{665}{x} \text{ hours} \quad \ldots \quad \ldots \quad \ldots \quad \text{(i)}$$

$$\text{Increased speed} = (x + 1\tfrac{1}{2}) \text{ knots}$$

$$\text{Reduced time} = \frac{665}{(x + 1\frac{1}{2})} \text{ hours} \quad \ldots \quad \ldots \quad \text{(ii)}$$

The reduced time (ii) is less than the actual time (i) by 3 hours,

$$\therefore \frac{665}{x} - \frac{665}{(x + 1\frac{1}{2})} = 3$$

Multiplying throughout by $x(x + 1\frac{1}{2})$,

$$665(x + 1\tfrac{1}{2}) - 665x = 3x(x + 1\tfrac{1}{2})$$
$$665x + 997{\cdot}5 - 665x = 3x^2 + 4{\cdot}5x$$
$$3x^2 + 4{\cdot}5x - 997{\cdot}5 = 0$$
$$x^2 + 1{\cdot}5x - 332{\cdot}5 = 0$$

The solution of this quadratic gives $x = 17{\cdot}5$ or -19
The minus quantity is impracticable.

$$\therefore \text{speed of ship} = 17\tfrac{1}{2} \text{ knots. Ans. (i)}$$

$$\text{Voyage time} = \frac{665}{17{\cdot}5} = 38 \text{ hours. Ans. (ii)}$$

13. Let x = speed of slow ship, in knots.
Time to do 330 nautical miles

$$= \frac{330}{x} \text{ hours} \quad \ldots \quad \ldots \quad \ldots \quad \text{(i)}$$

$(x + 3\frac{1}{2})$ = speed of fast ship, in knots
Time to do 334 nautical miles

$$= \frac{334}{x + 3\frac{1}{2}} \text{ hours} \quad \ldots \quad \ldots \quad \text{(ii)}$$

Time (ii) is 5 hours less than time (i), therefore,

$$\frac{330}{x} - \frac{334}{x + 3\frac{1}{2}} = 5$$

$$330(x + 3\tfrac{1}{2}) - 334x = 5x(x + 3\tfrac{1}{2})$$
$$330x + 1155 - 334x = 5x^2 + 17{\cdot}5x$$
$$5x^2 + 21{\cdot}5x - 1155 = 0$$
$$x^2 + 4{\cdot}3x - 231 = 0$$

The solution of this quadratic gives $x = 13{\cdot}2$

$\therefore$ speed of slow ship $= 13{\cdot}2$ knots
speed of fast ship $= 13{\cdot}2 + 3{\cdot}5 = 16{\cdot}7$ knots } Ans.

14. Working in centimetres;

Let L = length, and B = breadth,

length × breadth = area, $\therefore L \times B = 76$ (i)
length + breadth = $\frac{1}{2}$ perimeter, $\therefore L + B = 17{\cdot}5$ (ii)

Substituting $B = \dfrac{76}{L}$ from (i) into (ii),

$$L + \frac{76}{L} = 17{\cdot}5$$

Multiplying throughout by L,

$$L^2 + 76 = 17{\cdot}5L$$
$$L^2 - 17{\cdot}5L + 76 = 0$$

The solution of this quadratic gives $L = 9{\cdot}5$ or 8

$\therefore$ length $= 9{\cdot}5$ cm $= 95$ mm
breadth $= 8$ cm $= 80$ mm } Ans.

15. Working in centimetres:

$$(\text{length})^2 + (\text{breadth})^2 = (\text{diagonal})^2$$
$$\therefore L^2 + B^2 = 7{\cdot}5^2 \quad \dots \quad \dots \quad \dots \quad \dots \quad \text{(i)}$$
$$\text{length} \times \text{breadth} = \text{area}$$
$$\therefore L \times B = 27 \quad \dots \quad \dots \quad \dots \quad \dots \quad \text{(ii)}$$

Multiplying (ii) by 2, adding and subtracting result from (i)

$$L^2 + B^2 = 56{\cdot}25$$
$$2LB = 54$$

$$L^2 + 2LB + B^2 = 110{\cdot}25 \text{ by addition} \quad \dots \quad \text{(iii)}$$
$$L^2 - 2LB + B^2 = 2{\cdot}25 \text{ by subtraction} \quad \dots \quad \text{(iv)}$$

Taking square roots of (iii) and (iv) and adding results,

$$L + B = 10{\cdot}5 \quad \dots \quad (v)$$
$$L - B = 1{\cdot}5 \quad \dots \quad (vi)$$

$$2L = 12$$
$$L = 6$$

From (v) $B = 10{\cdot}5 - 6$
$$= 4{\cdot}5$$

$\therefore$ length = 6cm = 60 mm, breadth = 4·5 cm = 45 mm. Ans.

An alternative solution is as follows,

$$L^2 + B^2 = 56{\cdot}25 \quad \dots \quad (i)$$
$$L \times B = 27 \quad \dots \quad (ii)$$

Substituting $B = \dfrac{27}{L}$ from (ii) into (i),

$$L^2 + \frac{27^2}{L^2} = 56{\cdot}25$$

Multiplying throughout by L^2,

$$L^4 + 27^2 = 56{\cdot}25L^2$$
$$L^4 - 56{\cdot}25L^2 + 729 = 0$$

Let x represent L^2,

$$x^2 - 56{\cdot}25x + 729 = 0$$

The solution of this quadratic gives $x = 36$ or $20{\cdot}25$

$x = L^2$, $\quad \therefore L = \sqrt{x}$
$$L = \sqrt{36} \text{ or } \sqrt{20{\cdot}25}$$
$$L = 6 \text{ or } 4{\cdot}5$$

$\therefore$ length = 6 cm and breadth = 4·5 cm as before.

16. $2x^3 + 3x^2 - 17x - 30 = 0$

Finding the first root by trial,

Try $x = 1$, $\quad 2 \times 1^3 + 3 \times 1^2 - 17 \times 1 - 30$
$$= 2 + 3 - 17 - 30$$
$$= -42$$

Try $x = 2$, $\quad 2 \times 2^3 + 3 \times 2^2 - 17 \times 2 - 30$
$$= 16 + 12 - 34 - 30$$
$$= -36$$

Try $x = 3$, $\quad 2 \times 3^3 + 3 \times 3^2 - 17 \times 3 - 30$

$$= 54 + 27 - 51 - 30$$
$$= 0$$

$x = 3$ satisfies the equation, this is one root and $x - 3$ must be a factor. Divide the cubic equation by this factor.

$$
\begin{array}{l}
x - 3)\,2x^3 + 3x^2 - 17x - 30\,(2x^2 + 9x + 10 \\
\quad\quad\;\; 2x^3 - 6x^2 \\
\hline
\quad\quad\quad\quad\;\; 9x^2 - 17x - 30 \\
\quad\quad\quad\quad\;\; 9x^2 - 27x \\
\hline
\quad\quad\quad\quad\quad\quad\;\; 10x - 30 \\
\quad\quad\quad\quad\quad\quad\;\; 10x - 30 \\
\hline
\end{array}
$$

Equating the resulting quadratic to zero and finding its roots by formula,

$$2x^2 + 9x + 10$$

$$x = \frac{-b \pm \sqrt{b^2 - 4ac}}{2a}$$

$$= \frac{-9 \pm \sqrt{9^2 - 4 \times 2 \times 10}}{2 \times 2}$$

$$= \frac{-9 \pm 1}{4}$$

$$= -2 \text{ or } -2{\cdot}5$$

The roots of the given equation are:

3, −2, and −2·5 Ans.

17. $$2x - \frac{16}{x}\left\{2 - \frac{3}{x}\right\} = 3$$

Multiplying terms within the brackets by the multiplier outside, to eliminate the brackets,

$$2x - \frac{32}{x} + \frac{48}{x^2} = 3$$

Multiply throughout by the least common denominator, which is x^2, to eliminate fractions,

$$2x^3 - 32x + 48 = 3x^2$$

Arrange all terms on left hand side in order of descending powers of x,

$$2x^3 - 3x^2 - 32x + 48 = 0$$

Proceed to solve this cubic equation by finding the first root by trial,

Try $x = 1$,
$$2 \times 1^3 - 3 \times 1^2 - 32 \times 1 + 48$$
$$= 2 - 3 - 32 + 48$$
$$= 15$$

Try $x = 2$,
$$2 \times 2^3 - 3 \times 2^2 - 32 \times 2 + 48$$
$$= 16 - 12 - 64 + 48$$
$$= -12$$

Try $x = 3$,
$$2 \times 3^3 - 3 \times 3^2 - 32 \times 3 + 48$$
$$= 54 - 27 - 96 + 48$$
$$= -21$$

Try $x = 4$,
$$2 \times 4^3 - 3 \times 4^2 - 32 \times 4 + 48$$
$$= 128 - 48 - 128 + 48$$
$$= 0$$

$x = 4$ is a root and $x - 4$ is a factor. Divide the cubic equation by this factor,

$$
\begin{array}{l}
x - 4)\,2x^3 - 3x^2 - 32x + 48\,(2x^2 + 5x - 12 \\
\quad\quad\;\; \underline{2x^3 - 8x^2} \\
\quad\quad\quad\quad\;\; 5x^2 - 32x + 48 \\
\quad\quad\quad\quad\;\; \underline{5x^2 - 20x} \\
\quad\quad\quad\quad\quad\quad -12x + 48 \\
\quad\quad\quad\quad\quad\quad \underline{-12x + 48} \\
\quad\quad\quad\quad\quad\quad\quad \cdot\;\;\cdot
\end{array}
$$

Finding the roots of the resulting quadratic by formula,

$$2x^2 + 5x - 12 = 0$$

$$x = \frac{-b \pm \sqrt{b^2 - 4ac}}{2a}$$

$$= \frac{-5 \pm \sqrt{5^2 - 4 \times 2 \times (-12)}}{2 \times 2}$$

$$= \frac{-5 \pm 11}{4}$$

$$= 1{\cdot}5 \text{ or } -4$$

The roots of the given equation are:

4, 1·5, and –4 Ans.

SOLUTIONS TO TEST EXAMPLES 7

1. Finding plotting points:

(i) $y = 2 + x,$

$$\begin{aligned} \text{when } x &= 0, & y &= 2 + 0 = 2 \\ \text{when } x &= 12, & y &= 2 + 12 = 14 \end{aligned}$$

(ii) $y = 12 - 1{\cdot}5x,$

$$\begin{aligned} \text{when } x &= 0, & y &= 12 - 0 = 12 \\ \text{when } x &= 12, & y &= 12 - 18 = -6 \end{aligned}$$

(iii) $y = -1 - 0{\cdot}5x,$

$$\begin{aligned} \text{when } x &= 0, & y &= -1 - 0 = -1 \\ \text{when } x &= 12, & y &= -1 - 6 = -7 \end{aligned}$$

(iv) $y = -4 + 1{\cdot}25x,$

$$\begin{aligned} \text{when } x &= 0, & y &= -4 + 0 = -4 \\ \text{when } x &= 12, & y &= -4 + 15 = 11 \end{aligned}$$

The plotted graphs are shown in Fig. 156.

2. Being a straight line graph it can be represented by the equation $y = a + bx$. The graph is drawn through the plotting points:

$$\begin{aligned} \text{when } x &= 2, & y &= 4 \\ \text{and when } x &= 10, & y &= 16 \end{aligned}$$

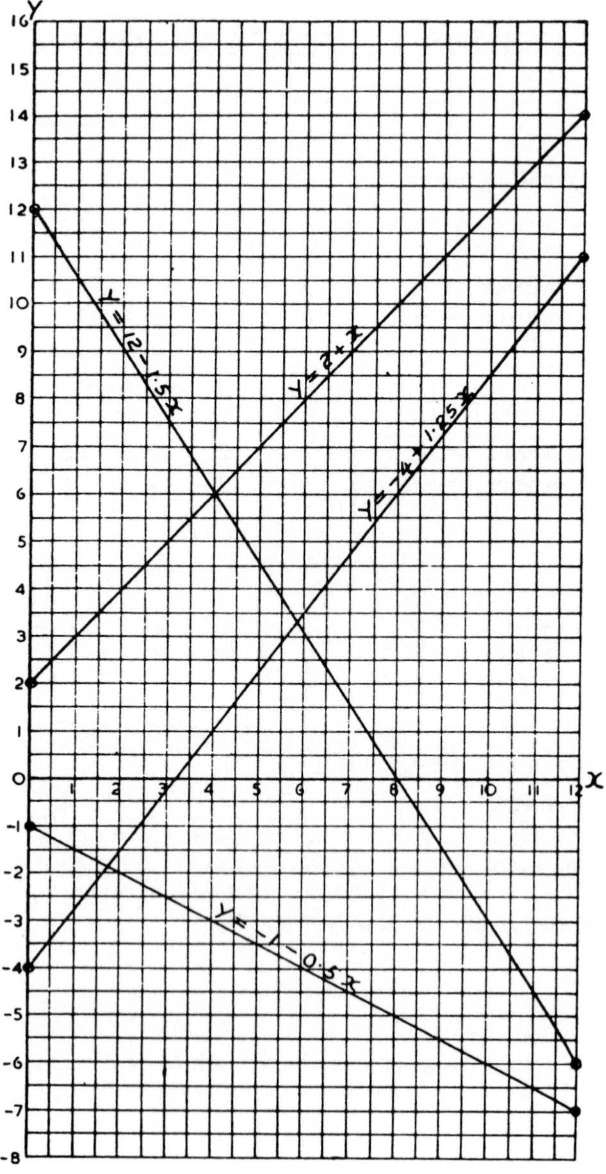
Y
16
15
14
13
12
11
10
9
8
7
6
5
4
3
2
1
0
-1
-2
-3
-4
-5
-6
-7
-8
1
2
3
4
5
6
7
8
9
10
11
12
X
Y = 12 − 1·5X
Y = 2 + X
Y = −4 + 1·25X
Y = −1 − 0·5X

Fig. 156

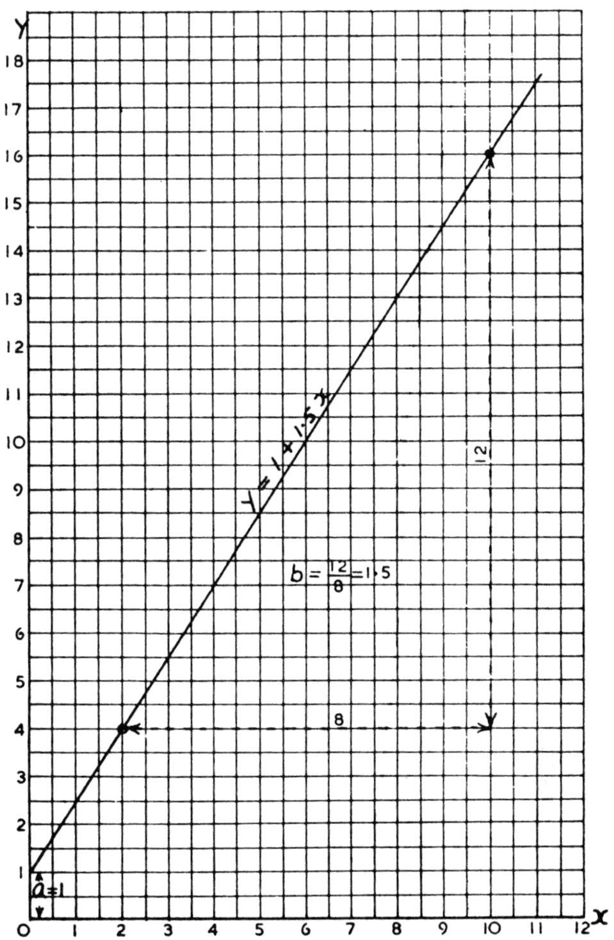

Fig. 157

Continued from page 349.

From the graph (see Fig. 157),

$$a = 1$$
$$b = 12 \div 8 = 1{\cdot}5$$

$\therefore$ the equation is $y = 1 + 1{\cdot}5x$ Ans.

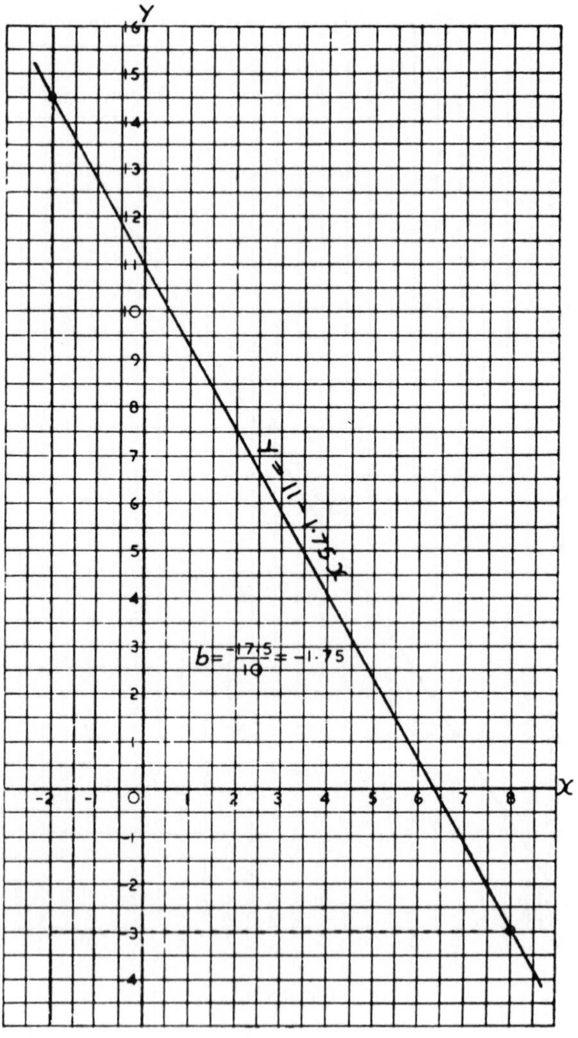

Fig. 158

3. From the graph in Fig. 158 we read:

$$a = 11$$
$$b = -17{\cdot}5 \div 10 = -1{\cdot}75$$

$\therefore$ the equation is $y = 11 - 1{\cdot}75x$ Ans.

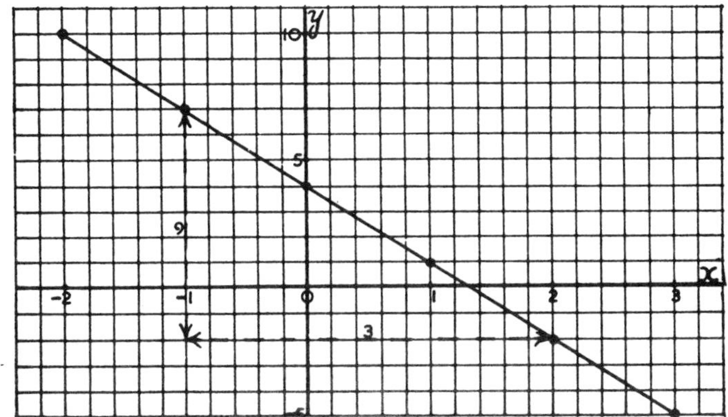

Fig. 159

4. From the plotted graph (see Fig. 159):

$$a = 4$$
$$b = -9 \div 3 = -3$$

$\therefore$ Law is $y = 4 - 3x$ Ans.

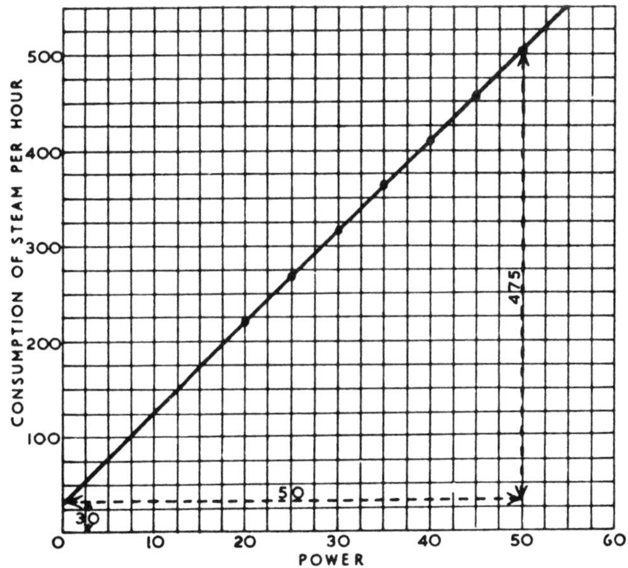

Fig. 160

5. Reading the graph (Fig. 160):

$$a = 30$$
$$b = 475 \div 50 = 9{\cdot}5$$

$\therefore$ Law is $m = 30 + 9{\cdot}5P$ Ans.

6. Expressing y in terms of the other quantities,

first equation,
$$3x + 5y = 23$$
$$5y = 23 - 3x$$
$$y = 4{\cdot}6 - 0{\cdot}6x \quad \ldots \quad \ldots \quad \text{(i)}$$

second equation,
$$5x - 2y = 12{\cdot}5$$
$$-2y = 12{\cdot}5 - 5x$$
$$2y = 5x - 12{\cdot}5$$
$$y = 2{\cdot}5x - 6{\cdot}25 \quad \ldots \quad \ldots \quad \text{(ii)}$$

Choosing two plotting points for the graphs of each equation,

equation (i), when $x = 0$, $y = 4{\cdot}6 - 0 = 4{\cdot}6$
,, $x = 5$, $y = 4{\cdot}6 - 3 = 1{\cdot}6$

equation (ii), when $x = 0$, $y = 0 - 6{\cdot}25 = -6{\cdot}25$
,, $x = 4$, $y = 10 - 6{\cdot}25 = 3{\cdot}75$

The two graphs are now drawn as in Fig. 161 from which we read the point of intersection,

$$\left.\begin{aligned} x &= 3{\cdot}5 \\ y &= 2{\cdot}5 \end{aligned}\right\} \text{Ans.}$$

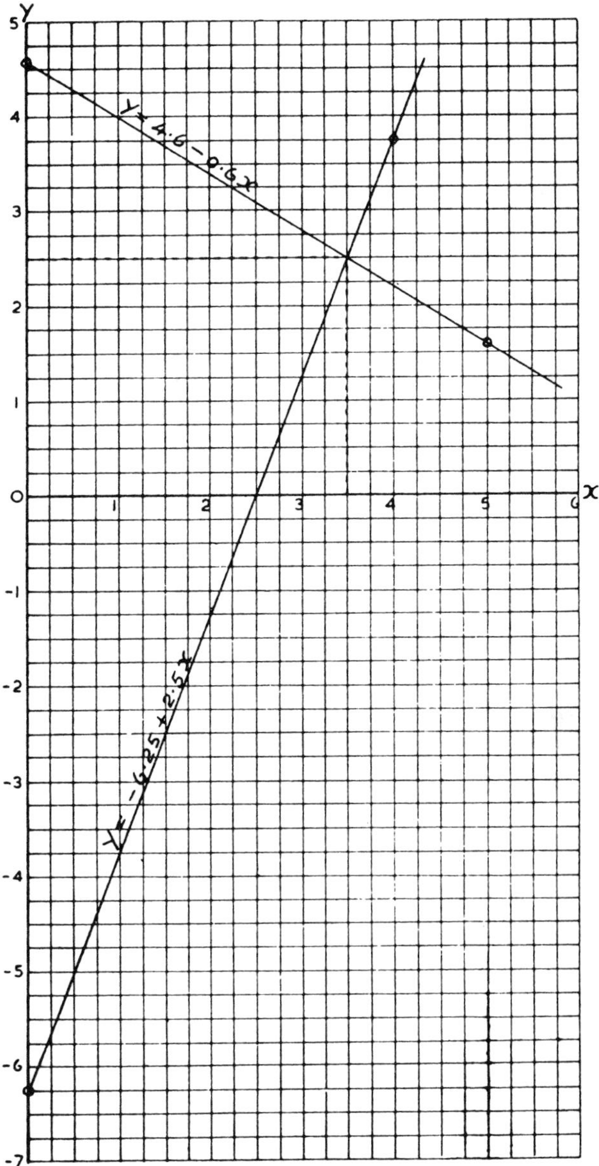

Fig. 161

7. Expressing p in each equation in terms of the other quantities,

$$5p - 2q = 5{\cdot}6$$
$$5p = 5{\cdot}6 + 2q$$
$$p = 1{\cdot}12 + 0{\cdot}4q \quad \ldots \quad \ldots \quad \text{(i)}$$
$$2p - 3q = -4{\cdot}8$$
$$2p = -4{\cdot}8 + 3q$$
$$p = -2{\cdot}4 + 1{\cdot}5q \quad \ldots \quad \ldots \quad \text{(ii)}$$

Plotting points for each equation:

(i) when $q = 0$, $p = 1{\cdot}12 + 0 = 1{\cdot}12$
,, $q = 5$, $p = 1{\cdot}12 + 2 = 3{\cdot}12$

(ii) when $q = 0$, $p = -2{\cdot}4 + 0 = -2{\cdot}4$
,, $q = 4$, $p = -2{\cdot}4 + 6 = 3{\cdot}6$

From the graphs (Fig. 162)

$$p = 2{\cdot}4 \text{ and } q = 3{\cdot}2 \text{ Ans.}$$

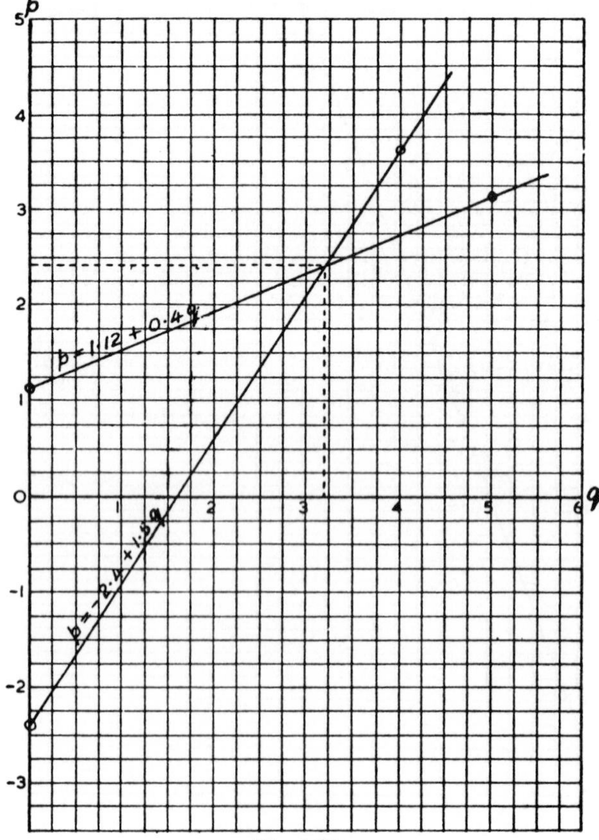

Fig. 162

8. $x^2 - 5x + 5\frac{1}{4} = 0$

Let $y = x^2 - 5x + 5\frac{1}{4}$

when $x = 0,\ y = 0 - 0 + 5\frac{1}{4} = 5\frac{1}{4}$

„ $x = 1,\ y = 1 - 5 + 5\frac{1}{4} = 1\frac{1}{4}$

„ $x = 2,\ y = 4 - 10 + 5\frac{1}{4} = -\frac{3}{4}$

„ $x = 3,\ y = 9 - 15 + 5\frac{1}{4} = -\frac{3}{4}$

„ $x = 4,\ y = 16 - 20 + 5\frac{1}{4} = 1\frac{1}{4}$

From the graph, Fig. 163, $y = 0$ when:

$$x = 1{\cdot}5 \text{ and } 3{\cdot}5$$

Therefore the roots of the equation are 1·5 and 3·5 Ans.

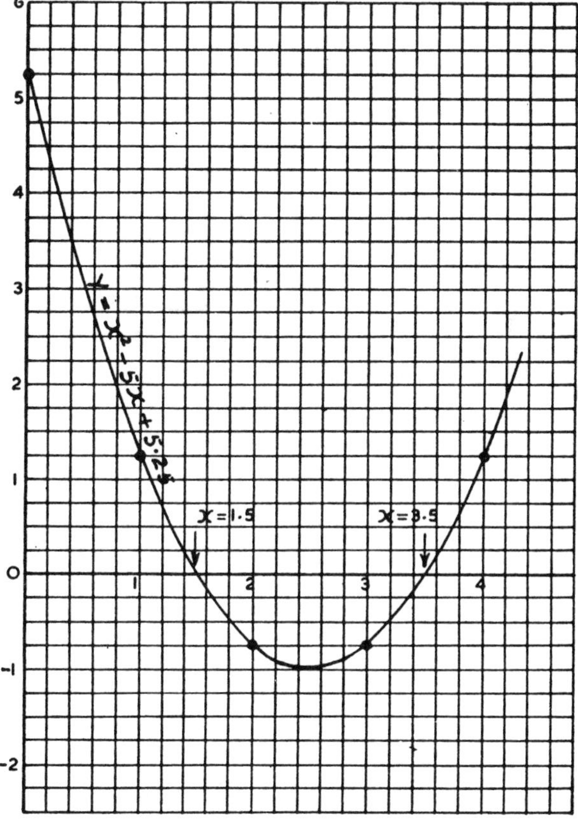

Fig. 163

9. Finding plotting points,

$$
\begin{aligned}
& & y &= 0{\cdot}5x^2 - 2x - 6 & & \\
x &= -4, & y &= 0{\cdot}5(-4)^2 - 2(-4) - 6 &&= 10 \\
x &= -3, & y &= 0{\cdot}5(-3)^2 - 2(-3) - 6 &&= 4{\cdot}5 \\
x &= -2, & y &= 0{\cdot}5(-2)^2 - 2(-2) - 6 &&= 0 \\
x &= -1, & y &= 0{\cdot}5(-1)^2 - 2(-1) - 6 &&= -3{\cdot}5 \\
x &= 0, & y &= 0 - 0 - 6 &&= -6 \\
x &= 1, & y &= 0{\cdot}5 - 2 - 6 &&= -7{\cdot}5 \\
x &= 2, & y &= 2 - 4 - 6 &&= -8 \\
x &= 3, & y &= 4{\cdot}5 - 6 - 6 &&= -7{\cdot}5 \\
x &= 4, & y &= 8 - 8 - 6 &&= -6 \\
x &= 5, & y &= 12{\cdot}5 - 10 - 6 &&= -3{\cdot}5 \\
x &= 6, & y &= 18 - 12 - 6 &&= 0 \\
x &= 7, & y &= 24{\cdot}5 - 14 - 6 &&= 4{\cdot}5 \\
x &= 8, & y &= 32 - 16 - 6 &&= 10
\end{aligned}
$$

The above points are plotted and the graph drawn as shown in Fig. 164

(i) $0{\cdot}5x^2 - 2x - 6 = 0$

The roots of this equation are where the graph crosses the base line $y = 0$, thus,

$$x = -2 \text{ and } +6 \quad \text{Ans. (i)}$$

(ii) The difference between $0{\cdot}5x^2 - 2x - 6$
and $0{\cdot}5x^2 - 2x - 4$ (subtract)

is -2

Thus, when $y = -2$ on the graph of $y = 0{\cdot}5x^2 - 2x - 6$ we read the roots of the equation $0{\cdot}5x^2 - 2x - 4 = 0$, that is,

$$x = -1{\cdot}5 \text{ and } +5{\cdot}5 \quad \text{Ans. (ii)}$$

(iii) The difference between $0{\cdot}5x^2 - 2x - 6$
and $0{\cdot}5x^2 - 2x - 1$ (subtract)

is -5

Therefore the values of x in this equation are where the graph crosses the line of $y = -5$, thus,

$$x = -0{\cdot}5 \text{ and } +4{\cdot}5 \quad \text{Ans. (iii)}$$

(iv) The difference between $0{\cdot}5x^2 - 2x - 6$

and $0{\cdot}5x^2 - 2x$ (subtract)

is -6

Hence, values of x are read where the graph crosses the line of $y = -6$,

$$x = 0 \text{ and } 4 \quad \text{Ans. (iv)}$$

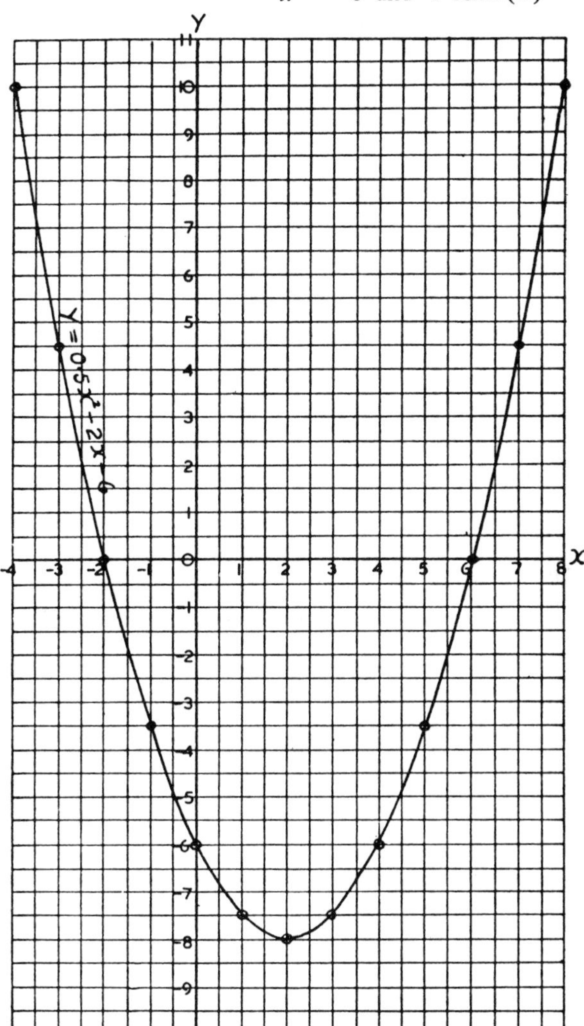

Fig. 164

(v) The difference between $0{\cdot}5x^2 - 2x - 6$
and $0{\cdot}5x^2 - 2x + 1$ (subtract)

is -7

Then read off the graph values of x when $y = -7$,

$$x = 0{\cdot}6 \text{ and } 3{\cdot}4 \text{ Ans. (v)}$$

10. The plotting points are calculated and the graphs drawn for each equation as in Fig. 165.

$$x^2 - 2{\cdot}5x - 3{\cdot}5 = 0$$

can be written $x^2 - (2{\cdot}5x + 3{\cdot}5) = 0$

and represented by $y_1 - y_2 = 0$

where $y_1 = x^2$, and $y_2 = 2{\cdot}5x + 3{\cdot}5$

Hence, where the graphs intersect we have the values of x where $y_1 - y_2 = 0$ and therefore the roots of the equation $x^2 - 2{\cdot}5x - 3{\cdot}5 = 0$.

$$\therefore x = -1 \text{ and } +3{\cdot}5 \text{ Ans.}$$

11. The graphs are plotted as in Fig. 166. The points of intersection produce the values of x and y as:

$$\left.\begin{array}{l} x = 1 \text{ and } y = -0{\cdot}6 \\ x = 10 \text{ and } y = 12 \end{array}\right\} \text{Ans.}$$

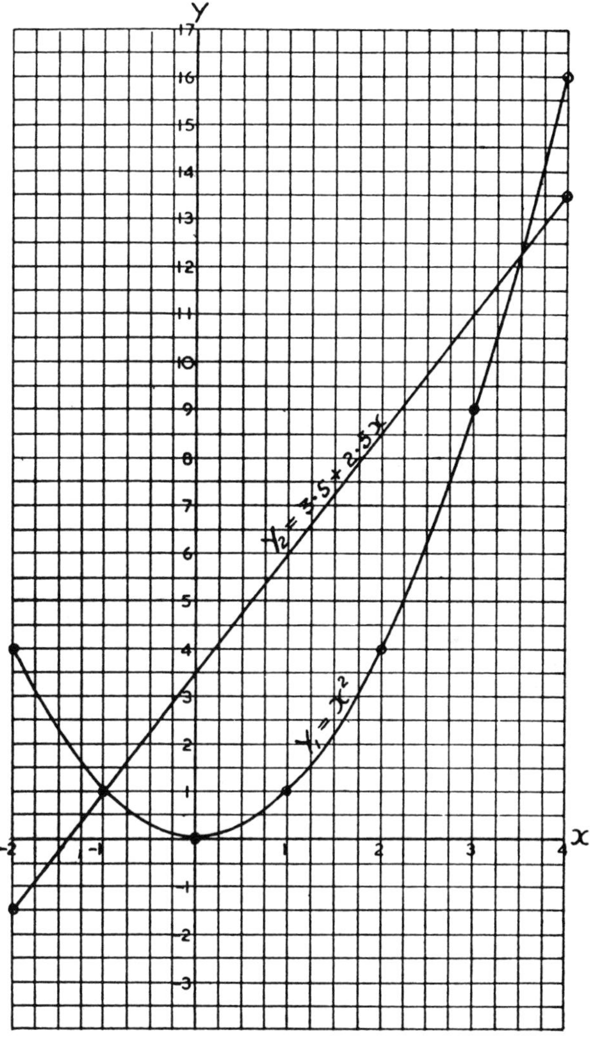

Fig. 165.

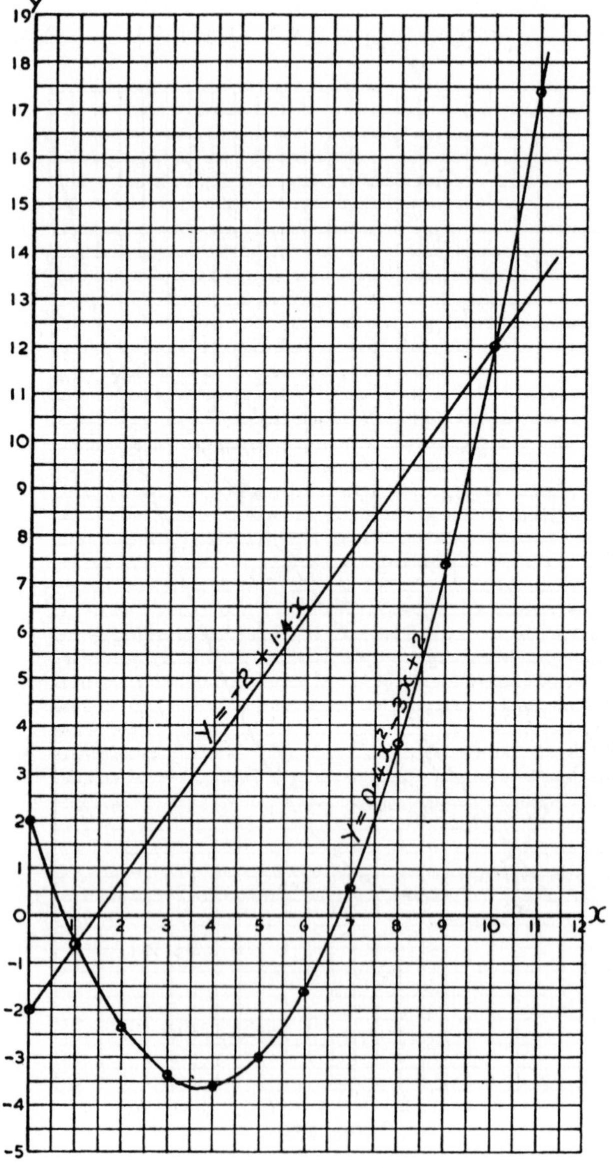
Y
X
19
18
17
16
15
14
13
12
11
10
9
8
7
6
5
4
3
2
1
0
-1
-2
-3
-4
-5
1
2
3
4
5
6
7
8
9
10
11
12
Y = -2 + 1·4X
Y = 0·4X² - 3X + 2

Fig. 166.

SOLUTIONS TO TEST EXAMPLES 8

1.

$$\text{One radian} = 57{\cdot}3^\circ = 57^\circ\ 18'$$
$$114^\circ\ 36' = 114{\cdot}6^\circ$$
$$114{\cdot}6 \div 57{\cdot}3 = 2 \text{ radians. Ans. (i)}$$
$$143^\circ\ 15' = 143{\cdot}25^\circ$$
$$143{\cdot}25 \div 57{\cdot}3 = 2{\cdot}5 \text{ radians. Ans. (ii)}$$
$$186^\circ\ 13{\cdot}5' = 186{\cdot}225^\circ$$
$$186{\cdot}225 \div 57{\cdot}3 = 3{\cdot}25 \text{ radians. Ans. (iii)}$$
$$286^\circ\ 30' = 286{\cdot}5^\circ$$
$$286{\cdot}5 \div 57{\cdot}3 = 5 \text{ radians. Ans. (iv)}$$

2. $\theta = \dfrac{\text{arc}}{\text{radius}}$ Radius of circle = 5 metres.

$$10 \div 5 = 2 \text{ radians Ans. (ai)}$$
$$21{\cdot}25 \div 5 = 4{\cdot}25 \text{ ,, ,, (aii)}$$
$$27{\cdot}5 \div 5 = 5{\cdot}5 \text{ ,, ,, (aiii)}$$
$$30{\cdot}4 \div 5 = 6{\cdot}08 \text{ ,, ,, (aiv)}$$

$$\text{Degrees} = \text{radians} \times 57{\cdot}3$$

$$2 \times 57{\cdot}3 = 114{\cdot}6^\circ = 114^\circ\ 36' \text{ Ans. (bi)}$$
$$4{\cdot}25 \times 57{\cdot}3 = 243{\cdot}525^\circ = 243^\circ\ 31{\cdot}5' \text{ ,, (bii)}$$
$$5{\cdot}5 \times 57{\cdot}3 = 315{\cdot}15^\circ = 315^\circ\ 9' \text{ ,, (biii)}$$
$$6{\cdot}08 \times 57{\cdot}3 = 348{\cdot}384^\circ = 348^\circ\ 23' \text{ ,, (biv)}$$

3. $\omega = \dfrac{2\pi \text{ rev/min}}{60}$ (value of π taken as 3·142)

$$\frac{2\pi \times 84}{60} = 8{\cdot}796 \text{ rad/s Ans. (i)}$$

$$\frac{2\pi \times 120}{60} = 12{\cdot}568 \text{ rad/s Ans. (ii)}$$

$$\frac{2\pi \times 350}{60} = 36{\cdot}65 \text{ rad/s Ans. (iii)}$$

$$\frac{2\pi \times 3000}{60} = 314{\cdot}2 \text{ rad/s Ans. (iv)}$$

4.

$$\text{rev/min} = \frac{\omega \times 60}{2\pi}$$

$$\frac{9{\cdot}25 \times 60}{2\pi} = 88{\cdot}33 \text{ rev/min} \quad \text{Ans. (i)}$$

$$\frac{12{\cdot}5 \times 60}{2\pi} = 119{\cdot}3 \text{ rev/min} \quad \text{Ans. (ii)}$$

$$\frac{37{\cdot}25 \times 60}{2\pi} = 355{\cdot}7 \text{ rev/min} \quad \text{Ans. (iii)}$$

$$\frac{500 \times 60}{2\pi} = 4775 \text{ rev/min} \quad \text{Ans. (iv)}$$

5.

$$v = \omega r$$

$$10{\cdot}52 \times 0{\cdot}1 = 1{\cdot}052 \text{ m/s} \quad \text{Ans. (i)}$$
$$10{\cdot}52 \times 0{\cdot}25 = 2{\cdot}63 \text{ m/s} \quad \text{,, (ii)}$$
$$10{\cdot}52 \times 0{\cdot}5 = 5{\cdot}26 \text{ m/s} \quad \text{,, (iii)}$$

6.

$$\text{Circumference} = \pi \times \text{diameter}$$

In one revolution, rim travels $\pi \times 2$ metres

,, ,, minute, ,, ,, $\pi \times 2 \times 125$ metres

,, ,, second, ,, ,, $\dfrac{\pi \times 2 \times 125}{60}$ metres

$$\text{Linear velocity} = 13{\cdot}09 \text{ m/s} \quad \text{Ans.}$$

7.

$$(AB)^2 = (AC)^2 + (BC)^2 \quad \text{(See Fig. 167)}$$
$$AB = \sqrt{36^2 + 27^2}$$
$$= \sqrt{2025} = 45 \text{ mm} \quad \text{Ans. (i)}$$

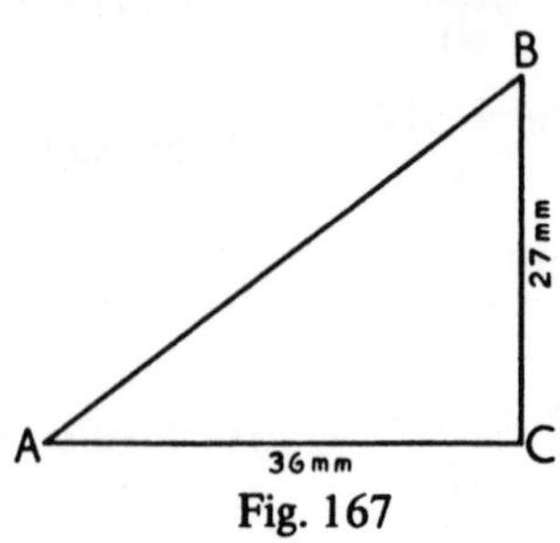

Fig. 167

$$\sin B = \frac{\text{opposite}}{\text{hypotenuse}} = \frac{36}{45} = 0{\cdot}8 \quad \text{Ans. (ii)}$$

$$\cos B = \frac{\text{adjacent}}{\text{hypotenuse}} = \frac{27}{45} = 0{\cdot}6 \quad \text{Ans. (iii)}$$

$$\tan B = \frac{\text{opposite}}{\text{adjacent}} = \frac{36}{27} = 1{\cdot}333 \quad \text{Ans. (iv)}$$

8.

$$\sin^2\theta + \cos^2\theta = 1$$

$$\sin^2\theta = 1 - \cos^2\theta$$

$$\sin\theta = \sqrt{1 - 0{\cdot}4924^2}$$

$$= \sqrt{0{\cdot}7575} = 0{\cdot}8704 \quad \text{Ans. (i)}$$

$$\tan\theta = \frac{\sin\theta}{\cos\theta} = \frac{0{\cdot}8704}{0{\cdot}4924}$$

$$= 1{\cdot}767 \quad \text{Ans. (ii)}$$

9.

ANGLE	SINE	COSINE
0	0	1
30	0·5	0·866
60	0·866	0·5
90	1	0
120	0·866	–0·5
150	0·5	–0·866
180	0	–1
210	–0·5	–0·866
240	–0·866	–0·5
270	–1	0
300	–0·866	0·5
330	–0·5	0·866
360	0	1

See Fig 168 for graphs

10.

ANGLE	SINE	COSINE	TANGENT
10° 33′	0·1831	0·9831	0·1862
46° 55′	0·7304	0·6831	1·0692
150° 47′	0·4882	–0·8728	–0·5593
201° 21′	–0·3641	–0·9314	0·3909
287° 14′	–0·9551	0·2963	–3·2235

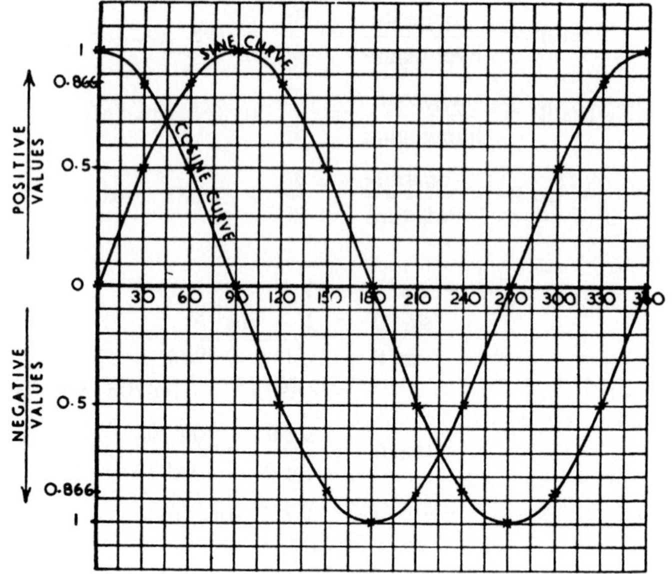

Fig. 168

11.

RATIO	VALUE	ANGLE
sine	0·3783	22° 14′ and 157° 46′
sine	−0·7005	224° 28′ and 315° 32′
cosine	0·9687	14° 22′ and 345° 38′
cosine	−0·8769	151° 16′ and 208° 44′
tangent	0·2010	11° 22′ and 191° 22′
tangent	−3·2006	107° 21′ and 287° 21′

12.

$$
\begin{aligned}
\sin 80^\circ &= 0{\cdot}9848 \\
\cos 80^\circ &= 0{\cdot}1736 \\
\sin 2\theta &= \sin 160^\circ = 0{\cdot}342 \\
\cos 2\theta &= \cos 160^\circ = -0{\cdot}9397 \\
\sin^2 \theta &= (0{\cdot}9848)^2 = 0{\cdot}9701 \\
\cos^2 \theta &= (0{\cdot}1736)^2 = 0{\cdot}03013
\end{aligned}
$$

13.
$$\sin^2\theta + \cos^2\theta = 1$$
$$\therefore \sin^2\theta = 1 - \cos^2\theta$$

Substituting this value of $\sin^2\theta$ into the given equation,

$$\cos\theta - \sin^2\theta = 0$$
$$\cos\theta - (1 - \cos^2\theta) = 0$$
$$\cos\theta - 1 + \cos^2\theta = 0$$

This is a quadratic equation, re-arrange in the usual way in descending powers of the unknown,

$$\cos^2\theta + \cos\theta - 1 = 0$$

Solve this quadratic equation either by formula or 'completing the square', if preferred, let x stand for $\cos\theta$,

$$x^2 + x - 1 = 0$$

Solution to this quadratic is 0·618 or −1·618

$$\therefore \cos\theta = 0{\cdot}618 \text{ (other value impossible)}$$

$$\left.\begin{array}{rl}\text{From tables, } \theta = & 51^\circ\ 50' \\ \text{or } (360^\circ - 51^\circ\ 50') = & 308^\circ\ 10'\end{array}\right\} \text{Ans.}$$

14.
$$\frac{a^2}{2} = \frac{2b^2}{5} - 4$$

$$\frac{2^2\sin^2\theta}{2} = \frac{2 \times 5^2\cos^2\theta}{5} - 4$$

$$2\sin^2\theta = 10\cos^2\theta - 4$$
$$\sin^2\theta = 5\cos^2\theta - 2$$
$$\sin^2\theta = 5(1 - \sin^2\theta) - 2$$
$$\sin^2\theta = 5 - 5\sin^2\theta - 2$$
$$6\sin^2\theta = 3$$
$$\sin^2\theta = 0{\cdot}5$$
$$\sin\theta = \pm\sqrt{0{\cdot}5} = \pm 0{\cdot}7071$$

For angles between 0 and 180°, only the positive value of the sine is applicable.

$$\sin\theta = 0{\cdot}7071$$
$$\therefore \theta = 45^\circ \text{ or } 135^\circ \text{ Ans.}$$

15. $$\operatorname{cosec}^2 A + \sec^2 A = \operatorname{cosec}^2 A . \sec^2 A$$

Simplifying left hand side of equation:

$$\operatorname{cosec}^2 A + \sec^2 A = \frac{1}{\sin^2 A} + \frac{1}{\cos^2 A}$$

$$= \frac{\cos^2 A + \sin^2 A}{\sin^2 A . \cos^2 A} = \frac{1}{\sin^2 A . \cos^2 A}$$

$$= \operatorname{cosec}^2 A . \sec^2 A = \text{right hand side of equation. Ans.}$$

16. $$\frac{1 - \sin A}{1 + \sin A} = (\sec A - \tan A)^2$$

Simplifying right hand side of equation:

$$(\sec A - \tan A)^2 = \left\{\frac{1}{\cos A} - \frac{\sin A}{\cos A}\right\}^2$$

$$= \left\{\frac{1 - \sin A}{\cos A}\right\}^2 = \frac{(1 - \sin A)^2}{\cos^2 A}$$

$$= \frac{(1 - \sin A)^2}{1 - \sin^2 A} = \frac{(1 - \sin A)(1 - \sin A)}{(1 + \sin A)(1 - \sin A)}$$

$$= \frac{1 - \sin A}{1 + \sin A} = \text{left hand side of equation. Ans.}$$

17. $$(\sec A - \cos A)(\operatorname{cosec} A - \sin A) = \frac{1}{\tan A + \cot A}$$

Simplifying left hand side of equation:

$(\sec A - \cos A)(\operatorname{cosec} A - \sin A)$

$$= \left\{\frac{1}{\cos A} - \cos A\right\}\left\{\frac{1}{\sin A} - \sin A\right\}$$

$$= \frac{1 - \cos^2 A}{\cos A} \times \frac{1 - \sin^2 A}{\sin A} = \frac{\sin^2 A}{\cos A} \times \frac{\cos^2 A}{\sin A}$$

$$= \sin A \cos A \quad \ldots \quad \ldots \quad \ldots \quad \ldots \quad \ldots \quad \ldots \quad \text{(i)}$$

Simplifying right hand side of equation:

$$\frac{1}{\tan A + \cot A} = \frac{1}{\frac{\sin A}{\cos A} + \frac{\cos A}{\sin A}}$$

$$= \frac{1}{\frac{\sin^2 A + \cos^2 A}{\sin A \cos A}} = \frac{\sin A \cos A}{\sin^2 A + \cos^2 A}$$

$$= \frac{\sin A \cos A}{1} = \sin A \cos A \quad \dots \quad \dots \quad \text{(ii)}$$

From (i) and (ii) both sides of the equation are equal to the same quantity, which proves the given identity. Ans.

18. $$\omega = \frac{2\pi \text{ rev/min}}{60} = \frac{2\pi \times 150}{60} = 15{\cdot}71 \text{ rad/sec.}$$

$$r = \text{half-stroke} = 0{\cdot}5 \text{ metre}$$

$$n = \frac{\text{connecting rod length}}{\text{crank length}} = \frac{2}{0{\cdot}5} = 4$$

$$\theta = 80^\circ,\ \sin 80^\circ = 0{\cdot}9848,\ \cos 80^\circ = 0{\cdot}1736$$

$$2\theta = 160^\circ,\ \sin 160^\circ = 0{\cdot}342,\ \cos 160^\circ = -0{\cdot}9397$$

$$v = \omega r \left[\sin\theta + \frac{\sin 2\theta}{2n}\right] \text{ m/s}$$

$$= 15{\cdot}71 \times 0{\cdot}5 \times \left[0{\cdot}9848 + \frac{0{\cdot}342}{2 \times 4}\right]$$

$$= 15{\cdot}71 \times 0{\cdot}5 \times 1{\cdot}02755$$

$$= 8{\cdot}072 \text{ m/s} \quad \text{Ans. (i)}$$

$$a = \omega^2 r \left[\cos\theta + \frac{\cos 2\theta}{n}\right]$$

$$= 15{\cdot}71^2 \times 0{\cdot}5 \times \left[0{\cdot}1736 + \frac{-0{\cdot}9397}{4}\right]$$

$$= 15{\cdot}71^2 \times 0{\cdot}5 \times (0{\cdot}1736 - 0{\cdot}2349)$$

$$= 15{\cdot}71^2 \times 0{\cdot}5 \times (-0{\cdot}0613)$$

$$= -7{\cdot}567 \text{ m/s}^2 \quad \text{Ans. (ii)}$$

SOLUTIONS TO TEST EXAMPLES 9

1.

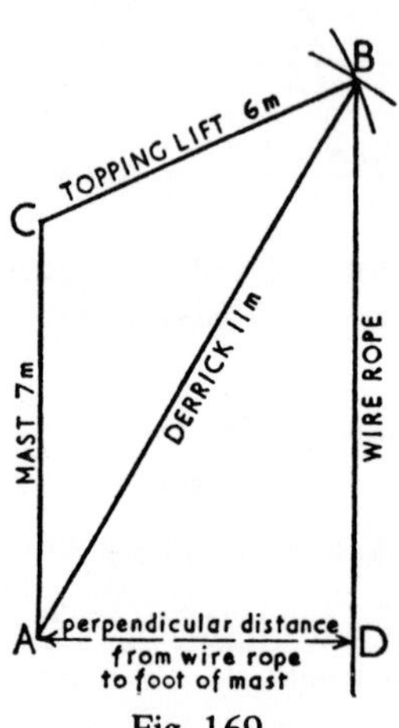

Fig. 169

By measurement (Fig. 169),

Angle A = $29\frac{1}{2}°$
,, B = $35°$
,, C = $115\frac{1}{2}°$
Perpendicular distance AD = 5·4 m Ans.

2.

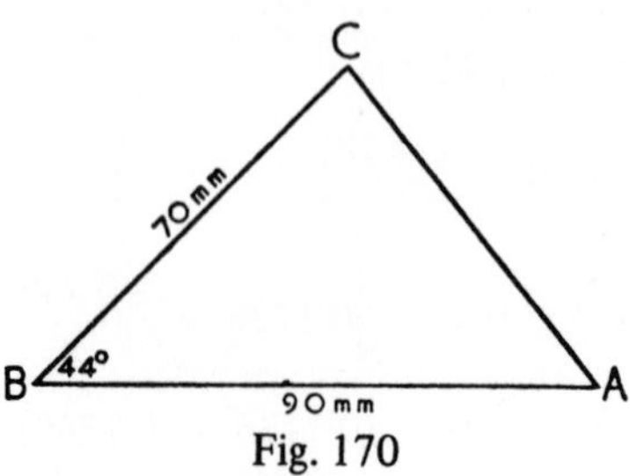

Fig. 170

By measurement (Fig. 170),

Remaining side, AC = 62·8 mm
,, angles A and C = 51° and 85° Ans.

3. There are two possible triangles which satisfy the conditions given as shown in Fig. 171.

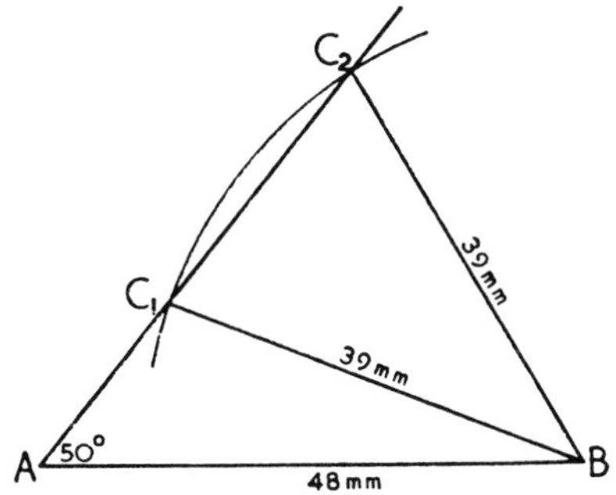

Fig. 171

One triangle:

Remaining side, AC_1 = 18 mm
,, angles, B and C_1 = $20\frac{1}{2}°$ and $109\frac{1}{2}°$ Ans.

Other triangle:

Remaining side, AC_2 = 44 mm
,, angles, B and C_2 = $59\frac{1}{2}°$ and $70\frac{1}{2}°$ Ans.

4.

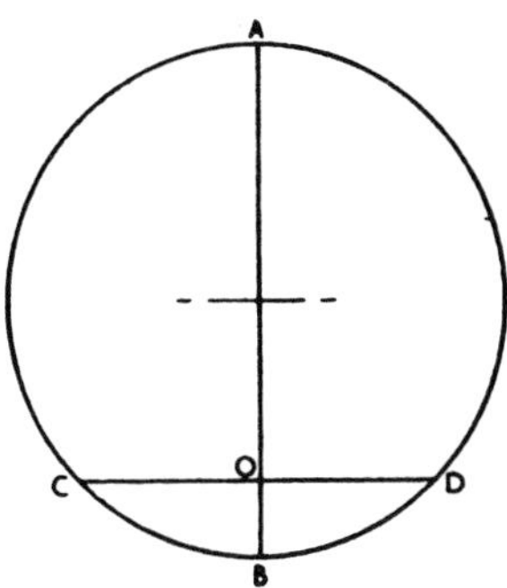

Fig. 172

Referring to Fig. 172,

$$AO \times BO = CO \times DO$$
$$AO \times 125 = 250 \times 250$$
$$AO = \frac{250 \times 250}{125} = 500 \text{ mm}$$

$$\text{Diameter of vessel} = AO + BO$$
$$= 500 + 125$$
$$= 625 \text{ mm Ans.}$$

5.

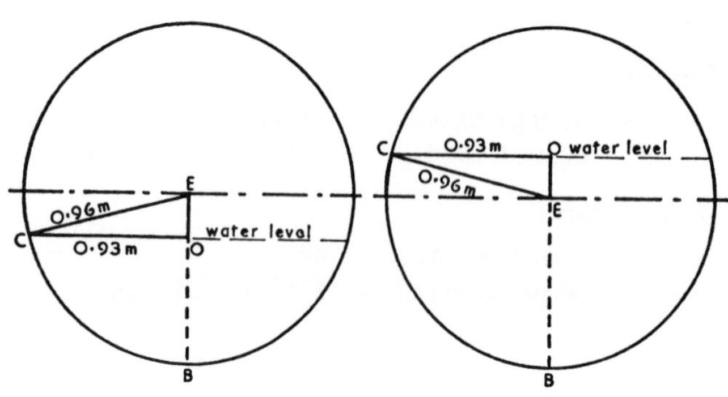

Fig. 173

Water-level width of 1·86 m can be below or above centre as shown in Fig. 173

$$\text{Radius of vessel} = 0{\cdot}96 \text{ m}$$
$$\text{Half width of water level} = 0{\cdot}93 \text{ m}$$
$$EO = \sqrt{0{\cdot}96^2 - 0{\cdot}93^2} = 0{\cdot}2381 \text{ m}$$

$$\text{Depth} = BO = EB - EO \text{ or } EB + EO$$
$$= 0{\cdot}96 - 0{\cdot}2381 \text{ or } 0{\cdot}96 + 0{\cdot}2381$$
$$= 0{\cdot}7219 \text{ or } 1{\cdot}1981 \text{ m}$$
$$= 721{\cdot}9 \text{ or } 1198{\cdot}1 \text{ mm Ans.}$$

6.

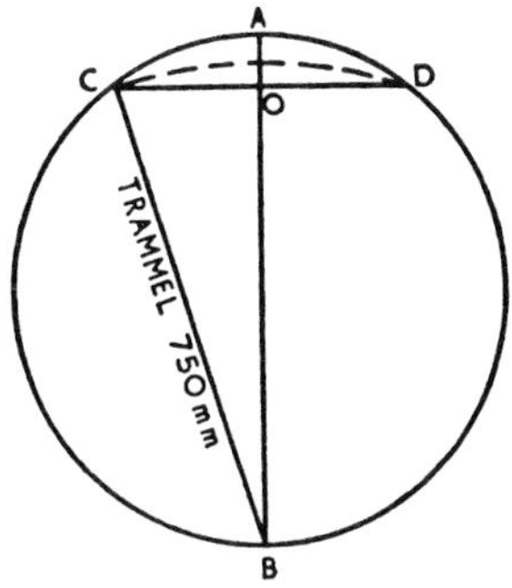

Fig. 174

Referring to Fig. 174,

$$\text{Chord CD} = 78 \text{ mm}$$
$$CO = DO = 39 \text{ mm}$$
$$(BO)^2 = (BC)^2 - (CO)^2$$
$$BO = \sqrt{750^2 - 39^2} = 749 \text{ mm}$$
$$AO \times BO = CO \times DO$$
$$AO \times 749 = 39 \times 39$$
$$AO = \frac{39 \times 39}{749} = 2{\cdot}031 \text{ mm}$$
$$\text{Diameter} = AO + BO$$
$$= 2{\cdot}031 + 749 = 751{\cdot}031 \text{ mm}$$
$$\text{Increase} = 751{\cdot}031 - 750$$
$$= 1{\cdot}031 \text{ mm Ans.}$$

7.

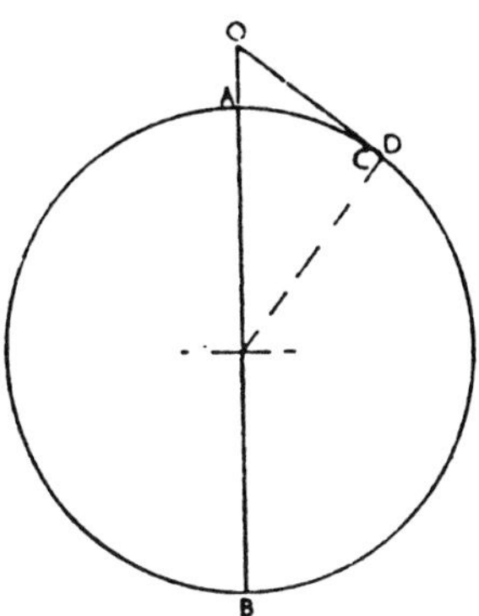

Fig. 175

$$AO = 30 \text{ metres} = 0{\cdot}03 \text{ kilometre}$$
$$BO = 12750 + 0{\cdot}03 = 12750{\cdot}03 \text{ km},$$

this can be taken as 12750 km

$$CO = DO = \text{observation distance.}$$

$$AO \times BO = CO \times DO$$
$$0{\cdot}03 \times 12750 = (CO)^2$$
$$CO = 19{\cdot}55 \text{ km Ans.}$$

8.

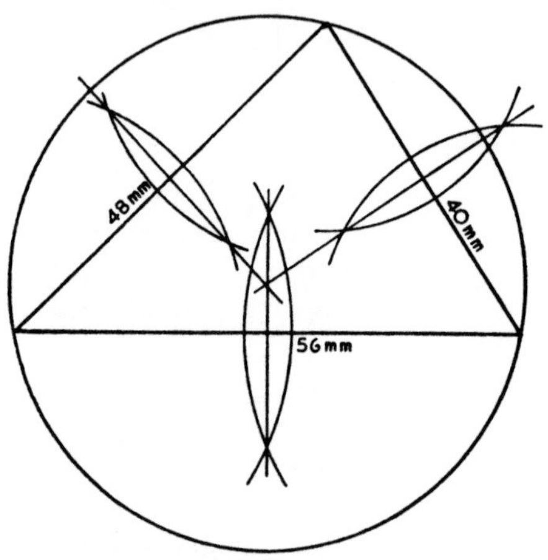

Fig. 176

Radius of circumscribing circle = 28·6 mm Ans.

9.

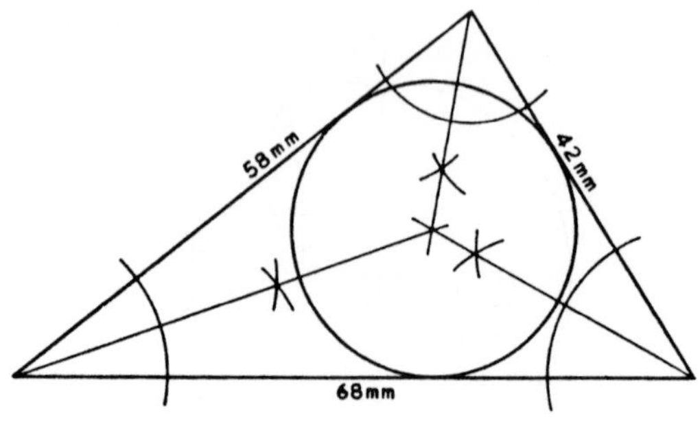

Fig. 177

Radius of inscribed circle = 14·4 mm Ans.

SOLUTIONS TO TEST EXAMPLES 10

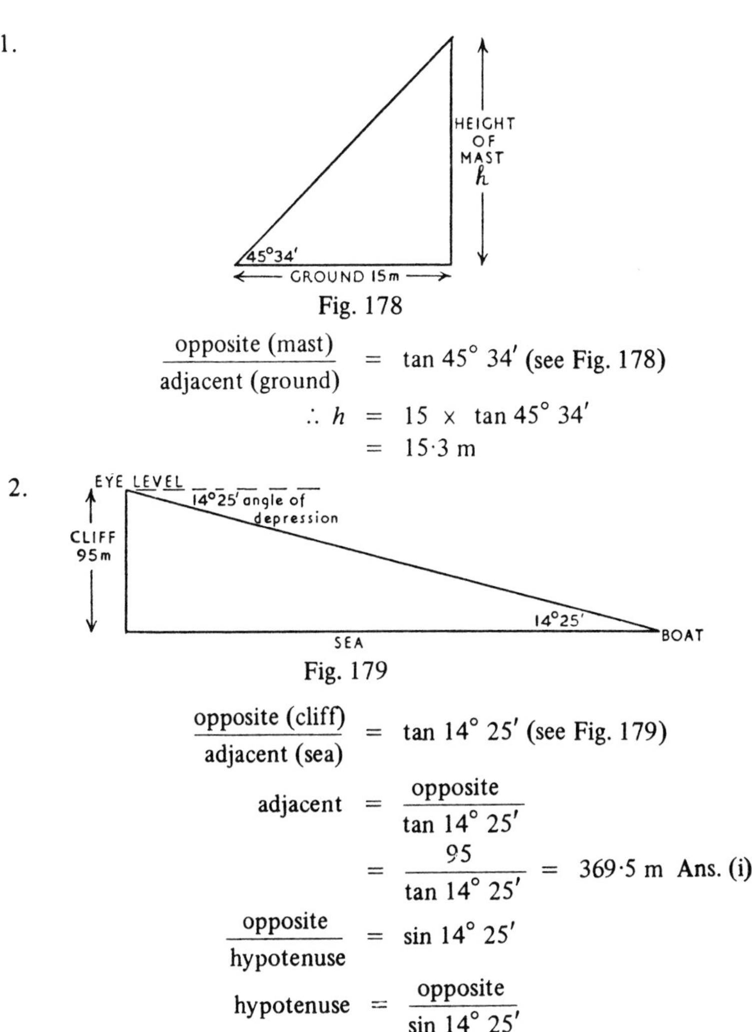

1.

Fig. 178

$$\frac{\text{opposite (mast)}}{\text{adjacent (ground)}} = \tan 45^\circ\ 34' \text{ (see Fig. 178)}$$

$$\therefore h = 15 \times \tan 45^\circ\ 34'$$

$$= 15{\cdot}3 \text{ m}$$

2.

Fig. 179

$$\frac{\text{opposite (cliff)}}{\text{adjacent (sea)}} = \tan 14^\circ\ 25' \text{ (see Fig. 179)}$$

$$\text{adjacent} = \frac{\text{opposite}}{\tan 14^\circ\ 25'}$$

$$= \frac{95}{\tan 14^\circ\ 25'} = 369{\cdot}5 \text{ m Ans. (i)}$$

$$\frac{\text{opposite}}{\text{hypotenuse}} = \sin 14^\circ\ 25'$$

$$\text{hypotenuse} = \frac{\text{opposite}}{\sin 14^\circ\ 25'}$$

$$= \frac{95}{\sin 14^\circ\ 25'} = 381{\cdot}5 \text{ m Ans. (ii)}$$

3.

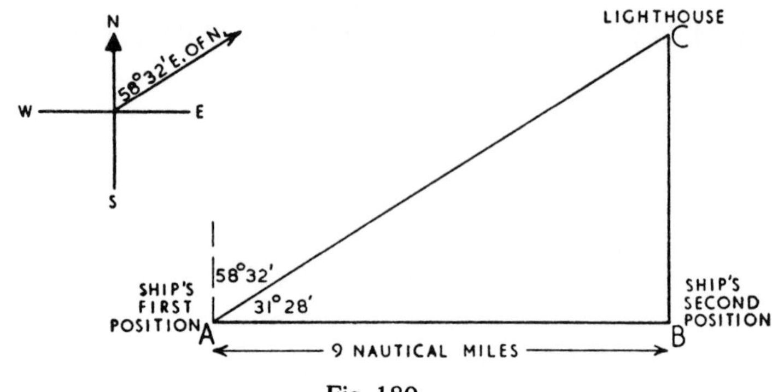

Fig. 180

Angle $A = 90° - 58° 32' = 31° 28'$ (See Fig. 180)

In half-an-hour, ship travels 9 naut. miles, = AB

$$\begin{aligned} BC &= AB \times \tan 31° 28' \\ &= 9 \times \tan 31° 28' \\ &= 5{\cdot}508 \text{ naut. miles. Ans.} \end{aligned}$$

4.

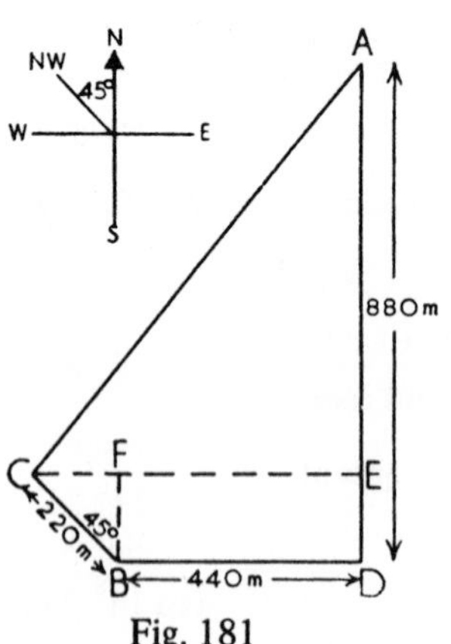

Fig. 181

Referring to Fig. 181,

$$\begin{aligned} CF &= 220 \times \sin 45° = 155{\cdot}6 \\ BF &= 220 \times \cos 45° = 155{\cdot}6 \\ AE &= 880 - 155{\cdot}6 = 724{\cdot}4 \\ CE &= 440 + 155{\cdot}6 = 595{\cdot}6 \end{aligned}$$

$$\tan A = \frac{\text{CE}}{\text{AE}} = \frac{595{\cdot}6}{724{\cdot}4} = 0{\cdot}8222$$

$$\therefore \text{ angle } A = 39^\circ\ 26'$$

$$\text{AC} = \frac{\text{CE}}{\sin A} = \frac{595{\cdot}6}{\sin 39^\circ\ 26'}$$

$$= 937{\cdot}6 \text{ metres.}$$

Point C lies 39° 26′ West of South from A, at a distance of 937·6 m **Ans.**

5.

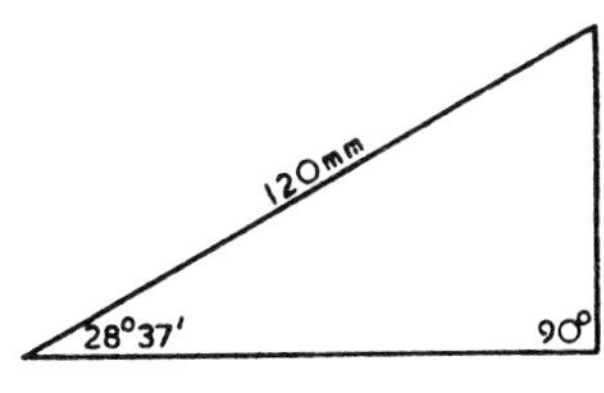

Fig. 182

$$\text{opposite} = 120 \times \sin 28^\circ\ 37' \text{ (see Fig. 182)}$$
$$= 57{\cdot}48 \text{ mm Ans. (i)}$$

$$\text{adjacent} = 120 \times \cos 28^\circ\ 37'$$
$$= 105{\cdot}3 \text{ mm Ans. (ii)}$$

$$\text{area} = \tfrac{1}{2}(\text{base} \times \text{perp. height})$$
$$= \frac{105{\cdot}3 \times 57{\cdot}48}{2}$$
$$= 3026 \text{ mm}^2 \text{ or } 30{\cdot}26 \text{ cm}^2 \text{ Ans. (iii)}$$

6.

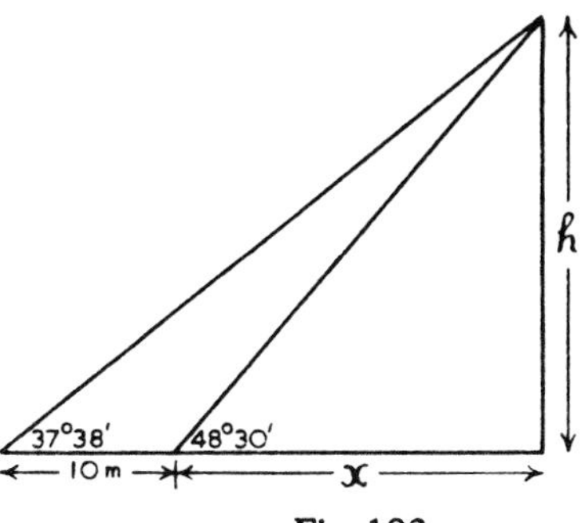

Fig. 183

$$h = x \times \tan 48° 30' \text{ (see Fig. 183)}$$
$$= x \times 1{\cdot}1303 \quad \dots \quad \dots \quad \dots \quad \text{(i)}$$

also,
$$h = (10 + x) \times \tan 37° 38'$$
$$= (10 + x) \times 0{\cdot}7710$$
$$= 7{\cdot}71 + 0{\cdot}771x \quad \dots \quad \dots \quad \dots \quad \text{(ii)}$$

From (i) and (ii),
$$1{\cdot}1303x = 7{\cdot}71 + 0{\cdot}771x$$
$$0{\cdot}3593x = 7{\cdot}71$$
$$x = 21{\cdot}46 \text{ m} \quad \text{Ans. (a)}$$

From (i),
$$h = x \times 1{\cdot}1303$$
$$= 21{\cdot}46 \times 1{\cdot}1303$$
$$= 24{\cdot}25 \text{ m} \quad \text{Ans. (b)}$$

7.

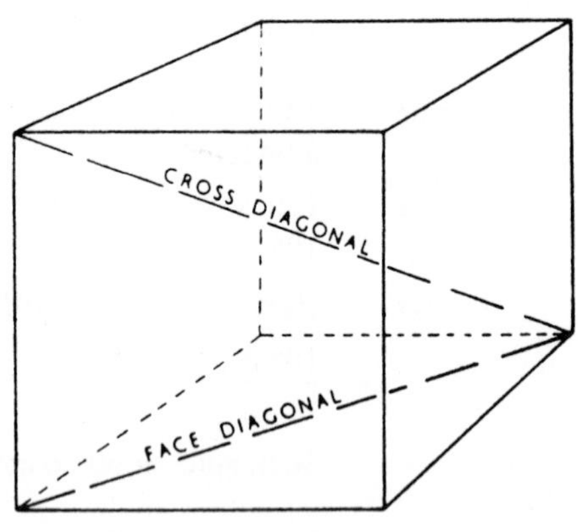

Fig. 184

By Pythagoras, referring to Fig. 184,

$$\text{diagonal across face} = \sqrt{60^2 + 60^2} = \sqrt{7200}$$
$$= 84{\cdot}85 \text{ mm} \quad \text{Ans. (i)}$$

Cross diagonal to opposite corners, through centre,

$$= \sqrt{60^2 + 84{\cdot}85^2}$$
$$= \sqrt{60^2 + 60^2 + 60^2} = \sqrt{10\,800}$$
$$= 103{\cdot}9 \text{ mm} \quad \text{Ans. (ii)}$$

8.

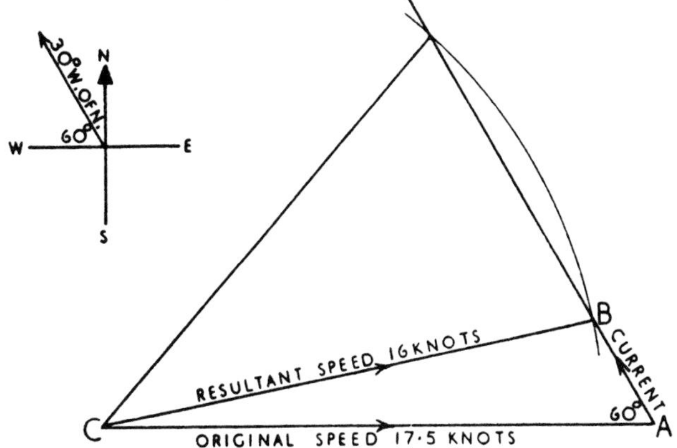

Fig. 185

By sine rule, referring to Fig. 185,

$$\frac{a}{\sin A} = \frac{b}{\sin B}$$

$$\sin B = \frac{b \times \sin A}{a}$$

$$= \frac{17{\cdot}5 \times \sin 60^\circ}{16} = 0{\cdot}9471$$

0·9471 is the sine of 71° 17′ or (180° − 71° 17′) which is 108° 43′

Referring to Fig. 185 it is seen that the probable value of B is 108° 43′, this gives the most likely speed of current and direction of ship.

$$C = 180^\circ - (108^\circ\ 43' + 60^\circ)$$
$$= 11^\circ\ 17'$$

$$\frac{a}{\sin A} = \frac{c}{\sin C}$$

$$c = \frac{a \times \sin C}{\sin A}$$

$$= \frac{16 \times \sin 11^\circ\ 17'}{\sin 60^\circ} = 3{\cdot}615$$

$\therefore$ Probable speed of current = 3·615 knots
Probable direction of ship = 11° 17′ North of East } Ans.

9.

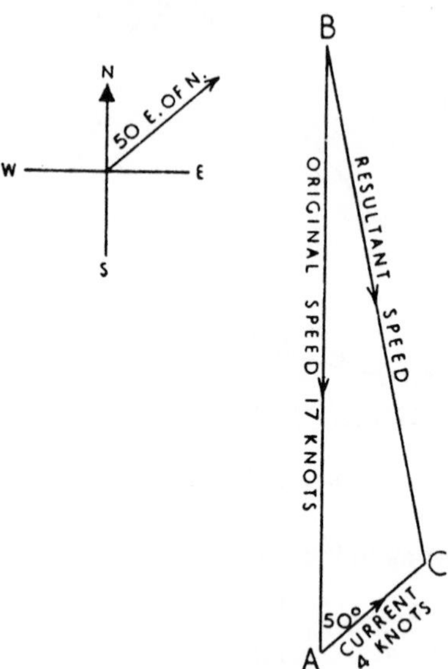

Fig. 186

By cosine rule, referring to Fig. 186,

$$a^2 = b^2 + c^2 - 2bc \cos A$$
$$= 4^2 + 17^2 - 2 \times 4 \times 17 \times \cos 50°$$
$$= 16 + 289 - 87{\cdot}42$$
$$a = \sqrt{217{\cdot}58} = 14{\cdot}75$$

By sine rule,

$$\frac{a}{\sin A} = \frac{b}{\sin B}$$

$$\sin B = \frac{b \times \sin A}{a}$$

$$= \frac{4 \times \sin 50°}{14{\cdot}75} = 0{\cdot}2078$$

$$\therefore B = 12°$$

$\therefore$ Resultant speed of ship = 14·75 knots
„ direction „ = 12° East of South } Ans.

10.

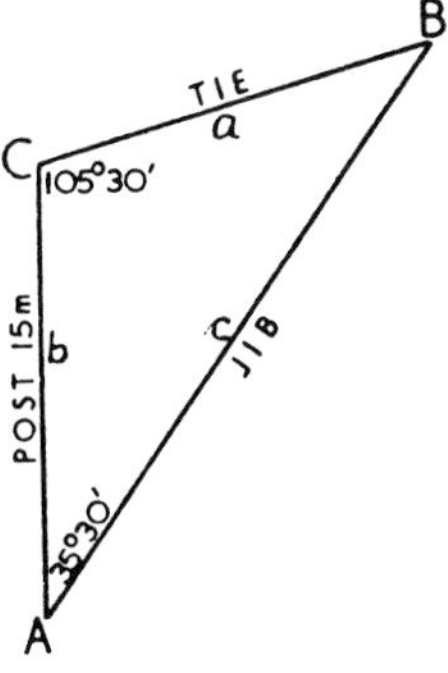

Fig. 187

$B = 180° - (105° 30' + 35° 30') = 39°$ (See Fig. 187).

By sine rule,

$$\frac{a}{\sin A} = \frac{b}{\sin B}$$

$$a = \frac{b \times \sin A}{\sin B}$$

$$= \frac{15 \times \sin 35° 30'}{\sin 39°} = 13{\cdot}84 \text{ m}$$

$$\frac{b}{\sin B} = \frac{c}{\sin C}$$

$$c = \frac{b \times \sin C}{\sin B}$$

$$= \frac{15 \times \sin 105° 30'}{\sin 39°} = 22{\cdot}97 \text{ m}$$

Length of jib = 22·97 m
,, ,, tie = 13·84 m } Ans.

11.

By sine rule, referring to Fig. 188,

$$\frac{a}{\sin A} = \frac{b}{\sin B}$$

$$\sin B = \frac{b \times \sin A}{a}$$

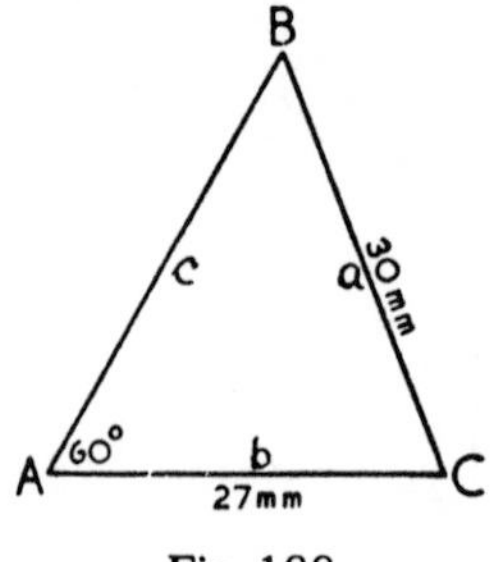

Fig. 188

$$= \frac{27 \times \sin 60^\circ}{30} = 0{\cdot}7794$$

$$\therefore \text{Angle } B = 51^\circ\, 12' \text{ Ans. (i)}$$

$$\text{,, } C = 180^\circ - (60 + 51^\circ\, 12')$$

$$= 68^\circ\, 48' \text{ Ans. (ii)}$$

$$\frac{a}{\sin A} = \frac{c}{\sin C}$$

$$c = \frac{a \times \sin C}{\sin A}$$

$$= \frac{30 \times \sin 68^\circ\, 48'}{\sin 60^\circ}$$

$$= 32{\cdot}29 \text{ mm Ans. (iii).}$$

12.

In Fig. 189, x is the distance the crosshead has moved from the top of its stroke when the crank is 35° past top centre.

Crank length = $\frac{1}{2}$ stroke = 400 mm

When crank is on top dead centre,
distance from shaft centre to crosshead

$$= 1600 + 400 = 2000 \text{ mm}$$

When crank is 35° past top dead centre, distance from shaft centre to crosshead is now to be calculated:

$$\frac{a}{\sin A} = \frac{c}{\sin C}$$

$$\sin A = \frac{a \times \sin C}{c}$$

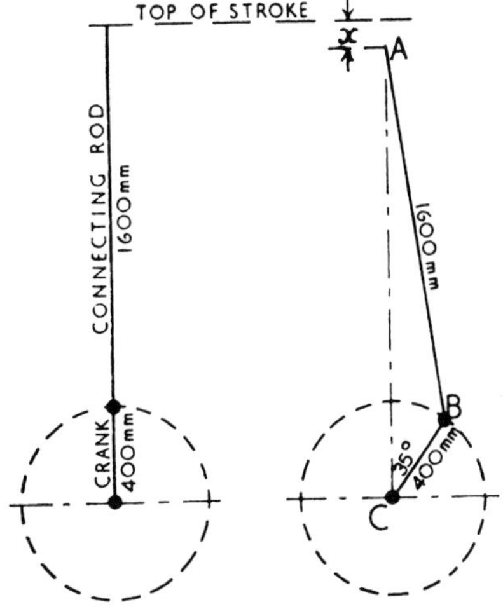

Fig. 189

$$= \frac{400 \times \sin 35°}{1600} = 0{\cdot}1434$$

$$\therefore A = 8° 15'$$

$$B = 180° - (35° + 8° 15') = 136° 45'$$

$$\frac{b}{\sin B} = \frac{c}{\sin C}$$

$$b = \frac{c \times \sin B}{\sin C}$$

$$= \frac{1600 \times \sin 136° 45'}{\sin 35°} = 1911 \text{ mm}$$

$$x = 2000 - 1911 = 89 \text{ mm Ans.}$$

13.

Let stroke = 1, crank length = 0·5, con rod length = 2.
(See Fig. 190).
When crank is on top dead centre,
distance from shaft centre to crosshead = 0·5 + 2 = 2·5
When crosshead has moved down 0·1 of its stroke,
distance from shaft centre to crosshead = 2·5 − 0·1 = 2·4,

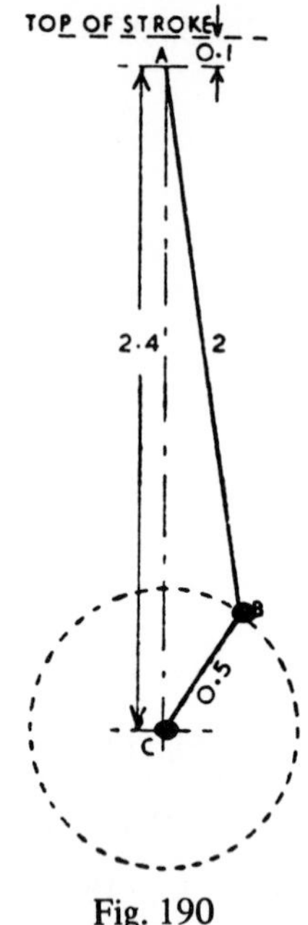

Fig. 190

By cosine rule,

$$\cos C = \frac{a^2 + b^2 - c^2}{2ab}$$

$$= \frac{0{\cdot}5^2 + 2{\cdot}4^2 - 2^2}{2 \times 0{\cdot}5 \times 2{\cdot}4} = 0{\cdot}8375$$

$$\therefore C = 33^\circ\, 7' \text{ Ans.}$$

14.

Crank length $= \frac{1}{2}$ stroke $= 125$ mm. Referring to Fig. 191,

$$\frac{a}{\sin A} = \frac{c}{\sin C}$$

$$\sin A = \frac{a \times \sin C}{c}$$

$$= \frac{125 \times \sin 116^\circ}{480} = 0{\cdot}2341$$

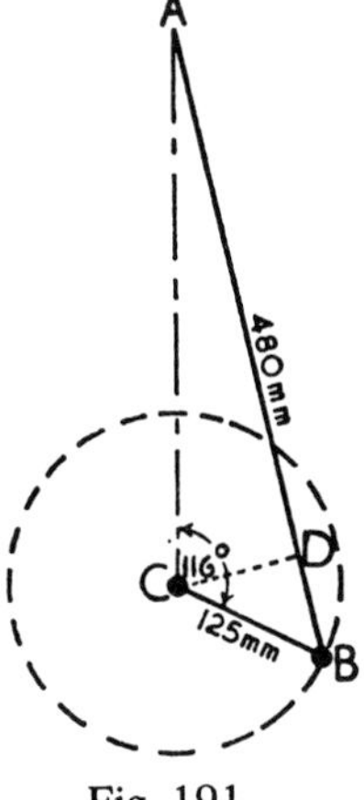

Fig. 191

$$A = 13° \; 32'$$
$$B = 180° - (116° + 13° \; 32') = 50° \; 28'$$

Perpendicular distance from con. rod to crank centre = CD

$$CD = 125 \times \sin 50° \; 28'$$
$$= 96{\cdot}4 \text{ mm Ans.}$$

15.

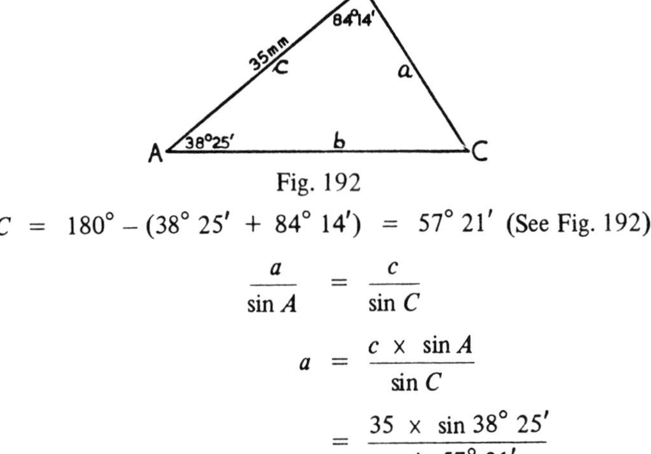

Fig. 192

$C = 180° - (38° \; 25' + 84° \; 14') = 57° \; 21'$ (See Fig. 192)

$$\frac{a}{\sin A} = \frac{c}{\sin C}$$

$$a = \frac{c \times \sin A}{\sin C}$$

$$= \frac{35 \times \sin 38° \; 25'}{\sin 57° \; 21'}$$

$$= 25{\cdot}83 \text{ mm} \quad \text{Ans. (i)}$$

$$\frac{b}{\sin B} = \frac{c}{\sin C}$$

$$b = \frac{c \times \sin B}{\sin C}$$

$$= \frac{35 \times \sin 84^\circ\, 14'}{\sin 57^\circ\, 21'}$$

$$= 41{\cdot}36 \text{ mm} \quad \text{Ans. (ii)}$$

16.

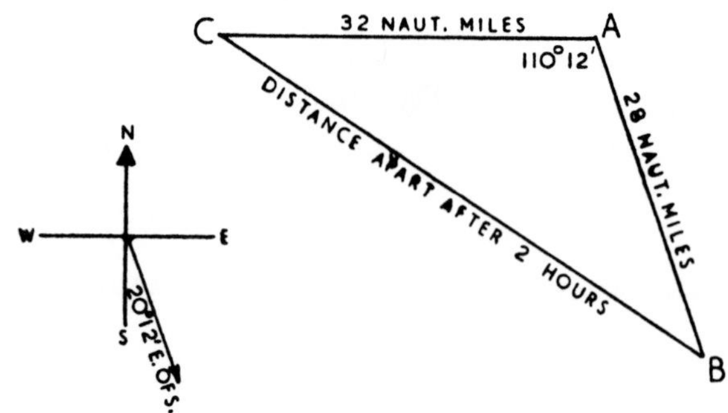

Fig. 193

Distance travelled by 14 knot ship $= 14 \times 2 = 28$ n. miles
,, ,, ,, 16 ,, ,, $= 16 \times 2 = 32$,,

By cosine rule, referring to Fig. 193,

$$\begin{aligned} a^2 &= b^2 + c^2 - 2bc \cos A \\ &= 32^2 + 28^2 - 2 \times 32 \times 28 \times \cos 110^\circ\, 12' \\ &= 1024 + 784 + 618{\cdot}9 \\ a &= \sqrt{2426{\cdot}9} \\ &= 49{\cdot}25 \text{ naut. miles.} \quad \text{Ans.} \end{aligned}$$

Note:

$$\begin{aligned} \cos 110^\circ\, 12' &= -\cos(180^\circ - 110^\circ\, 12') \\ &= -\cos 69^\circ\, 48' \end{aligned}$$

then, $-2 \times 32 \times 28 \times (-0{\cdot}3453)$

$$= +618{\cdot}9$$

Such questions as No. 16 can be worked out for a steaming time of one hour, then distance apart after x hours is x times distance apart after one hour, this avoids dealing with very large figures.

17.

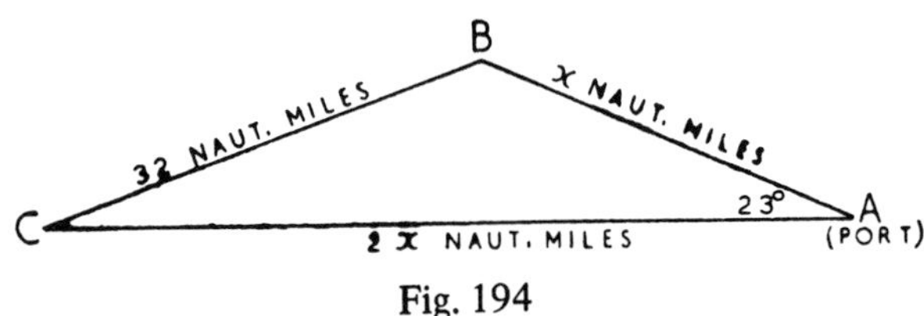

Fig. 194

$$\begin{aligned}
\text{Let } x &= \text{nearest ship's distance from port} \\
\text{then } 2x &= \text{other ship's distance from port. See Fig. 194} \\
a^2 &= b^2 + c^2 - 2bc\cos A \\
32^2 &= (2x)^2 + x^2 - 2 \times 2x \times x \times \cos 23^\circ \\
1024 &= 4x^2 + x^2 - 3{\cdot}682x^2 \\
1024 &= 1{\cdot}318x^2
\end{aligned}$$

$$x = \sqrt{\frac{1024}{1{\cdot}318}} = 27{\cdot}87$$

$\therefore$ distances from port $= 27{\cdot}87$ naut. miles
and $27{\cdot}87 \times 2 = 55{\cdot}74$ „ „ } Ans.

18.

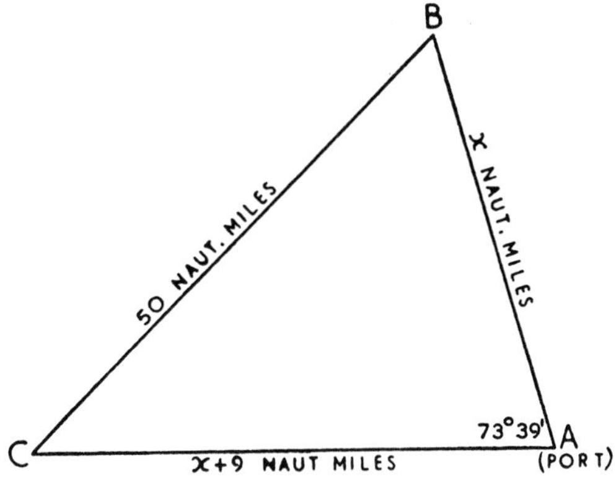

Fig. 195

Half-an-hour's steaming at 18 knots represents a distance of 9 nautical miles.

Let x naut. miles = distance of first ship from port
then $(x + 9)$,, ,, = ,, second ,, ,, ,,

By cosine rule, referring to Fig. 195

$$a^2 = b^2 + c^2 - 2bc \cos A$$
$$50^2 = (x + 9)^2 + x^2 - 2(x + 9) \times x \times \cos 73^\circ\ 39'$$
$$2500 = x^2 + 18x + 81 + x^2 - 2 \times x \times 0{\cdot}2815(x + 9)$$
$$2500 = 2x^2 + 18x + 81 - 0{\cdot}563x^2 - 5{\cdot}067x$$
$$2419 = 1{\cdot}437x^2 + 12{\cdot}933x$$

or, $1{\cdot}437x^2 + 12{\cdot}933x - 2419 = 0$

Solving this quadratic we get,

$$x = 36{\cdot}8$$

$\therefore$ Distance from port = 36·8 naut. miles
and $(36{\cdot}8 + 9)$ = 45·8 ,, ,, } Ans.

19.

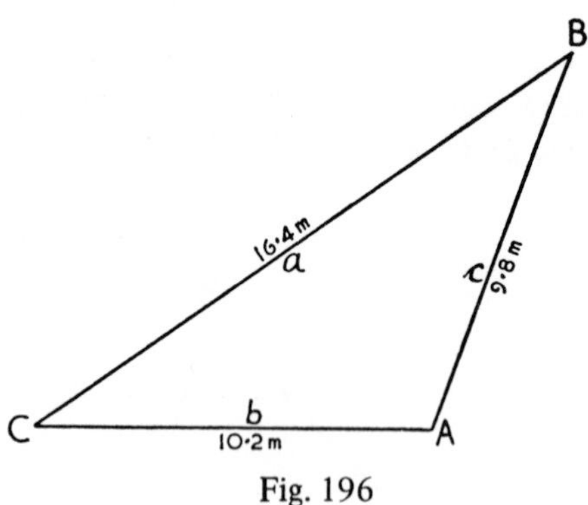

Fig. 196

By cosine rule, referring to Fig. 196,

$$\cos A = \frac{b^2 + c^2 - a^2}{2bc}$$

$$= \frac{10{\cdot}2^2 + 9{\cdot}8^2 - 16{\cdot}4^2}{2 \times 10{\cdot}2 \times 9{\cdot}8}$$

$$= \frac{104 + 96{\cdot}04 - 268{\cdot}96}{2 \times 10{\cdot}2 \times 9{\cdot}8}$$

$$= \frac{-68{\cdot}92}{2 \times 10{\cdot}2 \times 9{\cdot}8} = -0{\cdot}3447$$

$$\therefore \text{Angle } A = 180^\circ - 69^\circ\ 50' = 110^\circ\ 10'$$

By sine rule,

$$\frac{a}{\sin A} = \frac{b}{\sin B}$$

$$\sin B = \frac{b \times \sin A}{a}$$

$$= \frac{10{\cdot}2 \times \sin 110^\circ\ 10'}{16{\cdot}4} = 0{\cdot}5838$$

$$\therefore \text{Angle } B = 35^\circ\ 43'$$

$$\text{Angle } C = 180^\circ - (110^\circ\ 10' + 35^\circ\ 43')$$

$$= 34^\circ\ 7'$$

The three angles are,

$110^\circ\ 10'$, $35^\circ\ 43'$ and $34^\circ\ 7'$ Ans.

20. $$\text{Area} = \frac{ab \sin C}{2}$$

$$= \frac{6{\cdot}5 \times 7{\cdot}5 \times \sin 46^\circ\ 51'}{2}$$

$$= 17{\cdot}78 \text{ m}^2 \quad \text{Ans.}$$

21. $$\text{Area} = \sqrt{s(s-a)(s-b)(s-c)}$$

working in centimetres:

$$s = \frac{7{\cdot}1 + 4{\cdot}2 + 5{\cdot}3}{2} = 8{\cdot}3$$

$$s - a = 8{\cdot}3 - 7{\cdot}1 = 1{\cdot}2$$

$$s - b = 8{\cdot}3 - 4{\cdot}2 = 4{\cdot}1$$

$$s - c = 8{\cdot}3 - 5{\cdot}3 = 3{\cdot}0$$

$$\text{Area} = \sqrt{8{\cdot}3 \times 1{\cdot}2 \times 4{\cdot}1 \times 3}$$

$$= 11{\cdot}07 \text{ cm}^2 \quad \text{Ans.}$$

22. $$\text{Area} = 0{\cdot}433 \times \text{side}^2$$

$$\text{side} = \sqrt{\frac{\text{area}}{0{\cdot}433}} = \sqrt{\frac{57{\cdot}27}{0{\cdot}433}}$$

$$= 11{\cdot}5 \text{ cm} \quad \text{Ans.}$$

23.
$$\sin^2\theta + \cos^2 = 1$$
$$\therefore \cos^2\theta = 1 - \sin^2\theta$$

Substituting this value of $\cos^2\theta$ into the equation:

$$\cos 2\theta = \cos^2\theta - \sin^2\theta$$
$$= 1 - \sin^2\theta - \sin^2\theta$$
$$\therefore \cos 2\theta = 1 - 2\sin^2\theta \quad \text{Ans. (i)}$$

$$\sin^2\theta + \cos^2\theta = 1$$
$$\therefore \sin^2\theta = 1 - \cos^2\theta$$

Substituting this value of $\sin^2\theta$ into the equation:

$$\cos 2\theta = \cos^2\theta - \sin^2\theta$$
$$= \cos^2\theta - (1 - \cos^2\theta)$$
$$= \cos^2\theta - 1 + \cos^2\theta$$
$$\therefore \cos 2\theta = 2\cos^2\theta - 1 \quad \text{Ans. (ii)}$$

$$(1 + \cot^2\theta)(1 - \cos 2\theta)$$
$$= \left\{1 + \frac{\cos^2\theta}{\sin^2\theta}\right\} \times \{1 - (1 - 2\sin^2\theta)\}$$
$$= \left\{\frac{\sin^2\theta + \cos^2\theta}{\sin^2\theta}\right\} \times \{1 - 1 + 2\sin^2\theta\}$$
$$= \frac{1}{\sin^2\theta} \times 2\sin^2\theta = \frac{2\sin^2\theta}{\sin^2\theta}$$
$$= 2 \quad \text{Ans.}$$

24.
$$\sin 2\theta = \frac{2\tan\theta}{1 + \tan^2\theta}$$

Simplifying right hand side of equation:

$$\frac{2\tan\theta}{1 + \tan^2\theta} = \frac{2\tan\theta}{\sec^2\theta} = \frac{2\sin\theta}{\cos\theta} \times \frac{\cos^2\theta}{1}$$

$= 2\sin\theta\cos\theta = \sin 2\theta =$ left hand side of equation. Ans. (i)

$$\cos 2\theta = \frac{1 - \tan^2\theta}{1 + \tan^2\theta}$$

Simplifying right hand side of equation:

$$\frac{1 - \tan^2\theta}{1 + \tan^2\theta} = \frac{1 - \frac{\sin^2\theta}{\cos^2\theta}}{\sec^2\theta}$$

$$\frac{\cos^2\theta - \sin^2\theta}{\cos^2\theta} \times \frac{\cos^2\theta}{1} = \cos^2\theta - \sin^2\theta$$

$$= \cos 2\theta = \text{left hand side of equation. Ans. (ii)}$$

$$\sin 2\theta + 2\cos 2\theta = 1$$

$$\frac{2\tan\theta}{1+\tan^2\theta} + \frac{2(1-\tan^2\theta)}{1+\tan^2\theta} = 1$$

$$2\tan\theta + 2 - 2\tan^2\theta = 1 + \tan^2\theta$$
$$3\tan^2\theta - 2\tan\theta - 1 = 0$$

Positive solution of this quadratic for angle between 0 and 90°, $\tan\theta = 1$,

$$\therefore\ \theta = 45° \text{ Ans. (iii)}$$

25. Sin 3θ may be written as a compound angle,

$$\sin 3\theta = \sin(2\theta + \theta)$$

Using the compound angle formula,

$$\sin(A + B) = \sin A\cos B + \cos A\sin B$$

Let $A = 2\theta$, and $B = \theta$, then,

$$\sin(2\theta + \theta) = \sin 2\theta\cos\theta + \cos 2\theta\sin\theta$$

Substitute for $\sin 2\theta$ and $\cos 2\theta$ from the identities,

$$\sin 2\theta = 2\sin\theta\cos\theta$$
and
$$\cos 2\theta = 1 - 2\sin^2\theta$$

thus,

$$\begin{aligned}\sin(2\theta + \theta) &= \sin 2\theta\cos\theta + \cos 2\theta\sin\theta \\ &= (2\sin\theta\cos\theta) \times \cos\theta + (1 - 2\sin^2\theta) \times \sin\theta \\ &= 2\sin\theta\cos^2\theta + \sin\theta - 2\sin^3\theta\end{aligned}$$

Substitute for $\cos^2\theta$ from the identity,

$$\cos^2\theta = 1 - \sin^2\theta$$

thus,

$$\begin{aligned}\sin(2\theta + \theta) &= 2\sin\theta\cos^2\theta + \sin\theta - 2\sin^3\theta \\ &= 2\sin\theta(1 - \sin^2\theta) + \sin\theta - 2\sin^3\theta \\ &= 2\sin\theta - 2\sin^3\theta + \sin\theta - 2\sin^3\theta \\ \sin 3\theta &= 3\sin\theta - 4\sin^3\theta \text{ Ans.}\end{aligned}$$

SOLUTIONS TO TEST EXAMPLES 11

1.

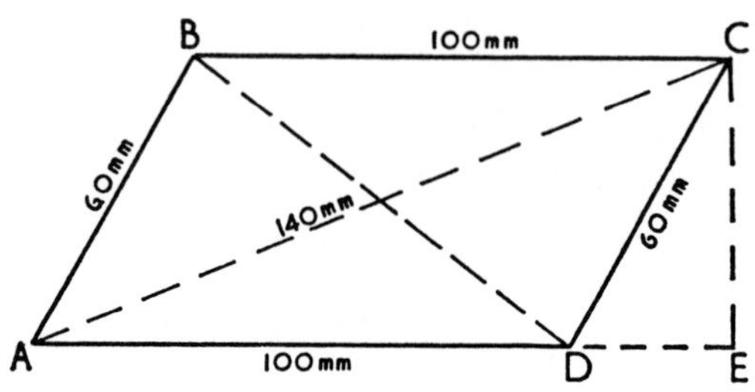

Fig. 197

By cosine rule, referring to Fig. 197, working in centimetres:

$$\cos ADC = \frac{(AD)^2 + (CD)^2 - (AC)^2}{2 \times (AD) \times (CD)}$$

$$= \frac{10^2 + 6^2 - 14^2}{2 \times 10 \times 6}$$

$$= -0{\cdot}5$$

$$\therefore \text{Angle } ADC = 180° - 60° = 120°$$

Obtuse angles are each 120°
Acute angles = 180° − 120° = 60° } Ans. (i)

$$(BD)^2 = (AD)^2 + (AB)^2 - 2 \times (AD) \times (AB) \times \cos BAD$$
$$= 10^2 + 6^2 - 2 \times 10 \times 6 \times \cos 60°$$
$$BD = \sqrt{76}$$

short diagonal = 8·718 cm = 87·18 mm Ans. (ii)

Angle CDE = 60°

Perpendicular height CE = 6 × sin 60°
= 5·196 cm = 51·96 mm Ans. (iii)

Area = base × perpendicular height
= 10 × 5·196
= 51·96 cm² or 5196 mm² Ans. (iv)

2.

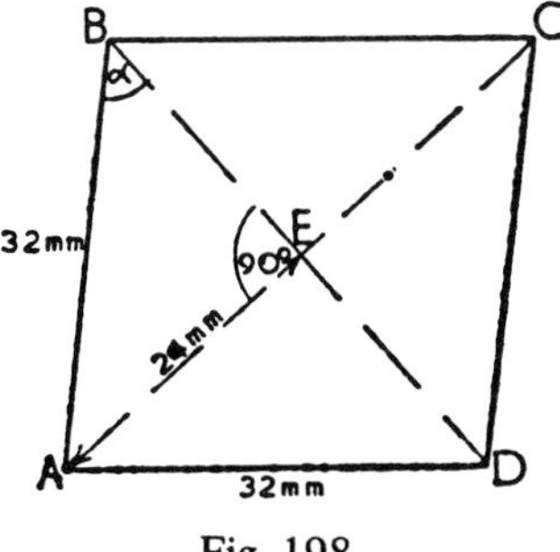

Fig. 198

Referring to Fig. 198

$$\sin \alpha = \frac{24}{32} = 0{\cdot}75$$

$$\therefore \alpha = 48^\circ\ 35'$$

$\therefore$ each obtuse angle $= 2 \times 48^\circ\ 35' = 97^\circ\ 10'$
,, acute ,, $= 180^\circ - 97^\circ\ 10' = 82^\circ\ 50'$ } Ans. (i)

$\frac{1}{2}$ length of short diagonal $= BE = 32 \times \cos 48^\circ\ 35'$
$= 21{\cdot}16$

$\therefore$ short diagonal $= 21{\cdot}16 \times 2 = 42{\cdot}32$ mm Ans. (ii)

Area $= \frac{1}{2}$ product of diagonals

$$= \frac{48 \times 42{\cdot}32}{2}$$

$= 1016\ \text{mm}^2$ or $10{\cdot}16\ \text{cm}^2$ Ans. (iii)

3.

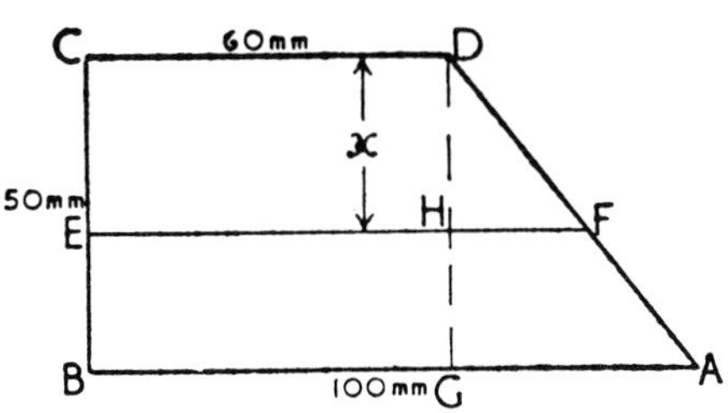

Fig. 199

Referring to Fig 199, working in centimetres:

$$\text{Mean length} = \tfrac{1}{2}(10 + 6) = 8 \text{ cm}$$

$$\begin{aligned} \text{Area} &= \text{mean length} \times \text{perp. height} \\ &= 8 \times 5 \\ &= 40 \text{ cm}^2 \quad \text{Ans. (i)} \end{aligned}$$

Each half area is to be 20 cm^2,
let dividing line EF be at x from CD

$$AG = 10 - 6 = 4 \text{ cm}$$

By similar triangles,

$$\frac{GA}{DG} = \frac{HF}{DH}$$

$$\frac{4}{5} = \frac{HF}{x}$$

$$HF = \frac{4x}{5} = 0{\cdot}8x$$

$$EF = 6 + 0{\cdot}8x$$

Mean length of half area FECD

$$\begin{aligned} &= \tfrac{1}{2}(CD + EF) \\ &= \tfrac{1}{2}(6 + 6 + 0{\cdot}8x) \\ &= 6 + 0{\cdot}4x \end{aligned}$$

$$\begin{aligned} \text{Area} &= \text{mean length} \times \text{perp. height} \\ 20 &= (6 + 0{\cdot}4x) \times x \\ 20 &= 6x + 0{\cdot}4x^2 \\ 0{\cdot}4x^2 + 6x - 20 &= 0 \\ x^2 + 15x - 50 &= 0 \\ \text{Solving this quadratic, } x &= 2{\cdot}81 \end{aligned}$$

$\therefore$ Dividing line should be at 2·81 cm = 28·1 mm from CD Ans. (ii)

4.

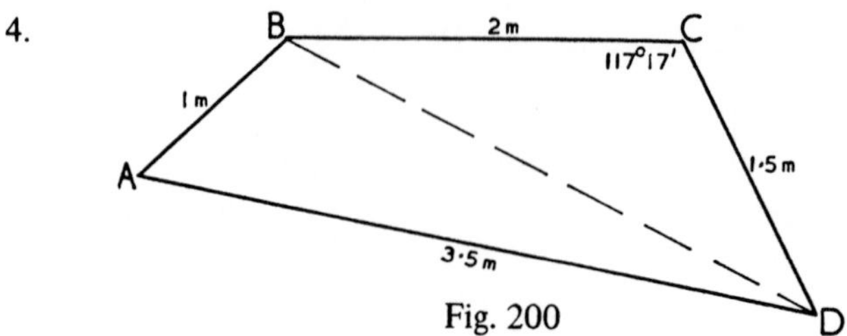

Fig. 200

By cosine rule, referring to Fig. 200

$$(BD)^2 = (BC)^2 + (CD)^2 - 2 \times (BC) \times (CD) \times \cos BCD$$
$$= 2^2 + 1{\cdot}5^2 - 2 \times 2 \times 1{\cdot}5 \times \cos 117^\circ\ 17'$$
$$= 4 + 2{\cdot}25 + 2{\cdot}75$$
$$BD = \sqrt{9} = 3 \text{ m}$$

Semisum of sides of triangle BCD

$$= \tfrac{1}{2}(2 + 1{\cdot}5 + 3) = 3{\cdot}25 \text{ m}$$
$$\text{Area of BCD} = \sqrt{s(s-a)(s-b)(s-c)}$$
$$= \sqrt{3{\cdot}25 \times 1{\cdot}25 \times 1{\cdot}75 \times 0{\cdot}25}$$
$$= 1{\cdot}333 \text{ m}^2$$

Alternatively the area could be found by $\frac{1}{2}(ab \sin C)$.

Semisum of sides of triangle ABD

$$= \tfrac{1}{2}(3 + 3{\cdot}5 + 1) = 3{\cdot}75 \text{ m}$$
$$\text{Area ABD} = \sqrt{3{\cdot}75 \times 0{\cdot}75 \times 0{\cdot}25 \times 2{\cdot}75}$$
$$= 1{\cdot}391 \text{ m}^2$$

$$\text{Area of quadrilateral} = \text{sum of areas of the two triangles}$$
$$= 1{\cdot}333 + 1{\cdot}391$$
$$= 2{\cdot}724 \text{ m}^2 \quad \text{Ans.}$$

5.

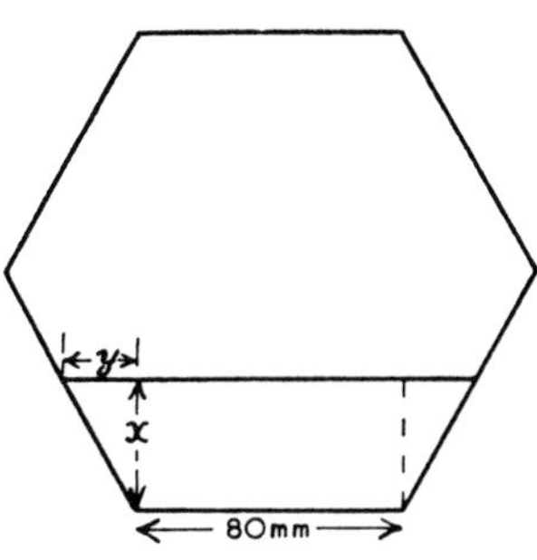

Fig. 201

Let x mm = thickness of piece cut off (see Fig. 201). The piece cut off is a trapezium, consisting of a rectangle 80 × x and two equal triangles; the height of each triangle being x, then the side perpendicular to this, marked y in the sketch, is $x \tan 30^\circ = 0{\cdot}5774x$, hence,

Area of piece cut off,

$$= (80 \times x) + (2 \times \tfrac{1}{2} \times 0{\cdot}5774x \times x)$$
$$= 80x + 0{\cdot}5774x^2 \quad \ldots \quad \ldots \quad \ldots \quad \text{(i)}$$

$$\text{Area of hexagon} = 6 \times 0{\cdot}433 \text{ side}^2 = 2{\cdot}598 \text{ side}^2$$

$$\text{Area of piece cut off} = 0{\cdot}1 \times 2{\cdot}598 \times 80^2$$
$$= 1663 \text{ mm}^2 \quad \ldots \quad \ldots \quad \text{(ii)}$$

$$\therefore 80x + 0{\cdot}5774x^2 = 1663$$
$$0{\cdot}5774x^2 + 80x - 1663 = 0$$
$$\text{or, } x^2 + 138{\cdot}6x - 2880 = 0$$

Solving this quadratic, $x = 18{\cdot}35$ mm Ans.

6. Considering one of the eight constituent triangles,

$$\text{Apex angle} = 360 \div 8 = 45°$$
$$\text{Each base angle} = \tfrac{1}{2}(180 - 45) = 67{\cdot}5°$$
$$\text{Perpendicular height} = 15 \times \tan 67° \, 30'$$
$$= 36{\cdot}21 \text{ mm}$$
$$\text{Area of each triangle} = \tfrac{1}{2}(\text{base} \times \text{perp. height})$$
$$= \tfrac{1}{2} \times 30 \times 36{\cdot}21$$
$$\text{Area of octagon} = 8 \times \tfrac{1}{2} \times 30 \times 36{\cdot}21$$
$$= 4346 \text{ mm}^2$$
$$\text{Area of hole} = 0{\cdot}7854 \times 50^2$$
$$= 1963 \text{ mm}^2$$
$$\text{Area of plate} = 4346 - 1963$$
$$= 2383 \text{ mm}^2 \text{ Ans.}$$

7.

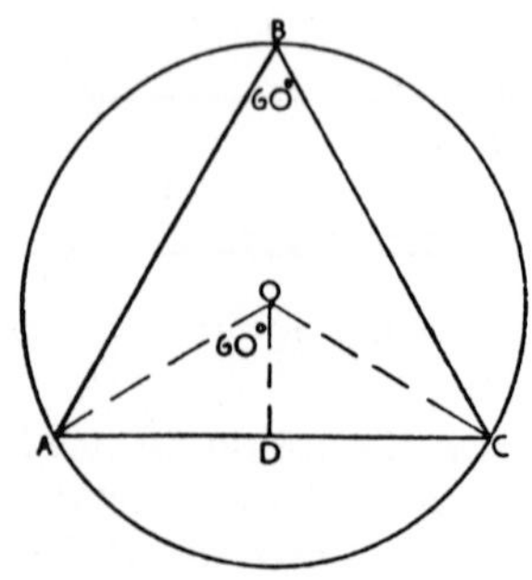

Fig. 202

7. Radius = AO = 60 mm (see Fig. 202)

Being an equilateral triangle, all angles are 60°

Angle at centre AOC is twice the angle at the apex.

$$\therefore \text{AOC} = 120°, \text{ and AOD} = 60°$$
$$\text{AD} = \text{AO} \times \sin 60$$
$$= 60 \times 0{\cdot}866 = 51{\cdot}96 \text{ mm}$$
$$\text{length of sides} = \text{AC} = 2 \times \text{AD}$$
$$= 103{\cdot}92 \text{ mm Ans. (i)}$$
$$\text{OD} = \text{AO} \times \cos 60$$
$$= 60 \times 0{\cdot}5 = 30 \text{ mm}$$
$$\text{BD} = \text{BO} + \text{OD}$$
$$= 60 + 30 = 90 \text{ mm}$$
$$\text{Area} = \tfrac{1}{2}(\text{base} \times \text{perp. height})$$
$$= \tfrac{1}{2} \times 103{\cdot}92 \times 90$$
$$= 4676 \text{ mm}^2 \text{ Ans. (ii)}$$
$$(\text{or formula, Area} = 0{\cdot}433 \times \text{side}^2 \text{ could be used})$$

8.
$$\text{Area of collar} = 0{\cdot}7854\,(D^2 - d^2)$$
$$= 0{\cdot}7854(D + d)(D - d)$$
$$= 0{\cdot}7854(755 + 415)(755 - 415)$$
$$= 0{\cdot}7854 \times 1170 \times 340$$
$$\text{Effective area} = 0{\cdot}7 \times 0{\cdot}7854 \times 1170 \times 340 \text{ mm}^2$$
$$1 \text{ m} = 10^3 \text{ mm}$$
$$1 \text{ m}^2 = (10^3)^2 \text{ mm}^2 = 10^6 \text{ mm}^2$$
$$\therefore \text{Area} = 0{\cdot}7 \times 0{\cdot}7854 \times 1170 \times 340 \times 10^{-6} \text{ m}^2$$
$$= 0{\cdot}2187 \text{ m}^2 \text{ Ans. (i)}$$

At a pressure of 2000 kN/m²:

$$\text{Total force} = 0{\cdot}2187 \times 2000$$
$$= 437{\cdot}4 \text{ kN Ans. (ii)}$$

9. Working in centimetres:

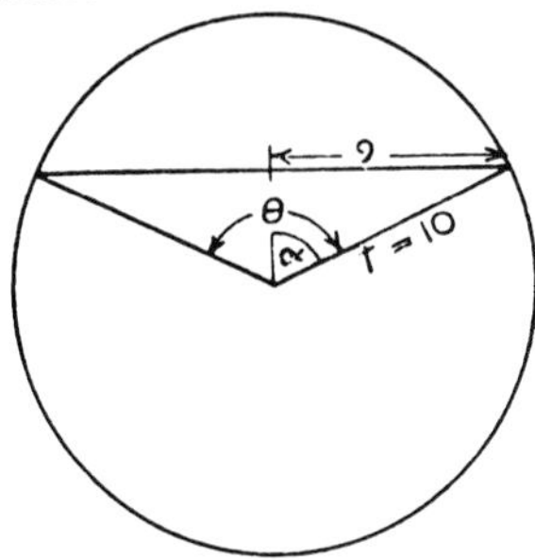

Fig. 203

Referring to Fig. 203,

$$\sin \alpha = \frac{9}{10} = 0{\cdot}9,$$

$$\therefore \alpha = 64^\circ\ 9'$$

$$\theta = 2\alpha = 128^\circ\ 18'$$

$$\sin 128^\circ\ 18' = \sin(180^\circ - 128^\circ\ 18') = \sin 51^\circ\ 42' = 0{\cdot}7848$$

$$\frac{128{\cdot}3}{57{\cdot}3} = 2{\cdot}239 \text{ radians}$$

$$\text{Area of segment} = \frac{r^2}{2}[\theta - \sin\theta]$$

$$= \frac{10^2}{2}(2{\cdot}239 - 0{\cdot}7848)$$

$$= 50 \times 1{\cdot}4542$$

$$= 72{\cdot}71 \text{ cm}^2 \quad \text{Ans.}$$

10. Diameter = 762 mm = 0·762 m

Length = 1270 mm = 1·27 m

$$\text{Surface area} = \text{circumference} \times \text{height}$$

$$= \pi d \times h$$

$$= \pi \times 0{\cdot}762 \times 1{\cdot}27$$

$$= 3{\cdot}041 \text{ m}^2 \quad \text{Ans.}$$

11. Curved surface area of sphere = $\pi d \times d$

$$\text{,, ,, hemisphere} = \frac{\pi d^2}{2}$$

$$\text{Surface area of circular base} = \frac{\pi d^2}{4}$$

$$\text{Total surface area} = \frac{\pi d^2}{2} + \frac{\pi d^2}{4}$$

$$\therefore 58{\cdot}9 = \pi d^2(\tfrac{1}{2} + \tfrac{1}{4})$$

$$58{\cdot}9 = \pi d^2 \times \tfrac{3}{4}$$

$$d = \sqrt{\frac{58{\cdot}9 \times 4}{\pi \times 3}} = 4{\cdot}999$$

say 5 cm or 50 mm diameter. Ans.

12. By crossed chords, reference to Fig. 204,

$$AO \times BO = CO \times DO$$

Let d mm = depth of indentation = CO

DO = ball diameter − CO = $10 - d$

AO = BO = $\frac{1}{2}$ surface diameter = 2·5 mm

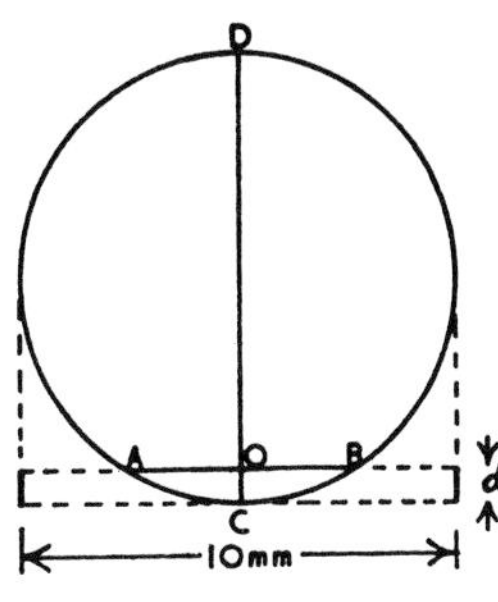

Fig. 204

$$AO \times BO = CO \times DO$$

$$2{\cdot}5 \times 2{\cdot}5 = d \times (10 - d)$$

$$6{\cdot}25 = 10d - d^2$$

$$d^2 - 10d + 6{\cdot}25 = 0$$

Solving this quadratic, d = 9·33 or 0·67

∴ depth = 0·67 mm Ans. (i)

Curved surface area of indentation is equal to the curved surface area of a slice of same depth off a circumscribing cylinder of same diameter as the ball,

$= \pi \times$ ball diameter $\times$ depth of indent

$= \pi \times 10 \times 0{\cdot}67$

$= 21{\cdot}05$ mm^2 Ans. (ii)

13.

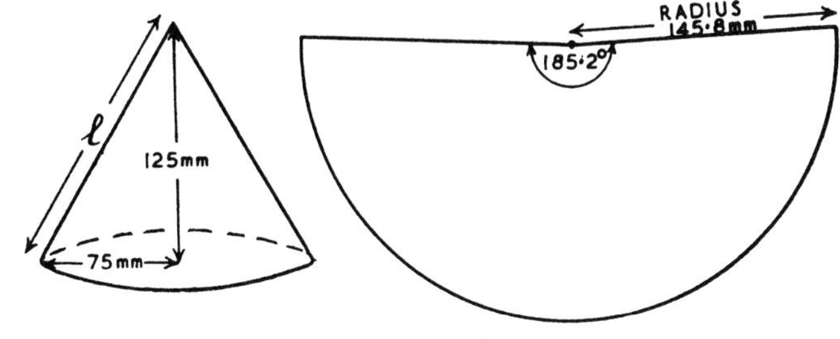

Fig. 205

Referring to Fig. 205,

$$\text{Slant height} = \sqrt{125^2 + 75^2} = 145{\cdot}8 \text{ mm}$$

this is the radius of the sector.

$$\text{Arc of sector} = \text{circumference of base of cone}$$
$$= \pi \times 150 = 471{\cdot}3 \text{ mm}$$

$$\text{Angle of sector} = \frac{\text{arc}}{\text{radius}} \text{ (in radians)}$$
$$= \frac{471{\cdot}3}{145{\cdot}8} = 3{\cdot}232 \text{ radians}$$

$$3{\cdot}232 \times \frac{360}{2\pi} = 185{\cdot}2 \text{ degrees}$$

∴ dimensions of sector are:

$$\left.\begin{aligned}\text{Radius} &= 145{\cdot}8 \text{ mm}\\ \text{Angle at centre} &= 185{\cdot}2^\circ\end{aligned}\right\} \text{Ans.}$$

14.

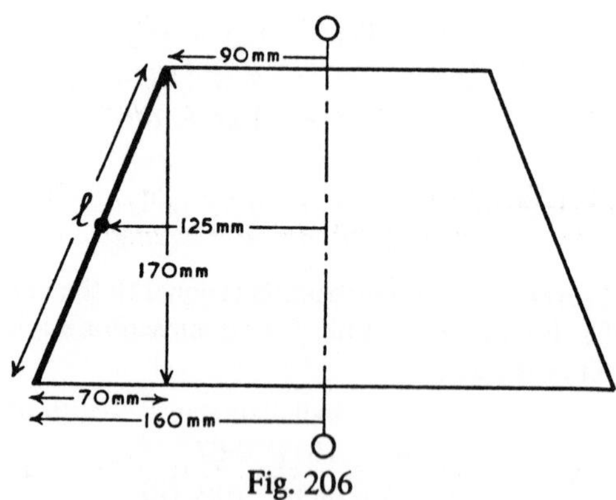

Fig. 206

Referring to Fig. 206,

$$l = \sqrt{170^2 + 70^2} = 183{\cdot}8 \text{ mm}$$

By theorem of Pappus, considering l as the line rotated one revolution about the fixed axis $o\ o$

$$\text{c.g. of line from } o\ o = \tfrac{1}{2}(90 + 160) = 125 \text{ mm}$$
$$\text{Area swept out} = \text{length of line} \times \text{distance c.g. moves}$$
$$\therefore \text{Curved surface area} = 183{\cdot}8 \times 2\pi \times 125$$
$$= 1{\cdot}444 \times 10^5 \text{ mm}^2$$

or 1444 cm² Ans.

15.

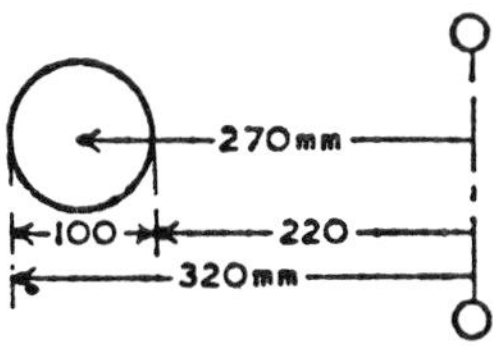

Fig. 207

Referring to Fig. 207,

Diameter of bar = difference in outer and inner radii
= 320 − 220 = 100 mm

Circumference of bar = $\pi \times 100$ mm
Position of c.g. from centre axis *o o* = 220 + 50 = 270 mm
By theorem of Pappus,

Area swept out = length of line × distance c.g. moves

∴ Curved surface area = $\pi \times 100 \times 2\pi \times 270$
= $5{\cdot}331 \times 10^5$ mm^2
or 5331 cm^2 **Ans.**

16. **Areas of similar figures vary as the square of their corresponding dimensions.**

Ratio of corresponding dimensions = 125 : 175
Dividing each by 125,

Ratio of dimensions = 1 : 1·4
Ratio of areas = 1^2 : $1{\cdot}4^2$
= 1 : 1·96
= 100 : 196

∴ Area of larger triangle is 96% greater than smaller Ans.

17.

SEMI-ORDINATES	SIMPSON'S MULTIPLIERS	PRODUCTS
0·1	1	0·1
3·0	4	12·0
5·85	2	11·7
7·2	4	28·8
8·1	2	16·2
8·4	4	33·6
8·4	2	16·8
8·25	4	33·0
8·1	2	16·2
7·5	4	30·0
6·3	2	12·6
3·75	4	15·0
0·5	1	0·5
	sum =	226·5

Number of ordinates = 13

∴ „ „ spaces = 12

Common interval = length ÷ no. of spaces

= 150 ÷ 12

= 12·5 mm

Area = sum of products × $\frac{1}{3}$ common interval

$$= \frac{226{\cdot}5 \times 12{\cdot}5}{3} = 943{\cdot}75 \text{ m}^2$$

As *semi*-ordinates have been used then this is the *half*-area of the water-plane.

Total area = 2 × 943·75

= 1887·5 m² Ans.

18. Plotting points for the two curves:

x	$y = x^2 + 3x + 6$	$y = 2x^2 - x + 1$
0	6	1
1	10	2
2	16	7
3	24	16
4	34	29

The plotted curves are shown in Fig. 208.

Finding area between curves by Simpson's rule:

ORDINATES BETWEEN CURVES	SIMPSON'S MULTIPLIERS	PRODUCTS
6 − 1 = 5	1	5
10 − 2 = 8	4	32
16 − 7 = 9	2	18
24 − 16 = 8	4	32
34 − 29 = 5	1	5
	sum =	92

Common interval between ordinates = 1

$$\text{Area} = \text{sum of products} \times \tfrac{1}{3} \text{ common interval}$$
$$= 92 \times \tfrac{1}{3} \times 1$$
$$= 30\tfrac{2}{3} \text{ units}^2 \quad \text{Ans.}$$

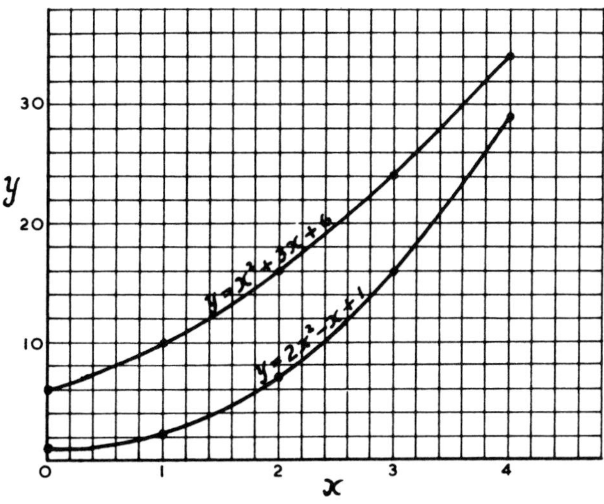

Fig. 208

19. Sum of mid-ordinates = 82·5 mm

Mean height = 82·5 ÷ 10 = 8·25 mm Ans. (i)

Mean effective pressure = 8·25 × 80

= 660 kN/m^2 Ans. (ii)

SOLUTIONS TO TEST EXAMPLES 12

1.

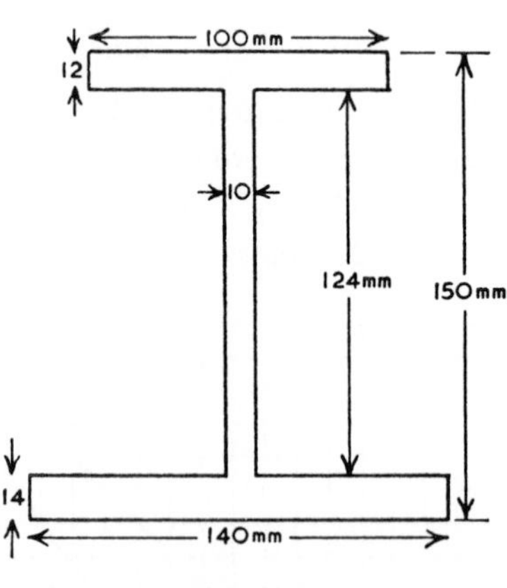

Fig. 209

Working in centimetres:

$$
\begin{aligned}
\text{Area of top flange} &= 10 \times 1{\cdot}2 = 12\ \text{cm}^2 \\
\text{Area of bottom flange} &= 14 \times 1{\cdot}4 = 19{\cdot}6\ \text{cm}^2 \\
\text{Area of centre web} &= 12{\cdot}4 \times 1 = 12{\cdot}4\ \text{cm}^2 \\
\text{Total area} &= 44{\cdot}0\ \text{cm}^2
\end{aligned}
$$

Volume [cm^3] of 1 metre length

$$
\begin{aligned}
&= \text{area [cm}^2] \times \text{length [cm]} \\
&= 44 \times 100 = 4400\ \text{cm}^3 \\
\text{Mass (g)} &= \text{volume [cm}^3] \times \text{density [g/cm}^3] \\
&= 4400 \times 7{\cdot}86 \\
&= 3{\cdot}458 \times 10^4\text{g} \\
&= 34{\cdot}58\ \text{kg per metre run. Ans.}
\end{aligned}
$$

2. Working in metres:

Volume of shaft body = area of end × length

$= 0{\cdot}7854\,(0{\cdot}4^2 - 0{\cdot}2^2) \times 6$

$= 0{\cdot}5654 \text{ m}^3$

Volume of two couplings

$= 0{\cdot}7854\,(0{\cdot}76^2 - 0{\cdot}4^2) \times 0{\cdot}075 \times 2$

$= 0{\cdot}0492 \text{ m}^3$

Total volume $= 0{\cdot}5654 + 0{\cdot}0492 = 0{\cdot}6146 \text{ m}^3$

Mass (tonne) = volume (m^3) × density (tonne/m^3)

$= 0{\cdot}6146 \times 7{\cdot}86$

$= 4{\cdot}831$ tonne Ans.

3.

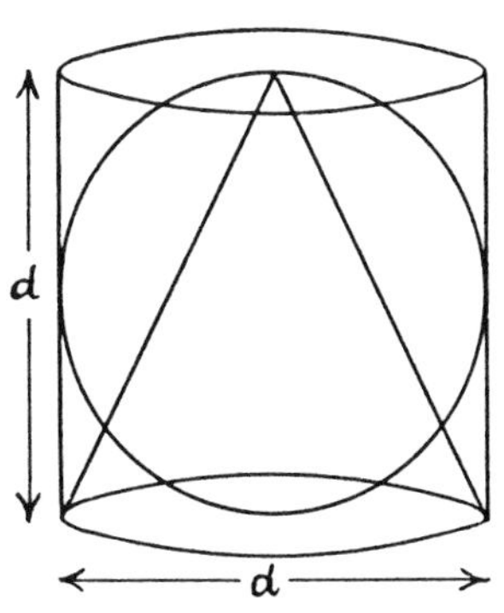

Fig. 210

Referring to Fig. 210,

Vol. of cylinder : Vol. of sphere : Vol. of cone

$$= \frac{\pi}{4} d^2 \times d \quad : \frac{\pi}{6} d^3 \quad : \frac{1}{3} \times \frac{\pi}{4} d^2 \times d$$

dividing throughout by $\frac{\pi}{12} d^3$

$= 3 : 2 : 1$ Ans.

4. Working in metres:

$$\text{Area of sector} = \frac{240}{360} \times \text{area of full circle}$$

$$= \tfrac{2}{3} \times \pi \times 0{\cdot}18^2 \text{ m}^2$$

$$\text{Mass of material} = \tfrac{2}{3} \times \pi \times 0{\cdot}18^2 \times 6{\cdot}5$$

$$= 0{\cdot}4412 \text{ kg} \quad \text{Ans. (i)}$$

Working in millimetres:

$$\text{Circumference of cone base} = \text{arc of sector}$$

$$\pi \times d = \frac{240}{360} \times 2\pi \times 180$$

$$d = \tfrac{2}{3} \times 2 \times 180$$

$$\therefore \text{diameter} = 240 \text{ mm} \quad \text{Ans. (ii)}$$

$$\text{Slant height of cone} = \text{radius of sector} = 180 \text{ mm}$$

$$\text{Radius of cone base} = 120 \text{ mm}$$

$$\therefore \text{Perp. height of cone} = \sqrt{180^2 - 120^2}$$

$$= 134{\cdot}2 \text{ mm} \quad \text{Ans. (iii)}$$

$$\text{Volume of cone} = \tfrac{1}{3} \times \text{area of base} \times \text{perp. ht.}$$

$$= \tfrac{1}{3} \times \pi r^2 \times h$$

$$= \tfrac{1}{3} \times \pi \times 120^2 \times 134{\cdot}2$$

$$= 2{\cdot}024 \times 10^6 \text{ mm}^3$$

$$1 \text{ litre} = 1 \text{ dm}^3 = 10^6 \text{ mm}^3$$

$$\therefore \text{Capacity} = 2{\cdot}024 \text{ litres} \quad \text{Ans. (iv)}$$

5. Working in centimetres:

$$\text{Volume of hemisphere} = \frac{1}{2} \times \frac{\pi}{6} d^3$$

$$= \frac{\pi}{12} \times 6^3 \text{ cm}^3$$

$$\text{Volume of cone} = \frac{1}{3} \times \frac{\pi}{4} d^2 \times h$$

$$= \frac{\pi}{12} \times 6^2 \times 5 \text{ cm}^3$$

$$\text{Total Volume} = \frac{\pi}{12} \times 6^3 + \frac{\pi}{12} \times 6^2 \times 5$$

$$= \frac{\pi}{12} \times 6^2(6 + 5)$$

$$= \frac{\pi}{12} \times 36 \times 11 \text{ cm}^3$$

$$\text{Mass} = \frac{\pi}{12} \times 36 \times 11 \times 8{\cdot}4$$

$$= 871 \text{ grammes Ans.}$$

6. 3 kilogrammes = 3000 grammes

Mass of 1 cm^3 of water is one gramme,

$\therefore$ Mass of 1 cm^3 of lead is 11·4 grammes

$\therefore$ Volume of lead in hollow sphere $= \frac{3000}{11{\cdot}4}$ cm^3

Let D = outside diameter, in cm, then

inside diameter = $(D - 2)$ cm

$$\frac{\pi}{6}(D^3 - d^2) = \frac{3000}{11{\cdot}4}$$

$$D^3 - (D - 2)^3 = \frac{3000 \times 6}{11{\cdot}4 \times \pi}$$

$$D^3 - (D^3 - 6D^2 + 12D - 8) = 502{\cdot}5$$
$$6D^2 - 12D + 8 = 502{\cdot}5$$
$$6D^2 - 12D - 494{\cdot}5 = 0$$
$$\text{or, } D^2 - 2D - 82{\cdot}41 = 0$$

Solving this quadratic equation,

D = 10·135 cm or 101·35 mm Ans.

7.

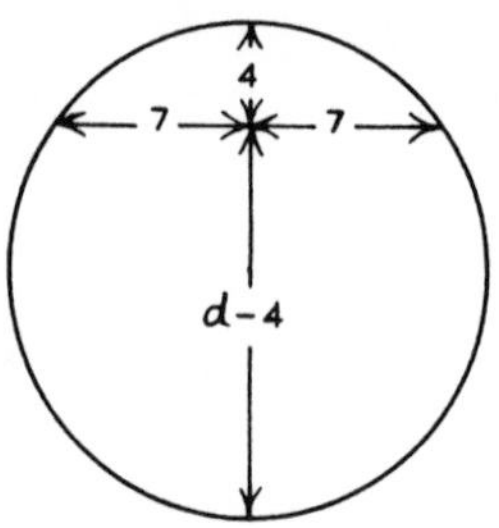

Fig. 211

Let d = diameter of sphere, by crossed chords:

$$4 \times (d - 4) = 7 \times 7$$

$$d = 16{\cdot}25 \text{ cm}$$

$$\text{Vol. of segment of sphere} = \frac{\pi}{6} h^2(3d - 2h)$$

$$= \frac{\pi}{6} \times 4^2(3 \times 16{\cdot}25 - 2 \times 4)$$

$$= \frac{\pi}{6} \times 16 \times 40{\cdot}75$$

$$= 341{\cdot}4 \text{ cm}^3 \quad \text{Ans.}$$

8.

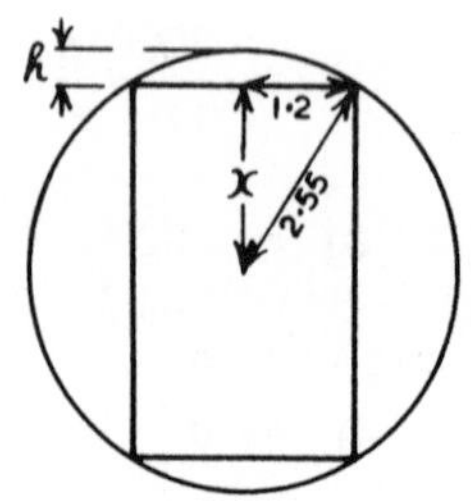

Fig. 212

Working in centimetres:

$$\text{radius of sphere} = \tfrac{1}{2} \times 5{\cdot}1 = 2{\cdot}55 \text{ cm}$$
$$\text{radius of hole} = \tfrac{1}{2} \times 2{\cdot}4 = 1{\cdot}2 \text{ cm}$$
$$\text{half depth of hole} = x \text{ cm}$$
$$x = \sqrt{2{\cdot}55^2 - 1{\cdot}2^2} = \sqrt{5{\cdot}0625} = 2{\cdot}25 \text{ cm}$$

Thickness of spherical segments at top and bottom of hole

$$= h \text{ cm}$$
$$h = \tfrac{1}{2}d - x = 2{\cdot}55 - 2{\cdot}25 = 0{\cdot}3 \text{ cm}$$

$$\text{Vol. of sphere} = \frac{\pi}{6} d^3$$
$$= \frac{\pi}{6} \times 5{\cdot}1^3 = 69{\cdot}465 \text{ cm}^3$$

$$\text{Vol. of cylindrical hole} = \frac{\pi}{4} d^2 l$$
$$= \frac{\pi}{4} \times 2{\cdot}4^2 \times (2 \times 2{\cdot}25) = 20{\cdot}36 \text{ cm}^3$$

$$\text{Volume of segment} = \frac{\pi}{6} h^2(3d - 2h)$$

Volume of the two end segments

$$= 2 \times \frac{\pi}{6} \times 0{\cdot}3^2(3 \times 5{\cdot}1 - 2 \times 0{\cdot}3)$$
$$= 2 \times \frac{\pi}{6} \times 0{\cdot}09 \times 14{\cdot}7$$
$$= 1{\cdot}386 \text{ cm}^3$$

Net vol. = vols. sphere – cyl. hole – 2 end segments
= $69{\cdot}465 - 20{\cdot}36 - 1{\cdot}386$
= $47{\cdot}72$ cm^3 Ans. (i)

Density = $7{\cdot}86 \times 10^3$ kg/m^3 = $7{\cdot}86$ g/cm^3
Mass [g] = volume [cm^3] × density [g/cm^3]
= $47{\cdot}72 \times 7{\cdot}86$
= $375{\cdot}1$ g Ans. (ii)

9.

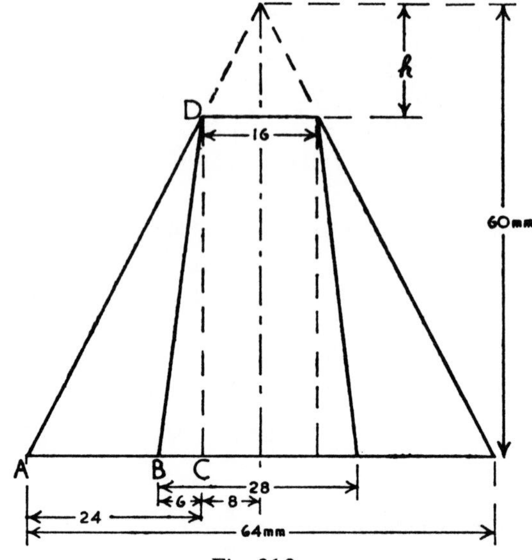

Fig. 213

By similar triangles (see Fig. 213),

$$\frac{h}{16} = \frac{60}{64} \qquad h = 15 \text{ mm}$$

height of hole $= 60 - 15 = 45$ mm

Volume of complete cone

$$= \tfrac{1}{3} \times \text{area of base} \times \text{perp. height}$$
$$= \tfrac{1}{3} \times 0{\cdot}7854 \times 64^2 \times 60 = 64\,340 \text{ mm}^3$$

Volume of top cone cut off

$$= \tfrac{1}{3} \times 0{\cdot}7854 \times 16^2 \times 15 = 1006 \text{ mm}^3$$

Volume of bored hole, this is a frustum of a cone,

$$= \tfrac{1}{12}\pi h(D^2 + Dd + d^2)$$
$$= \tfrac{1}{12} \times \pi \times 45(28^2 + 28 \times 16 + 16^2) = 17\,530 \text{ mm}^2$$

Net volume

$$= 64\,340 - 1006 - 17\,530$$
$$= 45\,804 \text{ mm}^3 \text{ or } 45{\cdot}804 \text{ cm}^3 \quad \text{Ans. (i)}$$

Mass [g]

$$= \text{volume [cm}^3] \times \text{density [g/cm}^3]$$
$$= 45{\cdot}804 \times 8{\cdot}4$$
$$= 384{\cdot}8 \text{ g} \quad \text{Ans. (ii)}$$ (see next page)

Alternatively, the volume can be obtained by the theorem of Pappus. Referring to Fig. 220 volume remaining after boring is equal to the volume swept out when triangular area ABD is swept through one complete revolution about the centre axis of the cone. This can be found by the difference between the volumes swept out by triangles ACD and BCD, the volumes in each case being area × distance c.g. moves.

Area of ACD $= \frac{1}{2} \times 24 \times 45 = 540 \text{ mm}^2$

Distance of its centroid from centre axis

$= \frac{1}{3} \times 24 + 8 = 16 \text{ mm}$

Area of BCD $= \frac{1}{2} \times 6 \times 45 = 135 \text{ mm}^2$

Distance of its centroid from centre axis

$= \frac{1}{3} \times 6 + 8 = 10 \text{ mm}$

Net volume $= 540 \times 2\pi \times 16 - 135 \times 2\pi \times 10$

$= 2\pi(8640 - 1350)$

$= 45\,810 \text{ mm}^3$ or $45{\cdot}81 \text{ cm}^3$ (as above)

10. Working in centimetres:

Area of hexagonal base $=$ six equilateral triangles

$= 6 \times 0{\cdot}433 \times \text{side}^2$

$= 2{\cdot}598 \times 2{\cdot}5^2$

Volume of pyramid $= \frac{1}{3} \times \text{area of base} \times \text{perp. ht.}$

$= \frac{1}{3} \times 2{\cdot}598 \times 2{\cdot}5^2 \times 6$

$= 32{\cdot}48 \text{ cm}^3$ Ans. (i)

Volumes of similar objects vary as the cube of their corresponding dimensions, therefore,

$$\frac{\text{vol. of top pyramid cut off}}{\text{vol. of whole pyramid}} = \frac{3^3}{6^3} = \frac{1}{8}$$

$\therefore$ vol. of top pyramid cut off $= \frac{1}{8}$ of whole pyramid

Volume of frustum $=$ the remaining $\frac{7}{8}$ of whole pyramid

$= \frac{7}{8} \times 32{\cdot}48$

$= 28{\cdot}42 \text{ cm}^3$ Ans. (ii)

11. Volumes of similar objects vary as the cube of their corresponding dimensions, therefore volume varies as depth³.

Let V = volume when depth is 7 cm

$$\frac{V}{200} = \frac{7^3}{5^3}$$

$$V = \frac{200 \times 7^3}{5^3} = 548{\cdot}8 \text{ cm}^3$$

Additional volume required $= 548{\cdot}8 - 200$

$= 348{\cdot}8 \text{ cm}^3$ Ans.

12.

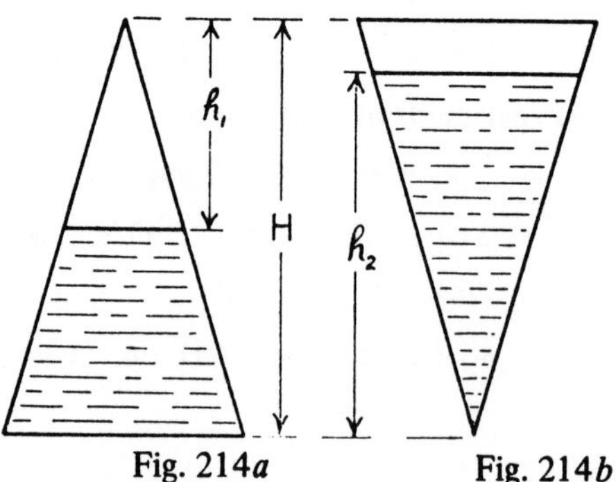

Fig. 214*a* Fig. 214*b*

Volumes of similar objects vary as the cube of their corresponding dimensions. Referring to Fig. 214*a*, comparing similar cones:

Let V = volume of whole cone
H = height of whole cone
v = volume of conical empty space
h_1 = height of conical empty space

$$\frac{V}{v} = \frac{H^3}{h_1^3} = \frac{500^3}{250^3} = \left\{\frac{500}{250}\right\}^3 = 2^3 = 8$$

$$\therefore V = 8 \times v \text{ or } v = \tfrac{1}{8}V$$

thus, the volume of the empty space is one-eighth of the whole volume, and the volume of the water is seven-eighths (= 0·875) of the whole volume.

Referring to Fig. 214*b*, comparing similar cones:

$$\frac{V}{0{\cdot}875\,V} = \frac{H^3}{h_2^3}$$

$$h_2 = 500 \times \sqrt[3]{0{\cdot}875} = 478{\cdot}3 \text{ mm Ans.}$$

13. Ratio of surface areas $= 1{\cdot}5 : 1$

" " diameters $= \sqrt{1{\cdot}5} : \sqrt{1}$

$= 1{\cdot}225 : 1$

" " volumes $= 1{\cdot}225^3 : 1^3$

$= 1{\cdot}8375 : 1$

Let v = volume of smaller sphere, in cm^3,
then volume of larger sphere $= (v + 10)\ cm^3$

$$\frac{v + 10}{v} = \frac{1{\cdot}8375}{1}$$

$$1{\cdot}8375v = v + 10$$
$$0{\cdot}8375v = 10$$
$$v = 11{\cdot}94\ cm^3 \quad \text{Ans. (i)}$$

$$\text{volume} = \frac{\pi}{6} d^3$$

$$\text{diameter} = \sqrt[3]{\frac{11{\cdot}94 \times 6}{\pi}}$$

$$= 2{\cdot}836\ cm = 28{\cdot}36\ mm \quad \text{Ans. (ii)}$$

14. Working in decimetres,

SECTIONAL AREAS	SIMPSON'S MULTIPLIERS	PRODUCTS
$0{\cdot}7854 \times 3{\cdot}95^2$	1	$0{\cdot}7854 \times 15{\cdot}6$
" $4{\cdot}77^2$	4	" $91{\cdot}0$
" $5{\cdot}00^2$	2	" $50{\cdot}0$
" $4{\cdot}77^2$	4	" $91{\cdot}0$
" $3{\cdot}95^2$	1	" $15{\cdot}6$
	sum =	$0{\cdot}7854 \times 263{\cdot}2$

There are 5 ordinates therefore 4 spaces

Common interval

$$= 5{\cdot}81 \div 4$$

$$\text{Volume} = \text{sum of products} \times \tfrac{1}{3} \text{ common interval}$$

$$= \frac{0{\cdot}7854 \times 263{\cdot}2 \times 5{\cdot}81}{3 \times 4}$$

$$= 100{\cdot}1\ dm^2 = 100{\cdot}1 \text{ litres} \quad \text{Ans.}$$

15. Finding plotting points, $y = 5 + 4x - x^2$,

x	-1	0	1	2	3	4	5
y	0	5	8	9	8	5	0

The graph of these values is shown in Fig. 215. Sweeping the area bounded by this graph through one revolution, the y ordinates of the graph become the radii at regular intervals along the length of the solid generated. Putting the cross-sectional areas of the solid through Simpson's rule to find the volume:

RADII r	CROSS-SECT. AREAS πr^2	SIMPSON'S MULTIPLIERS	PRODUCTS
0	0	1	0
5	$\pi \times 25$	4	$\pi \times 100$
8	$\pi \times 64$	2	$\pi \times 128$
9	$\pi \times 81$	4	$\pi \times 324$
8	$\pi \times 64$	2	$\pi \times 128$
5	$\pi \times 25$	4	$\pi \times 100$
0	0	1	0
		sum =	$\pi \times 780$

Common interval between ordinates $= 1$

$$\text{Volume} = \pi \times 780 \times \tfrac{1}{3} \times 1$$
$$= 816{\cdot}9 \text{ units}^3 \quad \text{Ans.}$$

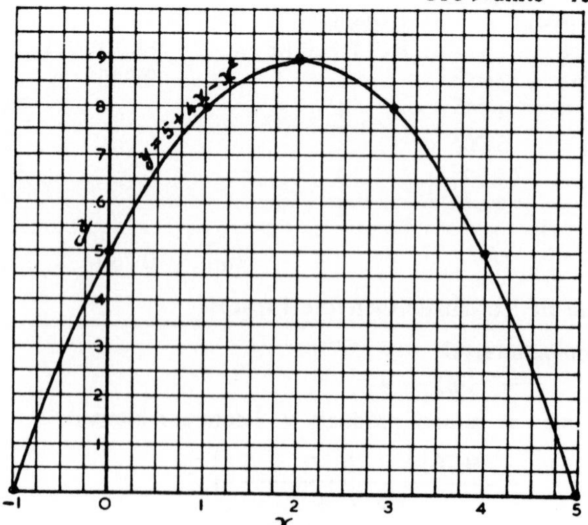

Fig. 215

16.

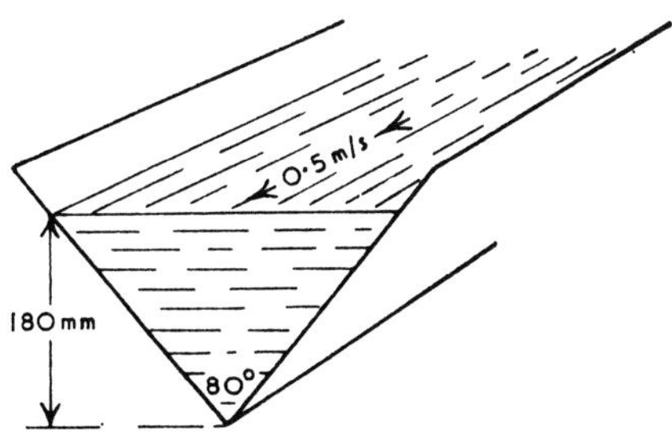

Fig. 216

Working in metres:

Half-breadth of water level $= 0{\cdot}18 \times \tan 40^\circ = 0{\cdot}151$ m

Area of cross-section $= \frac{1}{2}$(breadth $\times$ perp. ht.)

$= 0{\cdot}151 \times 0{\cdot}18\ \text{m}^2$

Volume flow [m^3/h] $=$ area [m^2] $\times$ velocity [m/h]

$= 0{\cdot}151 \times 0{\cdot}18 \times 0{\cdot}5 \times 3600$

$= 48{\cdot}94\ \text{m}^3/\text{h}$ Ans.

17.

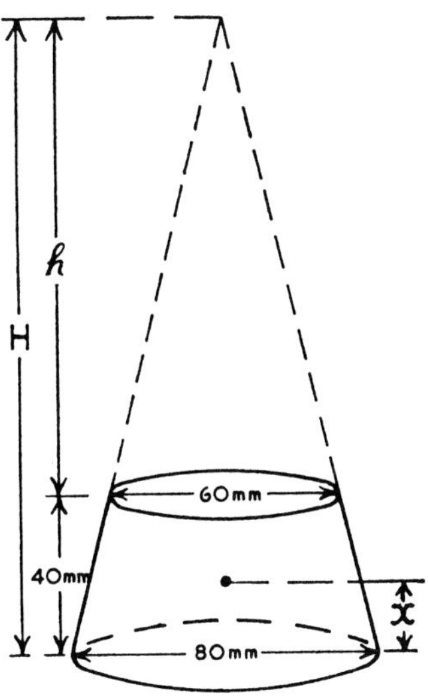

Fig. 217

Referring to Fig. 217 and working in centimetres:

By similar triangles, $\frac{h}{6} = \frac{h + 4}{8}$

$$\therefore h = 12 \text{ cm}$$

$$H = 12 + 4 = 16 \text{ cm}$$

c.g. of a cone is at $\frac{1}{4}$ height from base.

$$\text{Volume of whole cone} = \tfrac{1}{3} \times \pi \times 4^2 \times 16 \text{ cm}^3$$
$$\text{c.g. from base} = \tfrac{1}{4} \text{ of } 16 = 4 \text{ cm}$$
$$\text{Volume of top cone cut off} = \tfrac{1}{3} \times \pi \times 3^2 \times 12 \text{ cm}^3$$
$$\text{c.g. from base} = 4 + \tfrac{1}{4} \text{ of } 12 = 7 \text{ cm}$$

Moments about base:

$$x = \frac{\Sigma \text{ moments of volumes}}{\Sigma \text{ volumes}}$$

$$= \frac{\text{moment of whole cone} - \text{moment of top cone}}{\text{volume of whole cone} - \text{volume of top cone}}$$

$$= \frac{\tfrac{1}{3} \times \pi \times 4^2 \times 16 \times 4 - \tfrac{1}{3} \times \pi \times 3^2 \times 12 \times 7}{\tfrac{1}{3} \times \pi \times 4^2 \times 16 - \tfrac{1}{3} \times \pi \times 3^2 \times 12}$$

$\frac{1}{3}$, π and 4 cancel from every term,

$$x = \frac{4^2 \times 16 - 3^2 \times 3 \times 7}{4^2 \times 4 - 3^2 \times 3} = \frac{67}{37}$$

$$= 1{\cdot}811 \text{ cm} = 18{\cdot}11 \text{ mm from base Ans.}$$

18. Considering the plate as a rectangle 300 mm by 375 mm with a triangle 225 mm by 300 mm cut off one corner, working in centimetres:

Area of rectangle $= 30 \times 37{\cdot}5 = 1125 \text{ cm}^2$
Distance of its c.g. from 375 mm side $= 15$ cm
Distance of its c.g. from 300 mm side $= 18{\cdot}75$ cm
Area of triangle $= \frac{1}{2}(22{\cdot}5 \times 30) = 337{\cdot}5 \text{ cm}^2$
Distance of its c.g. from 375 mm side $= 30 - \frac{1}{3} \times 22{\cdot}5 = 22{\cdot}5$ cm
Distance of its c.g. from 300 mm side $= 37{\cdot}5 - \frac{1}{3} \times 30 = 27{\cdot}5$ cm
(c.g. of triangle is at $\frac{1}{3}$ height from its base)

Moments about 375 mm side:

$$x = \frac{\Sigma \text{ moments of areas}}{\Sigma \text{ areas}}$$

$$= \frac{1125 \times 15 - 337{\cdot}5 \times 22{\cdot}5}{1125 - 337{\cdot}5} = 11{\cdot}79 \text{ cm}$$

Moments about 300 mm side:

$$y = \frac{1125 \times 18{\cdot}75 - 337{\cdot}5 \times 27{\cdot}5}{1125 - 337{\cdot}5} = 15 \text{ cm}$$

$x = 117{\cdot}9$ mm, $y = 150$ mm Ans.

19.

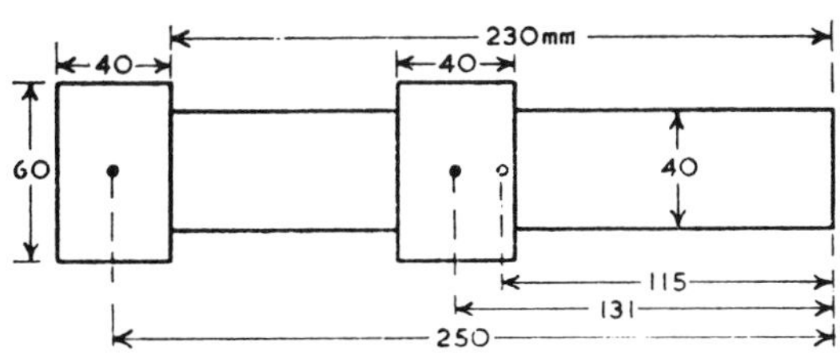

Fig. 218

Working in centimetres and taking moments about end of shank:

$$\text{c.g.} = \frac{\Sigma \text{ moments of volumes}}{\Sigma \text{ volumes}}$$

All parts being cylindrical of volume $0{\cdot}7854 \times d^2 \times l$, the multiplier 0·7854 is common to all terms and therefore cancels out.

$$\text{c.g.} = \frac{(6^2 \times 4 \times 25) + (4^2 \times 23 \times 11{\cdot}5) + (6^2 - 4^2) \times 4 \times 13{\cdot}1}{(6^2 \times 4) + (4^2 \times 23) + (6^2 - 4^2) \times 4}$$

$$= \frac{3600 + 4232 + 1048}{144 + 368 + 80} = \frac{8880}{592}$$

$= 15$ cm or 150 mm Ans.

SOLUTIONS TO TEST EXAMPLES 13

1. (a)

$$y = x^3 + 3x^2 - 9x + 4$$

$$\frac{dy}{dx} = 3x^2 + 6x - 9 \qquad \text{Answer (a).}$$

(b)

$$y = \frac{2x^3}{3} - \frac{7}{x^2} + x$$

$$\text{i.e. } y = \frac{2x^3}{3} - 7x^{-2} + x$$

$$\therefore \frac{dy}{dx} = 2x^2 + 14x^{-3} + 1$$

$$= 2x^2 + \frac{14}{x^3} + 1 \qquad \text{Answer (b).}$$

(c)

$$y = \sqrt[5]{x^3} + 1$$

$$\therefore y = x^{\frac{3}{5}} + 1$$

$$\frac{dy}{dx} = \frac{3x^{-\frac{2}{5}}}{5}$$

$$= \frac{3}{5x^{\frac{2}{5}}} \qquad \text{Answer (c).}$$

(d)

$$y = 5\cos x - 7\cos x + 2\sin x$$

$$\therefore \frac{dy}{dx} = -5\sin x + 7\sin x + 2\cos x \qquad \text{Answer (d).}$$

2.

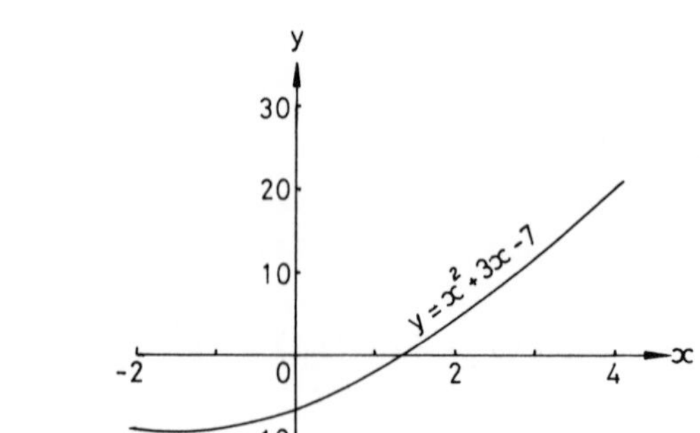

Fig. 219

$$y = x^2 + 3x - 7$$

$$\frac{dy}{dx} = 2x + 3$$

when $x = 3$,

$$\frac{dy}{dx} = 6 + 3$$

$$\text{i.e. gradient} = 9$$

when $x = -2$

$$\frac{dy}{dx} = -4 + 3$$

$$\text{i.e. gradient} = -1$$

Answer.

3. (a)

$$s = 20t - 5t^2 + 4$$

$$v = \frac{ds}{dt} = 20 - 10t \quad \ldots \quad \ldots \quad \ldots \quad \ldots \quad \text{(i)}$$

when $t = 1$,

$$v = 20 - 10$$

$$= 10 \text{ m/s}$$

Answer.

(b) when $v = 0$,

$$20 - 10t = 0$$

$$\therefore t = \frac{20}{10}$$

$$= 2 \text{ seconds.}$$

Answer.

(c) From equation (i),

$$v = 20 - 10t$$

$$a = \frac{dv}{dt}$$

$$\therefore a = -10 \text{ m/s}^2.$$

Answer.

4.

$$y = 4x + \frac{1}{x}$$

$$\text{i.e. } y = 4x + x^{-1}$$

$$\text{Gradient} = \frac{dy}{dx} = 4 - x^{-2}$$

∴ At any point on the curve,

$$\text{gradient} = 4 - \frac{1}{x^2}$$

When the gradient is zero,

$$\frac{dy}{dx} = 0$$

$$\therefore 4 - \frac{1}{x^2} = 0$$

$$\therefore \frac{1}{x^2} = 4$$

$$\therefore x = \sqrt{\frac{1}{4}}$$

$$\therefore x = \pm\frac{1}{2} \text{ (See Fig. 220)}$$

∴ Gradient is zero when $x = \frac{1}{2}$ and $x = -\frac{1}{2}$ **Answer.**

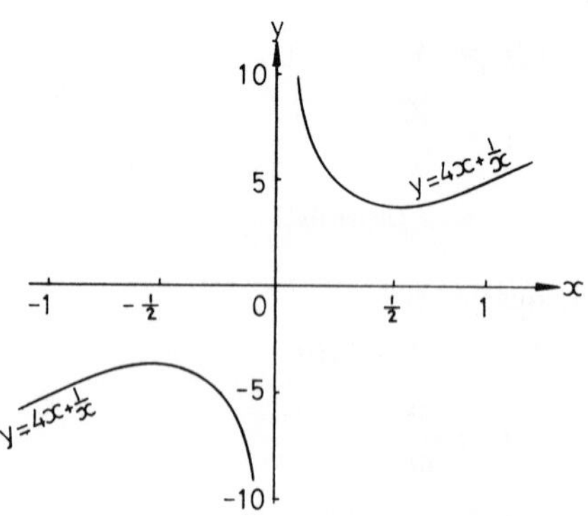

Fig. 220

5. $$y = 3x^2 - x + 2 \quad \ldots \quad \ldots \quad \ldots \quad \ldots \quad \text{(i)}$$

$$\text{Gradient} = \frac{dy}{dx} = 6x - 1$$

When the gradient is -7,

$$6x - 1 = -7$$
$$\therefore 6x = -6$$
$$\therefore x = -1$$

Substituting $x = -1$ in equation (i),

$$y = 3 + 1 + 2$$
$$= 6$$

$\therefore$ The gradient is -7 at the point $x = -1, y = 6$. Answer.

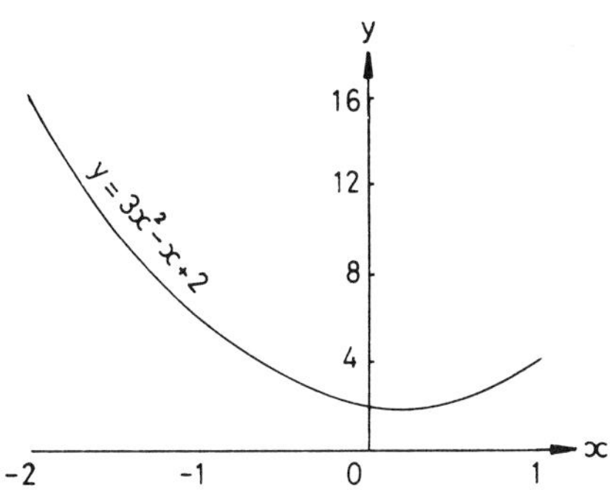

Fig. 221

6. $$\theta = 2 + 16t - \frac{t^2}{2}$$

$$\omega = \frac{d\theta}{dt} = 16 - t$$

when $t = 2$,

$$\omega = 16 - 2$$
$$= 14 \text{ rad/s}$$

when vel $= 0$

$$16 - t = 0$$
$$\therefore t = 16$$

i.e. shaft comes to rest after 16 seconds. Answer.

7. (a)

$$y = x^3 + 3x + \ln x$$

$$\frac{dy}{dx} = 3x^2 + 3 + \frac{1}{x}$$

$$= 3x^2 + 3 + x^{-1}$$

$$\frac{d^2y}{dx^2} = 6x - x^{-2}$$

$$= 6x - \frac{1}{x^2} \qquad \text{Answer.}$$

(b)

$$y = 3\cos\theta - 7\cos\theta + \theta$$

$$\frac{dy}{d\theta} = -3\sin\theta + 7\sin\theta + 1$$

$$\frac{d^2y}{d\theta^2} = -3\cos\theta + 7\cos\theta \qquad \text{Answer.}$$

8.

$$y = 12x + 3x^2 - 2x^3 \quad \dots \quad \dots \quad \dots \quad \text{(i)}$$

$$\frac{dy}{dx} = 12 + 6x - 6x^2$$

$$\frac{d^2y}{dx^2} = 6 - 12x \quad \dots \quad \dots \quad \dots \quad \dots \quad \dots \quad \text{(ii)}$$

For maximum or minimum values,

$$\frac{dy}{dx} = 0$$

i.e. $12 + 6x - 6x^2 = 0$

$\therefore 2 + x - x^2 = 0$

$\therefore x^2 - x - 2 = 0$

Solve for x:

$(x - 2)(x + 1) = 0$

$\therefore x = 2$ or $x = -1$

Substituting in equation (ii):

when $x = 2$,

$$\frac{d^2y}{dx^2} = 6 - 24 \text{ (i.e. negative)}$$

when $x = -1$,

$$\frac{d^2y}{dx^2} = 6 + 12 \text{ (positive)}$$

$\therefore$ A maximum value occurs when $x = 2$.

Substituting in equation (i):

$$\text{Maximum value of } y = 24 + 12 - 16$$
$$= 20$$

Answer.

9. $$M = \frac{wlx}{2} - \frac{wx^2}{2}$$

Differentiating with respect to x:

$$\frac{dM}{dx} = \frac{wl}{2} - wx \quad \dots \quad \dots \quad \dots \quad \dots \quad \dots \quad \text{(i)}$$

$$\frac{d^2M}{dx^2} = -w \text{ (i.e. negative)}$$

For maximum or minimum value,

$$\frac{dM}{dx} = 0$$

From equation (i):

$$\frac{wl}{2} - wx = 0$$

$$\therefore \frac{wl}{2} = wx$$

$$\therefore x = \frac{l}{2}$$

$\therefore$ Maximum bending moment occurs when

$$x = \frac{l}{2}$$

i.e. At the centre of the beam.

Answer.

10. (a) $$t = 2{\cdot}1 \sqrt[3]{\frac{1}{\theta^2}} - \frac{4}{\sqrt[5]{\theta}}$$

$$= 2{\cdot}1\,\theta^{-\frac{2}{3}} - 4\theta^{-\frac{1}{5}}$$

$$\frac{dt}{d\theta} = -1 \cdot 4\,\theta^{-\frac{5}{3}} + 0{\cdot}8\,\theta^{-\frac{6}{5}}$$

$$= \frac{0{\cdot}8}{\sqrt[5]{\theta^6}} - \frac{1{\cdot}4}{\sqrt[3]{\theta^5}} \qquad \text{Answer.}$$

(b)

$$z = \frac{au^n - 1}{C}$$

$$\frac{dz}{du} = \frac{an\,u^{n-1}}{C} \qquad \text{Answer.}$$

(c)

$$y = 3x(x^2 - 4)$$

$$= 3x^3 - 12x$$

$$\frac{dy}{dx} = 9x^2 - 12$$

$$= 3(3x^2 - 4) \qquad \text{Answer.}$$

(d)

$$y = 2\cos\theta + 5$$

$$\frac{dy}{d\theta} = -2\sin\theta \qquad \text{Answer.}$$

11.

$$f(x) = \frac{x^3}{3} - 2x^2 + 3x + 1 \quad \ldots \quad \ldots \quad \text{(i)}$$

$$f'(x) = x^2 - 4x + 3 \quad \ldots \quad \ldots \quad \ldots \quad \text{(ii)}$$

$$f''(x) = 2x - 4 \quad \ldots \quad \ldots \quad \ldots \quad \ldots \quad \text{(iii)}$$

$$f'(x) = 0 \text{ for a maximum or a minimum}$$

$$0 = x^2 - 4x + 3 \text{ from (ii)}$$
$$= (x-1)(x-3)$$
$$\therefore x = 1 \text{ or } x = 3$$

when $x = 1$ in (iii)

$$f''(x) = -2 \text{ i.e. a maximum}$$

when $x = 3$ in (iii)

$$f''(x) = 2 \text{ i.e. a minimum}$$

when $x = 3$ in (i)

$$f(x) = 9-18+9+1$$
$$= 1$$ Answer.

12. $$P = aV^b$$

$$\frac{dP}{dV} = ab\ V^{b-1}$$ Answer.

$$= 0{\cdot}5 \times 10^{-10} \times 6 \times 100^5$$
$$= 0{\cdot}5 \times 10^{-10} \times 6 \times 10^{10}$$
$$= 3 \text{ units}$$ Answer.

13. Let each of two sides be x long

Remaining two sides are $\frac{l}{2} - x$ long

i.e. $l = 2\left(\frac{l}{2} - x\right) + 2x$ Q.E.D.

$$\text{Area A} = x\left(\frac{l}{2} - x\right) \quad \ldots \quad \ldots \quad \ldots \quad \ldots \quad \text{(i)}$$

$$= \frac{xl}{2} - x^2$$

$$f'(A) = \frac{l}{2} - 2x \quad \ldots \quad \ldots \quad \ldots \quad \ldots \quad \text{(ii)}$$

$$f''(A) = -2 \text{ i.e. a maximum}$$

$$0 = \frac{l}{2} - 2x \text{ from (ii)}$$

$$x = \frac{l}{4}$$

Answer.

14. (a)

$$f(\theta) = \cos\theta - \text{In }\theta$$

$$f'(\theta) = -\sin\theta - \frac{1}{\theta}$$

$$f''(\theta) = \frac{1}{\theta^2} - \cos\theta$$

Answer.

(b)

$$f(t) = at^2 + 2\text{ In }t$$

$$f'(t) = 2at + \frac{2}{t}$$

$$f''(t) = 2a - \frac{2}{t^2}$$

$$= 2\left(a - \frac{1}{t^2}\right)$$

Answer.

(c)

$$f(x) = 5\,e^x$$

$$f''(x) = 5\,e^x$$

Answer.

15. $x = 0{\cdot}2\,t^2 + 10{\cdot}4$ ... ... (i)

$f'(x) = 0{\cdot}4\,t$ ... ... (ii)

$= 2\ m/s$ i.e. velocity at $t = 5$ seconds Answer.

$f''(x) = 0{\cdot}4\ m/s^2$ i.e. (constant) acceleration Answer.

$100 = 0{\cdot}2\,t^2 + 10{\cdot}4$ i.e. $x = 100$ in (i)

$500 = t^2 + 52$

$t = 21{\cdot}17\,s$

$f'(x) = 0{\cdot}4 \times 21{\cdot}17$ i.e. $t = 21{\cdot}17$ in (ii)

$= 8{\cdot}468\ m/s$ i.e. velocity Answer.

$f''(x) = 0{\cdot}4\ m/s^2$ i.e. acceleration Answer.

16. $\theta = 2{\cdot}1 - 3{\cdot}2\,t + 4{\cdot}8\,t^2$

$f'(t) = 9{\cdot}6\,t - 3{\cdot}2$

$= 9{\cdot}6 \times 1{\cdot}5 - 3{\cdot}2$ at $t = 1{\cdot}5\,s$

$= 11{\cdot}2$ rad/s angular velocity Answer.

$f''(t) = 9{\cdot}6$ rad/s^2 (constant) angular acceleration

Answer.

17. $f(x) = x^5 - 5x$ ... ... (i)

$f'(x) = 5x^4 - 5$ ... ... (ii)

$$0 = 5x^4 - 5 \text{ for maximum/minimum}$$

$$x^4 = 1$$

$$x = 1 \text{ or } x = -1$$

$$x = 1 \text{ in (i) gives } f(x) = -4$$

$$x = -1 \text{ in (i) gives } f(x) = 4$$

$$f''(x) = 20x^3 \qquad \text{(iii)}$$

$$= 20 \text{ i.e. a minimum when } x = 1$$ Answer.

$$= -20 \text{ i.e. a maximum when } x = -1$$ Answer.

$$\therefore x = 1, f(x) = -4, \text{ minimum}$$ Answer.

$$\therefore x = -1, f(x) = 4, \text{ maximum}$$ Answer.

SOLUTIONS TO TEST EXAMPLES 14

1\. (a) $\int(x^4 + x^2 - 8x + 5)dx = \frac{x^5}{5} + \frac{x^3}{3} - 4x^2 + 5x + C$

Answer.

(b) $\int\left(\frac{4}{x^2} + \frac{1}{x}\right)dx$

The expression above is re-arranged as shown:

$$\int(4x^{-2} + x^{-1})dx = -4x^{-1} + \ln x + C$$

$$= -\frac{4}{x} + \ln x + C \qquad \text{Answer.}$$

(c) $\int_1^3 (2x - 3x^2 - 1)dx = [x^2 - x^3 - x]_1^3$

$$= [9 - 27 - 3] - [1 - 1 - 1]$$

$$= [-21] - [-1]$$

$$= -21 + 1$$

$$= -20 \qquad \text{Answer.}$$

2\. Gradient $= \frac{dy}{dx} = x^2 + x - 2$

$$y = \int\frac{dy}{dx}.dx$$

$$\therefore y = \int(x^2 + x - 2)dx$$

$$y = \frac{x^3}{3} + \frac{x^2}{2} - 2x + C \quad \ldots \quad \text{(i)}$$

when $y = 5, \quad x = 2.$

$$\therefore 5 = \frac{(2)^3}{3} + \frac{(2)^2}{2} - 2 \times (2) + C$$

$$\therefore 5 = \frac{8}{3} + 2 - 4 + C$$

$$\therefore C = 4\tfrac{1}{3} \quad \ldots \quad \ldots \quad \ldots \quad \ldots \quad \text{(ii)}$$

From (i) and (ii):

$$y = \frac{x^3}{3} + \frac{x^2}{2} - 2x + 4\tfrac{1}{3} \quad \text{Answer.}$$

3. (a) $\int (3 \cos x - 2 \sin x + 4)dx = 3 \sin x + 2 \cos x + 4x + C$

Answer.

(b) $\int (4 \cos x - \cos x + x)dx = 4 \sin x - \sin x + \frac{x^2}{2} + C$

Answer.

(c)

$$\int_0^{\frac{\pi}{2}} \cos x \, dx = [\sin x]_0^{\frac{\pi}{2}}$$

$$= \left[\sin \frac{\pi}{2}\right] - [\sin 0]$$

$$= 1 - 0$$

$$= 1$$

Answer.

4.

$$y = x^3 - 4x^2 + 3x$$

$$\therefore \text{Area} = \int_1^3 (x^3 - 4x^2 + 3x)dx$$

$$= \left[\frac{x^4}{4} - \frac{4x^3}{3} + \frac{3x^2}{2}\right]_1^3$$

$$= \left[-\frac{27}{12}\right] - \left[\frac{5}{12}\right]$$

$$= -2\tfrac{2}{3}$$

Note: The negative sign indicates that the area is below the x-axis. This is confirmed by the sketch of the graph (Fig. 222).

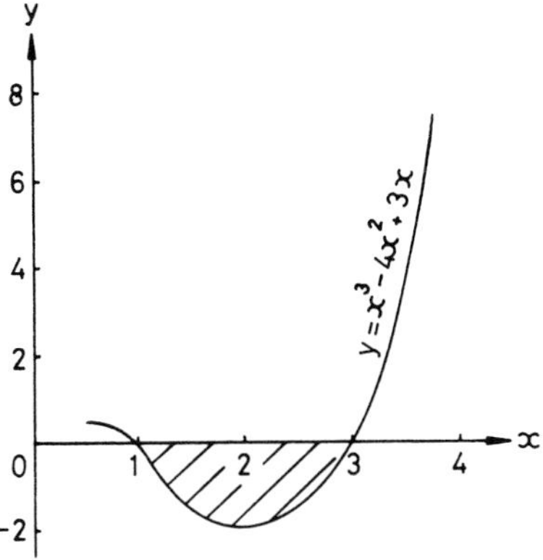

Fig. 222

$\therefore$ Area $= 2\frac{2}{3}$ square units. Answer.

5.

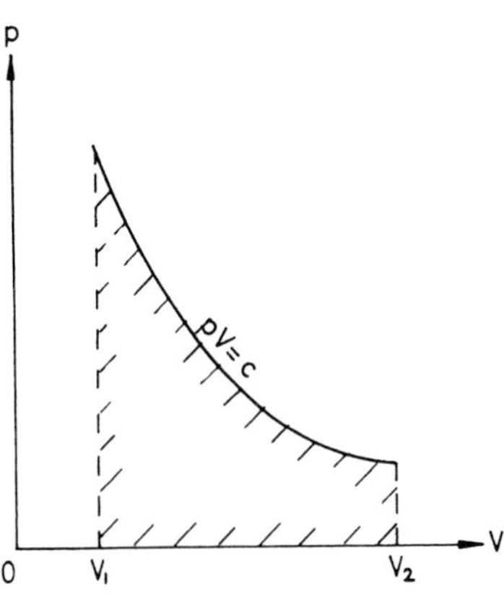

Fig. 223

The law of expansion is $pV = c$ ... ... (i)

$$\text{i.e. } p = \frac{c}{V}$$

The standard equation

$$\text{Area} = \int_a^b y\,dx$$

must be amended to suit the symbols in this problem:

$$\begin{aligned}
\text{Area under curve} &= \int_{V_1}^{V_2} p\,.\,dV \\
&= \int_{V_1}^{V_2} \frac{c}{V}\,.\,dV \\
&= c\int_{V_1}^{V_2} \frac{1}{V}\,.\,dV \\
&= c[\ln V]_{V_1}^{V_2} \\
&= c\{[\ln V_2] - [\ln V_1]\} \\
&= c \times \ln \frac{V_2}{V_1} \qquad \ldots \quad \ldots \quad \text{(ii)}
\end{aligned}$$

(Note: $\ln V_2 - \ln V_1$ is numerically equal to $\ln \frac{V_2}{V_1}$)

From equations (i) and (ii),

$$\text{Area under curve} = pV \ln \frac{V_2}{V_1} \qquad \text{Answer.}$$

6. (a)

$$\begin{aligned}
a &= 6t \\
\therefore v &= \int 6t\,dt \\
\therefore v &= 3t^2 + C \qquad \ldots \quad \ldots \quad \ldots \quad \text{(i)}
\end{aligned}$$

Substitute $v = 0$, $t = 0$ in equation (i),

$$\begin{aligned}
0 &= 0 + C \\
\therefore C &= 0 \\
\text{Hence, } v &= 3t^2 \qquad \text{Answer.}
\end{aligned}$$

(b) when t = 2, $v = 3(2)^2 = 12$

when $t = 3$, $v = 3(3)^2 = 27$

$$\therefore \text{Average velocity} = \frac{12 + 27}{2}$$

$$= 19{\cdot}5 \text{ m/s} \qquad \text{Answer.}$$

(c) when $t = 2{\cdot}5$, $v = 3(2{\cdot}5)^2$

i.e. Instantaneous velocity $= 18{\cdot}75$ m/s Answer.

7.
$$s = \int v\,dt$$
$$= \int (3t^2 + 8t + 12)dt$$
$$\therefore s = t^3 + 4t^2 + 12t + C \quad \ldots \quad \text{(i)}$$

Substitute $s = 10$, $t = 0$ in equation (i),

$$\therefore 10 = 0 + 0 + 0 + C$$
$$\therefore C = 10 \quad \ldots \quad \ldots \quad \ldots \quad \ldots \quad \text{(ii)}$$

From equations (i) and (ii),

$$s = t^3 + 4t^2 + 12t + 10$$

when $t = 10$,

$$s = 1000 + 400 + 120 + 10$$
$$= 1540 \text{ m} \qquad \text{Answer.}$$

8.

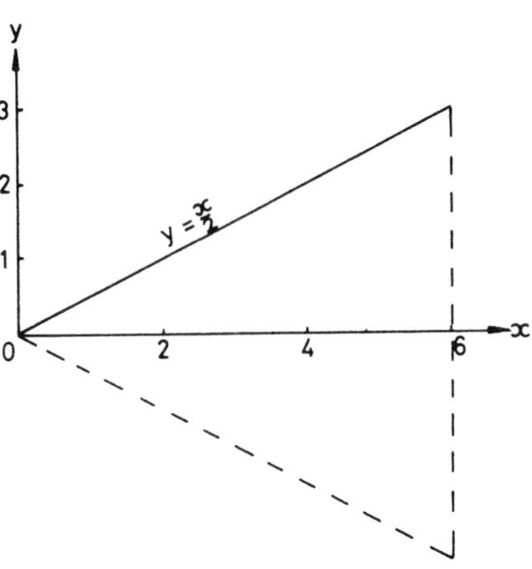

Fig. 224

$$\text{Volume of solid of revolution} = \pi \int_a^b y^2\,dx$$

$$= \pi \int_0^6 \left(\frac{x}{2}\right)^2 dx$$

$$= \pi \int_0^6 \frac{x^2}{4}\,dx$$

$$= \pi \left[\frac{x^3}{12}\right]_0^6$$

$$= \pi \left[\frac{6^3}{12}\right] - \pi\,[0]$$

$$= 56{\cdot}55 \text{ cubic units.}$$ **Answer.**

9. At the points where the graph intersects the x-axis, x has zero value.

$$\text{i.e. } x^2 - x = 0$$

hence, $x = 0$ or $x = 1$

$\therefore$ The graph intersects the x-axis at $x = 0$ and $x = 1$. (See Fig. 225)

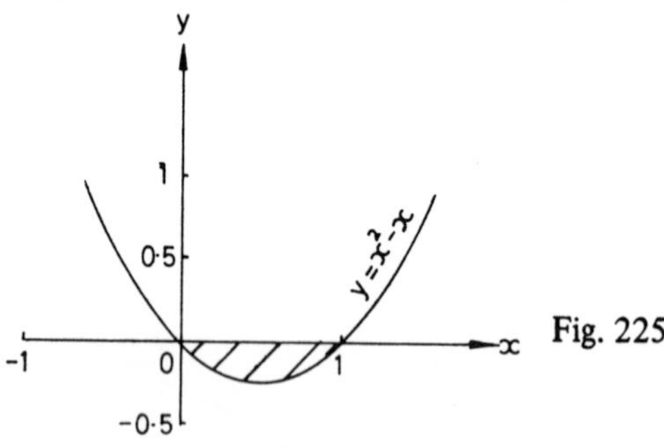

Fig. 225

$$\text{Volume of solid of revolution} = \pi \int_0^1 (x^2 - x)^2\,dx$$

$$= \pi \int_0^1 (x^4 - 2x^3 + x^2)\,dx$$

$$= \pi \left[\frac{x^5}{5} - \frac{x^4}{2} + \frac{x^3}{3}\right]_0^1$$

$$= \pi \left[\frac{1}{5} - \frac{1}{2} + \frac{1}{3}\right] - 0$$

$$= 0{\cdot}105 \text{ cubic units.}$$ **Answer.**

10. (a) $\int (x + 1)(x + 2)\,dx \quad = \quad \int (x^2 + 3x + 2)\,dx$

$$= \frac{x^3}{3} + \frac{3x^2}{2} + 2x + C$$

Answer.

(b) $\int \left(\frac{1}{x} - \frac{1}{x^2}\right) dx \quad = \quad \ln x + \frac{1}{x} + C$

Answer.

(c) $\int_1^2 (a + 2b)(x + 1)\,dx \quad = \quad a + 2b \int_1^2 (x + 1)\,dx$

$$= a + 2b \left[\frac{x^2}{2} + x\right]_1^2$$

$$= a + 2b \left[\left(2 + 2\right) - \left(\frac{1}{2} + 1\right)\right]$$

$$= \frac{5}{2}(a + 2b)$$ Answer.

(d) $\int 3e^{3x} dx \quad = \quad 3 \int e^x\,dx$

$$= 3e^x + C$$ Answer.

11. Area $= \int_6^8 (0{\cdot}06\,x^2 + 10)\,dx = \left[\frac{0{\cdot}06\,x^3}{3} + 10x\right]_6^8$

8

$$= \left[\frac{8^3 \times 0{\cdot}06}{3} + 80\right] - \left[\frac{6^3 \times 0{\cdot}06}{3} + 60\right]$$

$$= (0{\cdot}02 \times 512) + 80 - (0{\cdot}02 \times 216) - 60$$

$$= 0{\cdot}02 \times 296 + 20 = 5{\cdot}92 + 20$$

$$= 25{\cdot}92 \text{ sq. units.}$$ Answer.

12. Area $= \int_{V_1}^{V_2} p\,dV = \int_{V_1}^{V_2} \frac{C}{V^n}\,dV$

$$= C \int_{V_1}^{V_2} V^{-n}\,dV$$

$$= C\left[\frac{1}{1-n}\, V^{1-n}\right]_{V_1}^{V_2}$$

$$= C\,\frac{1}{1-n}\,(V_2^{1-n} - V_1^{1-n})$$

$$= \frac{-C}{n-1}\left(\frac{1}{V_2^{\,n-1}} - \frac{1}{V_1^{\,n-1}}\right)$$

Answer.

13. (a) $\int \left(\sin\theta - \frac{1}{2}\right)\frac{d\theta}{3} = \frac{1}{3}\int \sin d\theta - \frac{1}{3}\int \frac{d\theta}{2}$

$$= -\frac{1}{3}\left(\cos\theta + \frac{\theta}{2}\right) + C$$

Answer .

(b) $$\int_0^{\pi} \sin x \, dx = \Big[-\cos x\Big]_0^{\pi}$$

$$= (-\cos \pi) - (-\cos 0)$$

$$= 1 + 1$$

$$= 2$$ Answer.

(c) $$\int_0^{\frac{\pi}{2}} \left(5 \cos x + 3 \sin x - x\right) dx = \left[5 \sin x - 3 \cos x - \frac{x^2}{2}\right]_0^{\frac{\pi}{2}}$$

$$= \left(5 \sin \frac{\pi}{2} - 3 \cos \frac{\pi}{2} - \frac{\pi^2}{8}\right) - \left(5 \sin 0 - 3 \cos 0 - 0\right)$$

$$= \left(5 - 0 - \frac{\pi^2}{8}\right) - (0 - 3 - 0)$$

$$= 8 - \frac{\pi^2}{8}$$ Answer.

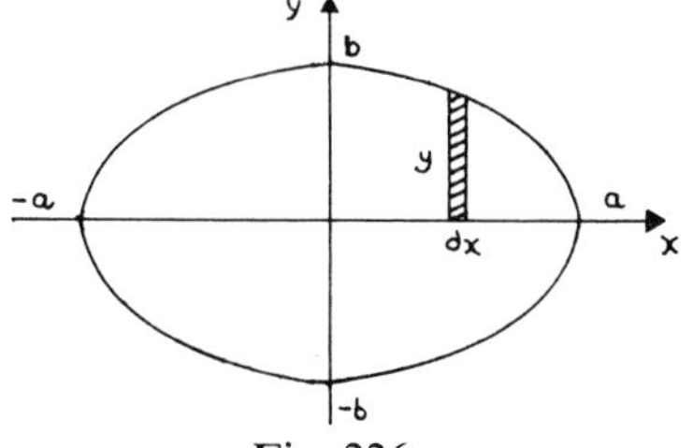

Fig. 226

14. $$V = \pi \int_{-a}^{a} y^2 \, dx$$

$$y^2 = b^2\left(1 - \frac{x^2}{a^2}\right)$$

$$V = \pi b^2 \int_{-a}^{a} \left(1 - \frac{x^2}{a^2}\right) dx$$

$$= \pi b^2 \left[x - \frac{x^3}{3a^2}\right]_{-a}^{a}$$

$$= \pi b^2 \left[\left(a - \frac{a^3}{3a^2}\right]\right. -$$

$$\pi b^2 \left[\left(-a - \frac{(-a)^3}{3(-a)^2}\right)\right]$$

$$= \pi b^2 \left(a - \frac{a}{3} + a - \frac{a}{3}\right)$$

$$\text{Volume V} = \frac{4}{3} ab^2\pi \qquad \text{Answer.}$$

15.

$$\frac{dv}{dt} = -kv$$

$$\frac{dv}{v} = -kt$$

$$\ln v = -kt + C$$

$$\text{when } t = 0 \; v = u \text{ so } C = \text{In } u$$

$$\ln v - \ln u = -kt$$

$$\ln \frac{v}{u} = -kt \qquad \text{Answer.}$$

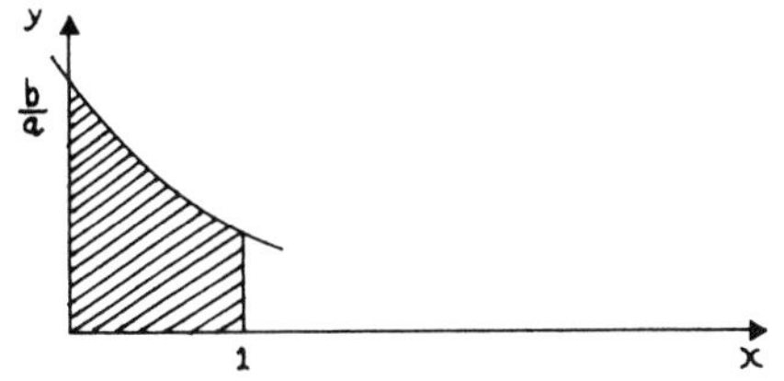

Fig. 227

16. $$\text{Area} = \int_0^1 y\,dx = \int_0^1 \frac{b}{x+a}\,dx$$

$$= b\int \frac{dx}{x+a} = b\Big[\text{In}(x+a)\Big]_0^1$$

$$= b\Big[\text{In}(1+a) - \text{In}(0+a)\Big]$$

$$= b\Big[\text{In}(1+a) - \text{In}\,a\Big]$$

$$= b\ \text{In}\ \frac{1+a}{a}$$

$$= 2\ \text{In}\ 2$$

$$= 1{\cdot}385 \text{ sq. units.} \qquad \text{Answer.}$$

17. (a) $$\int \frac{3}{x}\,dx = 3\int \frac{dx}{x}$$

$$= 3\ln x + C \qquad \text{Answer.}$$

(b) $$\int \frac{1}{a}\,x^3\,dx = \frac{1}{a}\int x^3\,dx$$

$$= \frac{x^4}{4a} + C \qquad \text{Answer.}$$

(c) $$\int (4z^3 + 3z^2 + 2z + 1)\,dz = z^4 + z^3 + z^2 + z + C \qquad \text{Answer.}$$

(d) $$\int (2\cos\theta - 5\sin\theta)\,d\theta = 2\sin\theta + 5\cos\theta + C \qquad \text{Answer.}$$

SELECTION OF EXAMINATION QUESTIONS

1. Determine the second differential coefficients of:

   (a) $y = x^4 - \dfrac{2}{x^2} + 7x - 9$

   (b) $y = \ln x + \sin x$

2. (a) Prove that x can have any value to satisfy the equation:

$$\left(x + \frac{1}{x}\right)^2 - \left(x - \frac{1}{x}\right)^2 = 4$$

   (b) Find the values of (i) $x^2 + \dfrac{1}{x^2}$, (ii) $x^4 + \dfrac{1}{x^4}$, and (iii) x, in the equation

$$\left\{x + \frac{1}{x}\right\}^2 = 9$$

3. Water flows along a horizontal pipe of 100 mm internal diameter, the depth of water in the pipe is 75 mm. Calculate the hydraulic mean depth d of the water from the formula:

$$d = \frac{\text{cross-sectional area of water stream}}{\text{perimeter of wetted pipe surface}}$$

4. Given $P = 120/V$, where P and V represent the pressure (N/m^2) and volume (m^3) of a given mass of gas at constant temperature, calculate by integration the work done when the gas expands from

$$V = 0{\cdot}5\text{m}^3 \text{ to } V = 6{\cdot}5\text{m}^3$$

5. The volumetric analysis of a certain mixture is 68% water, 18% alcohol and the remainder solids. Calculate (a) how much water should be added to each litre of this mixture to reduce the alcohol content to 15%, (b) the percentage analysis of the new mixture.

6. Differentiate the following with respect to x.

(a) $Rgx - \frac{Cx^3}{64L}$ (b) $ax^3 + \frac{b}{x^2}$

Integrate the following

(c) $\int_{+1}^{+2} (k^{2\cdot 6} - 1)\, dk$ with respect to k, and

(d) $\int [(2v^2 - 4)v]\, dv$ with respect to v.

7. A chord in a circle is 80 mm long. A tangent to the circle drawn through one end of the chord where it meets the circumference makes an angle of 60 degrees to the chord. Find the area of the smaller segment of the circle cut off by the chord.

8. Sketch the graph of $y = 11 + 7x - 2x^2$ and use differential calculus to determine the maximum value of y.

9. A component consists of a cone of base diameter 52·5 mm and perpendicular height 45 mm, standing concentrically on the top of a solid cylinder 60 mm diameter and 40 mm high. The base of this component is firmly fixed on to a flat base-plate. Calculate the total surface area of the component exposed above the base-plate.

10. 0·016 m³ of metal is used to cast two solid spheres. If one is 300 mm diameter, find the diameter and mass of the other taking the density of the metal as 7·21 g/cm³.

11. (a) The distance S metres travelled by a body in time t seconds is given by:

$$S = t^3 - 5{\cdot}5t^2 - 4t + 68{\cdot}5$$

Determine the time, distance travelled, and acceleration of the body when its velocity becomes zero.

(b) Integrate with respect to x:

(i) $3\sqrt{x} - 3x^2$

(ii) $\frac{2}{x^3} + 2$

12. A roof has the form of a square pyramid, the base is 24 metres square and the length of the ridge from corner of base to apex is 18 metres. Find (i) the perpendicular height from base to apex, (ii) the total surface area of the roof.

13. (i) Multiply $2x^3 - 3x^2 + 2x + 4$ by $x^2 - 3x + 5$
 (ii) Divide $6x^5 + x^4 + 2x^3 - 4x^2 - 7x + 4$
 by $3x^3 - x^2 + 3x - 4$
 (iii) Multiply $(2a^2b)^3$ by $(3ab^2)^2$
 (iv) Divide $(3ab^2)^3$ by $(ab^3)^2$

14. Find the values of angle θ between 0 and 360° which will satisfy the equation:

$$\tan\theta = \cos\theta$$

15. (a) Determine the second differential coefficients of:
 (i) $4\cos x + \ln x$
 (ii) $3x^3 + \dfrac{4}{x^2} + 6x - 2$

 (b) The expression $T = 1700 + 16000/x^2 + 0{\cdot}6x^2$ is used to calculate the hoop tensile strength T of certain rotating discs. If x is the disc radius, use calculus to determine the value of x for which T is a minimum and the corresponding value at T at that radius.

16. The section of a water trough is an isosceles triangle, the angle at the apex (bottom) being 90 degrees. Find the volume flow of water along the trough, in cubic metres per hour and in litres per minute, when the velocity of the water is 0·2 m/s and its depth is 90 mm.

17. The following table gives corresponding values of x and y. Plot y against x and, assuming they are connected by a straight line law, find the values of the constants and state the law.

x	−3·8	−2	+2·1	+4·1	+5·9	+8
y	−18	−13·7	−5·9	−2	+1·7	+6

18. A sector of a circle is cut out of a piece of sheet metal, the radius of the sector being 102 mm and the angle at the centre 150 degrees. The sector is then rolled into the shape of a cone. Find the surface area of the metal used, the diameter of the base of the cone and its perpendicular height.

19. Find the values of x and y in the simultaneous equations:

$$10y + 4b = -2x$$
$$3a + x = y$$

given that the same values of a and b in the above also satisfy the equations:

$$19a + 27b = 43$$
$$7a + 4b = -2$$

20. The rim of a flywheel is elliptical in cross-section with the major axis in the radial direction. The major and minor axes of the rim section measure 300 and 200 mm respectively, and the overall diameter of the wheel is 3·6 m. If the material is cast iron of density $7{\cdot}21 \times 10^3$ kg/m^3, find the mass of the rim.

21. Determine the gradient of a tangent to any point in the curve

$$y = \frac{x^3}{4} - \frac{x^2}{2}$$

and show that there are two points where the gradient is zero. Determine the point on the curve where y is a minimum.

22. Factorise completely:

(i) $2x^2 - 2$

(ii) $7x^2 + 9xy - 10y^2$

(iii) $14x^3y^2 + 22x^2y^3 - 12xy^4$

(iv) $3x^3 + 8x^2y - xy^2 - 10y^3$ given that $(x - y)$ is one of the factors.

23. A solid cone 40 mm diameter at the base and of perpendicular height 60 mm, has a 15 mm diameter hole bored through the centre axis from base to apex. Find the volume of material removed from the cone, in cm^3.

24. Using integral calculus determine the difference in area bounded by the curves $y = 10x^2 + x + 4$ and $y = x^2/2 - 1/x - 1/2$ and the x axis between the ordinates $x = 1$ and $x = 5$.

25. (a) Using integral calculus, determine the area bounded by the curve $y = 2 \cos x$ and the x-axis between the ordinates at $x = 0$ and $x = \pi/4$.

(b) A curve which passes through the point ($x = 1, y = -2$) has a gradient of $3x^2 - 4$. Determine the equation of the curve.

26. (i) Prove that the solution of a quadratic equation of the form,

$$ax^2 + bx + c = 0$$

can be obtained from,

$$x = \frac{-b \pm \sqrt{b^2 - 4ac}}{2a}$$

(ii) Solve the following quadratic equation by the method of "completing the square",

$$x^2 + 5x = 84$$

(iii) Find the value of x in the following quadratic equation by factorising,

$$2x^2 - 3ax - bx = 0$$

27. Determine the second differential coefficient of:

(a) $y = x^2 - 5x + 7 - \frac{1}{x}$

(b) $y = 2\cos x - 3\sin x$

28. A block of brass is 80 mm long and has a varying cross-section throughout its length. Commencing at one end the cross-sectional areas, at regular distances apart, are 0, 2·9, 3·8, 4·1, 3·7, 2·7 and 1·1 square centimetres respectively. Find the volume by Simpson's rule and the mass taking the density of brass as 8·4 g/cm³.

29. On a certain production machine, the cost C of manufacturing N articles is given by $C = a + bN$ where a and b are constants. If it costs £8·50 for 300 articles, and £7·50 for 200 (a) produce an equation to give the cost in pounds to manufacture any number of articles, (b) find the cost to make 150, (c) what is the minimum cost of running the machine?

30. Using integral calculus, determine the area bounded by the curve.

$$y = \sin x + \tfrac{1}{2}\cos x$$

and the values $x = 60°$ and $x = 30°$

31. A piece of wire 1 m long is cut into two pieces and each is bent into a form of a square.
Using differential calculus, determine the least possible sum of the two areas.

32. A flat mild steel plate of elliptical shape measures 420 mm across the major axis and 370 mm across the minor axis, and is 45 mm thick. There is a groove around the whole of the perimeter, 13 mm wide by 13 mm deep. Taking the density of steel as 7·86 g/cm^3, find the mass of the plate.

33. Two chords, 60 and 80 mm long respectively, are drawn parallel to each other inside a circle. If the chords are 10 mm apart, find the radius of the circle.

34. A railway cutting, in the form of a trapezium, is 10 metres wide at the bottom, 15 metres wide at the top, and 14 metres deep. The cutting sweeps around on a circular arc, the radius of the arc to the centre of the cutting is 805 metres, and the angle subtended is 6 degrees. Calculate the volume of earth removed, in cubic metres, in the making of the cutting.

35. In a triangle ABC the lengths of the sides AB and AC are 50 and 30 mm respectively and the angle BAC is 20 degrees. Calculate the length of the other side, the two remaining angles, and the area.

36. Draw graphs of the following simultaneous equations between the limits $x = -2$ and $x = +3$ and find the values of x and y,

$$x^2 - y = 1$$
$$2y - 3x = 4$$

37. Determine the area contained between the curves $P = \dfrac{78}{V^{1\cdot 2}}$, $P = \dfrac{7}{V^{1\cdot 15}}$ and the ordinates at $V = 2$ and $V = 5$.

38. A body moves s metres in t seconds according to the relationship

$$s = t^3 - 6t^2 + 9t + 3$$

Given that velocity $v = ds/dt$ and acceleration $a = dv/dt$, calculate

(i) the values of t at which the body is stationary

(ii) the values of the acceleration corresponding to the values of t determined in (i).

39. Find the value of x in the following equation:

$$\frac{x+2}{x-1} - \frac{x-1}{3x-4} = \frac{7}{2}$$

40. Calculate the area of the largest isoceles triangular plate with base angles of 75 degrees that can be cut out of a circular plate having an area of 804·2 square centimetres.

41. (a) Differentiate the following equations

(i) $y = (Ei - Ri^2)$ with respect to i

(ii) $y = \left[\frac{C}{1-n}\right]\left[v^{1-n} - p^{1-n}\right]$ with respect to v

(b) evaluate the following integrals

(iii) $\int_0^1 12x^7 dx$

(iv) $\frac{3}{2\pi r^3} \int_0^r \pi(r^2 - x^2)\,dx$

42. Draw a straight line graph through a pair of points $(-1, 5)$ and $(4, -10)$ and find the equation to the graph.

43. (i) Prove that $\sin^2\theta + \cos^2\theta = 1$

(ii) Find the values of θ between 0 and 360 degrees which will satisfy the equation:

$$2\sin^2\theta - 2\cos^2\theta - 2 = 0$$

44. Prove that the radius of a circle inscribed in a triangle is given by:

$$R = \frac{2 \times \text{area of triangle}}{\text{perimeter of triangle}}$$

and calculate the radius of the circle inscribed in a triangle of sides 50, 60 and 70 mm respectively.

45. Determine the value of x for which expression

$$y = 4x^3 - 3x^2 - 18x$$

is a maximum or minimum, and calculate the corresponding values of y stating whether it is a maximum or minimum.

46. A solid consists of a right circular cone of 100 mm base diameter mounted concentrically on the flat surface of a hemisphere of equal diameter, the perpendicular height of the conical part being also 100 mm.

Calculate the volume of the solid in cm^3 and the curved surface area in cm^2.

If another similar solid was made with all linear dimensions double those of the original, what would be its volume and curved surface area?

47. Plot the graph of the equation $y = x^2 - 8x + 7$ between $x = 0$ and $x = 9$, using intervals of unity. From the graph solve for x in each of the following equations:

(a) $x^2 - 7x + 6 = 0$

(b) $x^2 - 9x + 8 = 0$

Suggested scales: x axis 2 cm = 1
y axis 2 cm = 5

48. For a certain ship at deadweight displacement, an expression connecting the power of the main engines and the speed of the ship is given by:

$$P = V(a + bV^2)$$

where, P = power of the engines in kW.
V = speed of the ship in knots.
a and b = constants.

If the powers are 6024 and 13408 kW when the speeds are 12 and 16 knots respectively, calculate the probable power when the speed is 14 knots.

49. A wheel 1·5 m diameter stands on the ground on its rim and the rim is marked at the point where it touches the ground. Find the height of the mark above the ground after the wheel has been rolled along the ground for a distance of 1·5 m.

50. Obtain the second differential coefficients of:

(i) $y = \dfrac{x^3}{3} - \dfrac{3}{x^3} + 3x^2 - 2x$

(ii) $y = 3 \sin x + 4 \cos x$

51. A buoy is constructed by welding the base of a hollow cone to the flat base rim of a hollow hemisphere, the common diameter of their bases being 2·44 m. The greatest length of the buoy is 4·27 m. Calculate (i) the curved surface area of the cone, (ii) the angle subtended at the centre of the sector of a circle when the cone is developed, (iii) the curved surface area of the hemisphere, (iv) the external volume of the buoy.

52. Solve for a, b, c and d in the following four simultaneous equations:

$$\begin{aligned} 3a + 7b + 2c &= 25 \\ 2a + 6b + c &= 18 \\ a + b + c + d &= 14 \\ 2a + d &= 9 \end{aligned}$$

53. A lens is flat on one side and convex on the other side, the convex surface being part of a sphere. The diameter of the flat surface is 60 mm. The maximum thickness of the lens is 4 mm and the minimum thickness is nil. Calculate the curved surface area of the lens in cm^2.

54. (a) If $dy/dx = -\cos x - 2 \sin x$, determine an expression for y in terms of x, given that $y = 3$ when $x = 0$.

(b) The area bounded by the curve

$$y = \frac{x^2}{4}$$

the ordinate $x = 4$ and the x-axis between 0 and 4 is rotated once around the x-axis. Use integral calculus to determine the volume of the solid generated.

55. Transpose the terms in the following equation to make K the subject:

$$p = \left[\frac{wu^2}{g\left\{\dfrac{K + a}{K(2a + K) + a^2} + \dfrac{D}{Et}\right\}}\right]^{\frac{1}{2}}$$

56. (i) If x varies directly as z and inversely as y^2, calculate the percentage change in x when z is increased by 12 per cent and y is decreased by 20 per cent.

(ii) The time for one beat of a pendulum varies as the square root of its length. If a pendulum 174·4 mm long makes 105 beats in 44 seconds, calculate the length of a similar pendulum to beat exactly once every second.

57. Solve the following simultaneous equations to give the smallest values of the angles x and y:

$$3\cos x + 4\sin y = 2\sqrt{2} - 1{\cdot}5$$
$$5\cos x - \sqrt{2}\sin y = -3{\cdot}5$$

58. Show that the height of a frustum of a right circular cone is one-third the height of the complete cone when the volume of the frustum is 19/27 that of the complete cone.

59. (a) Determine the area bounded by the curve $y = 2\sin\theta$ and the θ axis between the ordinates $\theta = \pi/6$ radians and $\theta = 5\pi/6$ radians.

(b) Evaluate:

(i) $\displaystyle\int_{\frac{\pi}{2}}^{\pi} (3\cos\theta)d\theta$

(ii) $\displaystyle\int_{1}^{2} (x^2 - 6x)dx$

(iii) $\displaystyle\int_{1}^{3} (x^4 - 7x + 8)dx$

60. Express V in terms of v, f_1 and f_2 in the following expression where $K = f_1/f_2$ and find the values of V when $f_1 = 12$, $f_2 = 18$ and $v = 20$,

$$K = \left\{\frac{V + v}{V - v}\right\}^2$$

61. Plot the graph $y = x^2 + 2x - 8$ between the limits $x = -4$ and $x = +2$. If the area bounded by this curve and the x-axis is rotated through one revolution about the x-axis, calculate the volume generated, using Simpson's rule.

62. The total area of metal A needed to make a box of stated volume is given by the expression

$$A = x^2 + \frac{16000}{x}$$

where x is the side of the square base. Use calculus to find:

(i) the value of x for which A is a minimum.

(ii) the minimum value of A.

63. Three steel ball bearings, each 2 cm diameter, are placed inside a cylinder 6 cm diameter, equally spaced around the bottom, then another ball bearing 4 cm diameter is placed on top of these. Calculate the volume of water, in millilitres, required to pour into the cylinder so that the top ball is just submerged.
If all linear dimensions were half of those given above, what would be the volume of water required?

64. The distance from the centre to a corner of a regular pentagon is 3 cm, find the area of the pentagon. If a piece is sliced off by a straight cut from one corner at an angle of 30° to a side, find the ratio of the areas of the two pieces so formed.

65. (a) Find the angles between 0 and 360° that will satisfy the equation:

$$2 \sin^2 \theta - \cos \theta = 1$$

(b) Find the angles between 0 and 360° that will satisfy the equation:

$$3 \cos 2\theta + 2 \cos \theta = 0$$

66. The resistance R (newtons) to motion of a body and its velocity V (metres per second) is tabulated below:

V	5	10	15	20	25
R	10	25	50	85	130

Show that these values are connected by the law $R = k + cV^2$, find the values of the constants k and c and calculate the probable resistance to motion when the velocity is 40 metres per second.

67. Transpose the following to express k in terms of p and V, and find the value of k when $p = 9{\cdot}2$ and $V = 0{\cdot}42$,

$$p = V + \frac{k(p + V)}{k + p + V}$$

68. (a) Calculate the area of a trapezium inscribed in a semi-circle of radius 10 cm if the length of each non-parallel side of the trapezium is equal to the radius of the semi-circle.

(b) If this trapezium is rotated through one complete revolution about an axis passing through the centre of, and perpendicular to, its long side, calculate the volume of the solid frustum of a cone thus generated.

69. (a) Find the values of a, b and w in the simultaneous equations:

$$\begin{aligned} 4a + 3b + 2w &= 2 \\ 2a + 8b + 3w &= 11 \\ 3a + 5b - 6w &= -36 \end{aligned}$$

(b) Find the values of x and y in the simultaneous equations:

$$\begin{aligned} x^2 + y^2 &= 18{\cdot}5 \\ x - y &= 1 \end{aligned}$$

(c) Find the value of x in the following equation:

$$\frac{100}{x} + \frac{100}{x + 5} = \frac{50}{3}$$

70. (a) Prove, $\dfrac{1 - \tan^2 \theta}{1 + \tan^2 \theta} = \cos^2 \theta - \sin^2 \theta$

(b) Prove, $\sin \theta + \sin \theta \cot^2 \theta + \dfrac{\tan \theta}{\cos \theta} = \dfrac{1}{\sin \theta \cos^2 \theta}$

71. Draw the graphs of the following equations between the limits $x = -2$ and $x = +8$ and find the area of the triangle enclosed by the graphs,

$$\begin{aligned} 4y - x - 20 &= 0 \\ y - x + 1 &= 0 \\ 2y + x - 7 &= 0 \end{aligned}$$

72. A hexagon of unit side is to have its area reduced by 20% by cutting it parallel to one side. Calculate the thickness of the piece cut off and the distance from the centre of the original hexagon to the corner of the new face.

73. Using Simpsons rule, with eleven ordinates, calculate the area between the curve $y = e^{-2x}$ and the x axis between the limits $x = 0$ and $x = +2$ and hence determine the length of the mean ordinate for this section of the curve.

74. The metal from the melting down of a solid cone 60 cm base diameter and 50 cm slant height is to be cast into the form of a hollow sphere of 1·5 cm uniform thickness. Calculate the outer and inner diameters.

75. Transpose for h in the following expression and use the result to evaluate h when $a = 1{\cdot}5, c = 15, s = 16$ and $d = 17$

$$a = \frac{c}{2s}\left\{\frac{h^2}{d - h}\right\}$$

76. Three points A, B and C lie on a horizontal ground. C is due North of A, and B is due East of a point W which lies on the line joining A and C. A vertical mast stands on point B and the angle of elevation to the top of the mast from A is 20°, AB = 60 m, BC = 32 m, and AC = 68 m.

(a) Prove that the angle at B is a right angle.

(b) Find the angles at A and C.

(c) Calculate the height of the mast.

(d) Find the angle of elevation from point W to the top of the mast.

77. Draw the graph $y = x^2 + 1$ between the limits $x = -2$ and $x = +3$ and use this to solve the equation $x^2 - x = 2$.

78. (a) Prove that the area of a segment of a circle is equal to

$$\frac{r^2}{2}(\theta - \sin\theta)$$

where r = radius of circle
θ = angle subtended in radians

(b) A horizontal pipe is more than half full of water. The breadth of the water-level is 60 cm and the height from water-level to crown of pipe is 15 cm. Calculate (i) the diameter of the pipe, (ii) the cross-sectional area of the water, (iii) the litres of water contained over a length of 2 metres.

79. (a) Find the values of WZ, W^2Z, W^2/Z and Z^3/W when $W = x^{a+b}$ and $Z = x^{a-b}$

(b) Find the values of x and y in the simultaneous equations:

$$4x + y = 10$$

$$\frac{2}{x} + \frac{7}{y} = 3$$

80. In each of the following formulae, transpose and re-write in terms of k:

(a) $Q = \sqrt{\dfrac{2gh\,D\,k^2}{d(s^2 - k^2)}}$

(b) $F = \dfrac{1}{2\pi}\sqrt{\dfrac{1}{Lk}}$

(c) $L = \dfrac{\pi}{2}(k + b) + \dfrac{(k - b)}{4c} + 2C$

81. The equation $7x^2 - 28 + 7y^2 = 0$ represents the law of a circle. Use integral calculus to determine the volume of revolution generated by the middle part of the circle about its x axis between the limits $x = +1$ and $x = -1$.

Take each unit of x and y as representing 1 centimetre.

SOLUTIONS TO EXAMINATION QUESTIONS

1. (a)

$$y = f(x) = x^4 - \frac{2}{x^2} + 7x - 9$$

$$f(x) = x^4 - 2x^{-2} + 7x - 9$$

$$f'(x) = 4x^3 + 4x^{-3} + 7$$

$$f''(x) = 12x^2 - 12x^{-4}$$

$$f''(x) = 12\left(x^2 - \frac{1}{x^4}\right)$$ Answer.

(b)

$$y = f(x) = \text{In}\, x + \sin x$$

$$f'(x) = \frac{1}{x} + \cos x$$

$$f'(x) = x^{-1} + \cos x$$

$$f''(x) = -x^{-2} - \sin x$$

$$f''(x) = -\left(\sin x + \frac{1}{x^2}\right)$$ Answer.

2. $$\left\{x + \frac{1}{x}\right\}^2 - \left\{x - \frac{1}{x}\right\}^2 = 4$$

This is the 'difference between two squares', similar to $a^2 - b^2$.
Factorising $a^2 - b^2$ we get $(a + b)(a - b)$
Factorising the given expression we have:

$$\left[\left(x + \frac{1}{x}\right) + \left(x - \frac{1}{x}\right)\right] \times \left[\left(x + \frac{1}{x}\right) - \left(x - \frac{1}{x}\right)\right] = 4$$

$$\left[x + \frac{1}{x} + x - \frac{1}{x}\right] \times \left[x + \frac{1}{x} - x + \frac{1}{x}\right] = 4$$

$$2x \times \frac{2}{x} = 4$$

$$4 = 4$$

x cancels therefore it can have any value. Ans. (a)

$$\left\{x + \frac{1}{x}\right\}^2 = 9$$

$$x^2 + 2 + \frac{1}{x^2} = 9$$

$$\therefore x^2 + \frac{1}{x^2} = 7 \text{ Ans. b (i)}$$

Squaring both sides,

$$x^4 + 2 + \frac{1}{x^4} = 49$$

$$x^4 + \frac{1}{x^4} = 47 \text{ Ans. b (ii)}$$

$$\left\{x + \frac{1}{x}\right\}^2 = 9$$

Taking square root of both sides,

$$x + \frac{1}{x} = 3$$

Multiplying throughout by x,

$$x^2 + 1 = 3x$$

$$\text{or, } x^2 - 3x + 1 = 0$$

Solving this quadratic,

$$x = 2{\cdot}618 \text{ or } 0{\cdot}382 \quad \text{Ans. b (iii)}$$

3.

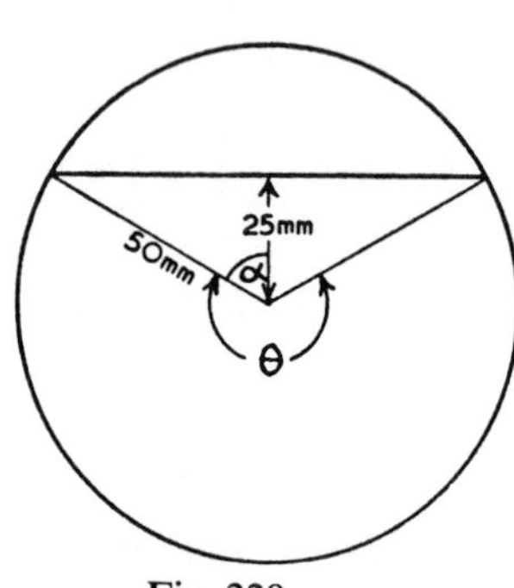

Fig. 228

Working in centimetres:

$$\cos\alpha = \frac{2{\cdot}5}{5} = 0{\cdot}5 \qquad \alpha = 60^\circ$$

$$\theta = 360 - (2 \times \alpha) = 240^\circ$$

θ in radians

$$= \frac{240}{360} \times 2\pi = 4{\cdot}19$$

$$\sin\theta = -\sin(240 - 180) = -\sin 60 = -0{\cdot}866$$

Area of segment of angle θ,

$$= 0{\cdot}5r^2(\theta - \sin\theta)$$

$$= 0{\cdot}5 \times 5^2[4{\cdot}19 - (-0{\cdot}866)]$$

$$= 12{\cdot}5 \times 5{\cdot}056 = 63{\cdot}2 \text{ cm}^2$$

Wetted perimeter of pipe surface

$$= \frac{240}{360} \times \pi \times 10 = 20{\cdot}94 \text{ cm}$$

Hydraulic mean depth

$$= \frac{63{\cdot}2}{20{\cdot}94} = 3{\cdot}017 \text{ cm or } 30{\cdot}17 \text{ mm} \quad \text{Ans.}$$

4. $$W = \int_{V_1}^{V_2} P\,dV \text{ where } W \text{ is work done}$$

$$= \int_{0\cdot5}^{6\cdot5} \frac{120\,dV}{V}$$

$$= 120\,[\ln V]_{0\cdot5}^{6\cdot5}$$

$$120[\ln 6\cdot5 - \ln 0\cdot5]$$

$$= 120 \ln \frac{6\cdot5}{0\cdot5}$$

$$= 120 \ln 13$$

$$= 120 \times 2\cdot565$$

$$W = 307\cdot8 \text{ Nm}$$ Answer.

5. By adding x litre of water, the original one litre of the mixture is increased to $(1 + x)$ litres; the original amount of alcohol was 18 per cent of one litre = 0·18 litre, and this is to be 15 per cent of the new mixture.

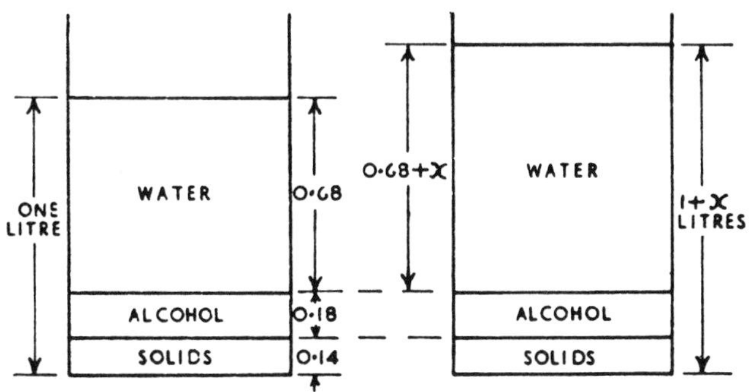

Fig. 229

$$\frac{15}{100}(1 + x) = 0{\cdot}18$$

$$15(1 + x) = 0{\cdot}18 \times 100$$
$$15 + 15x = 18$$
$$15x = 3$$
$$x = 0{\cdot}2 \text{ litre or } 200 \text{ ml} \quad \text{Ans. (i)}$$

$$\% \text{ water in new mixture} = \frac{0{\cdot}68 + 0{\cdot}2}{1 + 0{\cdot}2} \times 100$$
$$= \frac{0{\cdot}88 \times 100}{1{\cdot}2}$$
$$= 73\tfrac{1}{3}\%$$

$$\% \text{ solids} = \text{remainder} = 100 - (73\tfrac{1}{3} + 15)$$
$$= 11\tfrac{2}{3}\%$$

$\therefore$ New analysis is: Water $= 73\frac{1}{3}\%$, Alcohol $= 15\%$, Solids $= 11\frac{2}{3}\%$ } Ans.

6. (a)

$$y = Rg\,x - \frac{Cx^3}{64L}$$

$$\frac{dy}{dx} = Rg - \frac{3\,Cx^2}{64L} \qquad \text{Answer.}$$

$$y = ax^3 + \frac{b}{x^2}$$

$$y = ax^3 + bx^{-2}$$

$$\frac{dy}{dx} = 3ax^2 - 2bx^{-3}$$

$$= 3ax^2 - \frac{2b}{x^3} \qquad \text{Answer.}$$

(c) $$\int_{+1}^{+2} (k^{2 \cdot 6} - 1)\, dk$$

$$= \left[\frac{k^{3\cdot6}}{3\cdot6} - k\right]_1^2$$

$$= \left[\frac{2^{3\cdot6}}{3\cdot6} - 2\right] - \left[\frac{1^{3\cdot6}}{3\cdot6} - 1\right]$$

$$= \left[\frac{12\cdot13}{3\cdot6} - 2\right] - \left[\frac{1}{3\cdot6} - 1\right]$$

$$= [3\cdot369 - 2] - [0\cdot2778 - 1]$$

$$= 1\cdot169 + 0\cdot7222$$

$$= 1\cdot891$$ Answer.

(d) $$= \int [2v^2 - 4)v]\, dv$$

$$= \int [2v^3 - 4v)\, dv$$

$$= \frac{v^4}{2} - 2v^2 + C$$ Answer.

7.

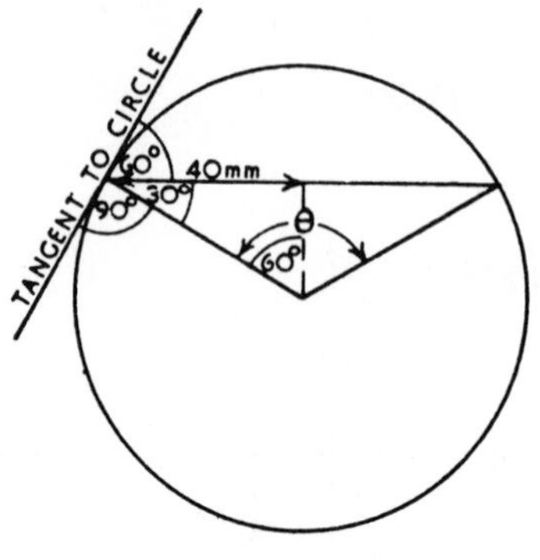

Fig. 230

Referring to Fig. 228, working in centimetres,

$$r = \frac{4}{\sin 60} = 4{\cdot}619 \text{ cm}$$

$$\theta = 2 \times 60 = 120° = 2{\cdot}0944 \text{ radians}$$

$$\sin 120° = 0{\cdot}866$$

$$\begin{aligned}\text{Area of segment} &= \tfrac{1}{2}r^2\,[\theta - \sin\theta] \\ &= \tfrac{1}{2} \times 4{\cdot}619^2\,[2{\cdot}0944 - 0{\cdot}866] \\ &= \tfrac{1}{2} \times 4{\cdot}619^2 \times 1{\cdot}2284 \\ &= 13{\cdot}1 \text{ cm}^2 \text{ Ans.}\end{aligned}$$

8.

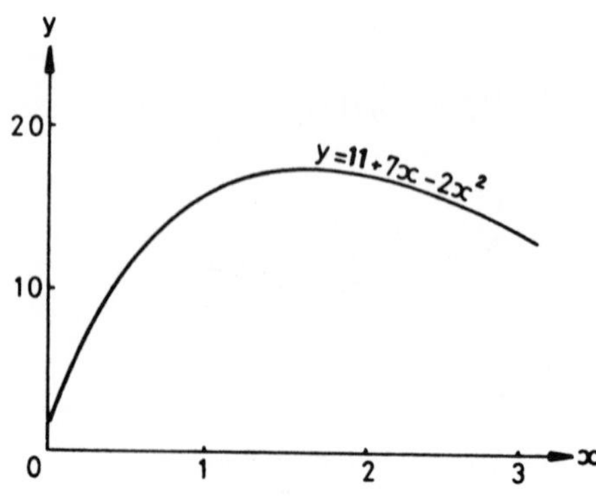

Fig. 231

$$y = 11 + 7x - 2x^2$$

$$\frac{dy}{dx} = 7 - 4x$$

For turning points,

$$7 - 4x = 0$$

$$\therefore x = \frac{7}{4}$$

From the graph, it is obvious that $x = 1\frac{3}{4}$ gives a *maximum* value of y.

$$\therefore y = 11 + 7(\tfrac{7}{4}) - 2(\tfrac{7}{4})^2$$

$$= 11 + \frac{49}{4} - \frac{98}{16}$$

$$= 17\tfrac{1}{8} \quad \text{Answer.}$$

9. Working in centimetres:

Radius of cone base $= \frac{1}{2} \times 5{\cdot}25 = 2{\cdot}625$ cm

Slant height of cone $= \sqrt{4{\cdot}5^2 + 2{\cdot}625^2} = 5{\cdot}21$ cm

Curved surface area of cone $= \pi r l$

$= \pi \times 2{\cdot}625 \times 5{\cdot}21$

$= 42{\cdot}96$ cm^2 ... ... (i)

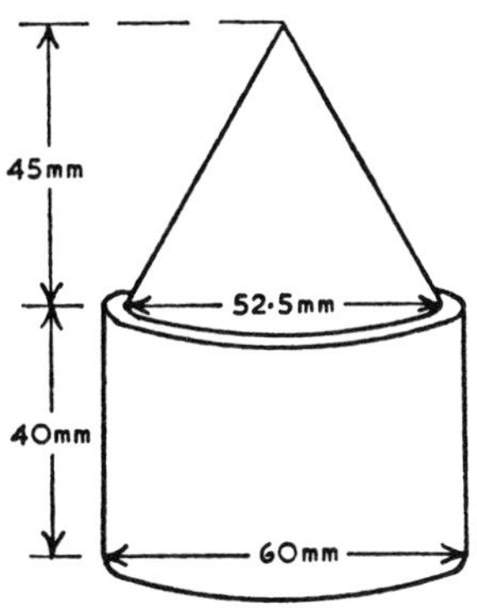

Fig. 232

Area of annulus between top of cylinder and base of cone,

$$= 0{\cdot}7854 \times (D^2 - d^2)$$
$$= 0{\cdot}7854(6 + 5{\cdot}25)(6 - 5{\cdot}25)$$
$$= 6{\cdot}628 \text{ cm}^2 \quad \ldots \quad \ldots \quad \text{(ii)}$$

Curved surface area of cylinder

$$= \pi dh$$
$$= \pi \times 6 \times 4$$
$$= 75{\cdot}41 \text{ cm}^2 \quad \ldots \quad \ldots \quad \text{(iii)}$$

Total surface area exposed $= 42{\cdot}96 + 6{\cdot}628 + 75{\cdot}41$

$= 124{\cdot}998$

say 125 cm^2 Ans.

10. Volume of sphere $= \dfrac{\pi}{6} d^3$

Vol. of 300 mm dia. sphere $= \dfrac{\pi}{6} \times 0{\cdot}3^3 = 0{\cdot}01414 \text{ m}^3$

Vol. of other sphere $= 0{\cdot}016 - 0{\cdot}01414 = 0{\cdot}00186 \text{ m}^3$

$$\therefore d = \sqrt[3]{\frac{0{\cdot}00186 \times 6}{\pi}}$$

$= 0{\cdot}1526 \text{ m} = 152{\cdot}6 \text{ mm}$ Ans. (i)

Mass [kg] $=$ volume [m^3] $\times$ density [kg/m^3]

$= 0{\cdot}00186 \times 7{\cdot}21 \times 10^3$

$= 13{\cdot}41$ kg ... ... Ans. (ii)

11. (a)

$$S = t^3 - 5{\cdot}5t^2 - 4t + 68{\cdot}5$$

$$v = \frac{dS}{dt} = 3t^2 - 11t - 4 = 0$$

$$0 = (3t + 1)(t - 4)$$

$t = 4s$ i.e. minus time inadmissable

$$S = 4^3 - 5{\cdot}5 \times 4^2 - 4 \times 4 + 68{\cdot}5$$

$$= 64 - 88 - 16 + 68{\cdot}5$$

$$= 28{\cdot}5\ m$$

$$f = \frac{dv}{dt} = 6t - 11$$

$$= 24 - 11$$

$$= 13\ m/s^2$$

Time $= 4s$
Distance travelled $= 28{\cdot}5\ m$ } when velocity is zero — Answer.
Acceleration $= 13\ m/s^2$

(b). (i)

$$\int (3\sqrt{x} - 3x^2)\,dx$$

$$= \int (3x^{1/2} - 3x^2)\,dx$$

$$= 2x^{3/2} - x^3 + C$$

$$= 2\sqrt{x^3} - x^3 + C \qquad \text{Answer}$$

(b). (ii)

$$\int \left(\frac{2}{x^3} + 2\right) dx$$

$$= \int (2x^{-3} + 2)\,dx$$

$$= -x^{-2} + 2x + C$$

$$= 2x - \frac{1}{x^2} + C \qquad \text{Answer.}$$

12.

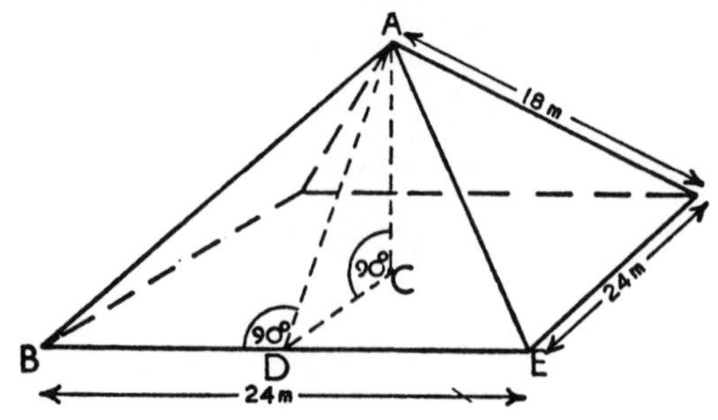

Fig. 233

AB = 18, BD = 12, DC = 12

$$(AD)^2 = (AB)^2 - (BD)^2$$
$$= 18^2 - 12^2$$
$$AD = \sqrt{180} = 13{\cdot}42$$
$$(AC)^2 = (AD)^2 - (DC)^2$$
$$AC = \sqrt{18^2 - 12^2 - 12^2}$$
$$= 6 \text{ m} \quad \text{Ans. (i)}$$

Area of one sloping triangular flat surface,

$$= \tfrac{1}{2} \times BE \times AD$$

$$\text{Total surface area of roof} = 4 \times \tfrac{1}{2} \times 24 \times 13{\cdot}42$$
$$= 644{\cdot}2 \text{ m}^2 \quad \text{Ans. (ii)}$$

13.

$$\begin{array}{rrrrrr} & & 2x^3 & -\ 3x^2 & +\ 2x & +\ 4 \\ & & & x^2 & -\ 3x & +\ 5 \\ \hline & & +\ 10x^3 & -\ 15x^2 & +\ 10x & +\ 20 \\ & -\ 6x^4 & +\ 9x^3 & -\ 6x^2 & -\ 12x & \\ +\ 2x^5 & -\ 3x^4 & +\ 2x^3 & +\ 4x^2 & & \\ \hline 2x^5 & -\ 9x^4 & +\ 21x^3 & -\ 17x^2 & -\ 2x & +\ 20 \end{array} \quad \text{Ans. (i)}$$

$$3x^3 - x^2 + 3x - 4)6x^5 + x^4 + 2x^3 - 4x^2 - 7x + 4(2x^2 + x - 1 \quad \text{Ans. (ii)}$$
$$6x^5 - 2x^4 + 6x^3 - 8x^2$$

$$3x^4 - 4x^3 + 4x^2 - 7x$$
$$3x^4 - x^3 + 3x^2 - 4x$$

$$-3x^3 + x^2 - 3x + 4$$
$$-3x^3 + x^2 - 3x + 4$$

. . . . .

$$\begin{aligned} &(2a^2b)^3 \times (3ab^2)^2 \\ =\ &8a^6b^3 \times 9a^2b^4 \\ =\ &72a^8b^7 \text{ Ans. (iii)} \end{aligned}$$

$$\begin{aligned} &(3ab^2)^3 \div (ab^3)^2 \\ =\ &27a^3b^6 \div a^2b^6 \\ =\ &\frac{27a^3b^6}{a^2b^6} \\ =\ &27a \text{ Ans. (iv)} \end{aligned}$$

14.

$$\begin{aligned} \tan\theta &= \cos\theta \\ \frac{\sin\theta}{\cos\theta} &= \cos\theta \\ \sin\theta &= \cos^2\theta \end{aligned}$$

From the standard identity,

$$\begin{aligned} \sin^2\theta + \cos^2\theta &= 1 \\ \cos^2\theta &= 1 - \sin^2\theta \end{aligned}$$

Substituting for $\cos^2\theta$ and simplifying:

$$\begin{aligned} \sin\theta &= 1 - \sin^2\theta \\ \sin^2\theta + \sin\theta - 1 &= 0 \end{aligned}$$

Solving this quadratic of $\sin\theta$,

$$\begin{aligned} \sin\theta &= 0{\cdot}6180 \text{ or } -1{\cdot}618 \\ \text{From } \sin\theta &= 0{\cdot}6180 \text{ (other value impossible)} \\ \theta &= 38^\circ\ 10' \\ \text{and } 180^\circ - 38^\circ\ 10' &= 141^\circ\ 50' \end{aligned} \Bigg\} \text{ Ans.}$$

15. (a). (i) $f(x) = 4\cos x + \ln x$

$$f'(x) = -4\sin x + \frac{1}{x}$$

$$= -4\sin x + x^{-1}$$

$$f''(x) = -4\cos x - \frac{1}{x^2}$$

$$= -\left(4\cos x + \frac{1}{x^2}\right)$$ Answer.

(ii) $f(x) = 3x^3 + \dfrac{4}{x^2} + 6x - 2$

$$= 3x^3 + 4x^{-2} + 6x - 2$$

$$f'(x) = 9x^2 - 8x^{-3} + 6$$

$$f''(x) = 18x + 24x^{-4}$$

$$= 6\left(\frac{4}{x^4} + 3x\right)$$ Answer.

(b) $T = 1700 + \dfrac{16000}{x^2} + 0{\cdot}6\,x^2$

$$= 1700 + 16000\,x^{-2} + 0{\cdot}6\,x^2$$

$$\frac{dT}{dx} = -16000 \times 2\,x^{-3} + 0{\cdot}6 \times 2x$$

$$= 1{\cdot}2\,x - \frac{32000}{x^3}$$

$$1 \cdot 2\,x = \frac{32000}{x^3} \text{ for maximum or minimum}$$

$$1 \cdot 2 = \frac{32000}{x^4}$$

$$x = \sqrt[4]{\frac{32000}{1 \cdot 2}}$$

$$= \sqrt[4]{\frac{26667}{1}}$$

$$= 12 \cdot 77$$ Answer.

$$\frac{dT}{dx} = 1 \cdot 2x - \frac{32000}{x^3}$$

$$= 1 \cdot 2\,x - 32000\,x^{-3}$$

$$\frac{d^2 T}{dx^2} = 1 \cdot 2 + 96000\,x^{-4}$$

$$= 1 \cdot 2 + \frac{96000}{x^4}$$ positive when $x = 12 \cdot 77$ i.e. a minimum

$$T = 1700 + \frac{16000}{x^2} + 0{\cdot}6\,x^2$$

$$= 1700 + \frac{16000}{12 \cdot 77^2} + 0{\cdot}6 \times 12{\cdot}77^2$$

$$= 1700 + \frac{16000}{163{\cdot}07} + 0{\cdot}6 \times 163{\cdot}07$$

$$= \;=\; 1700 + 98{\cdot}12 + 97{\cdot}84$$

$$= 1896$$ Answer.

16.

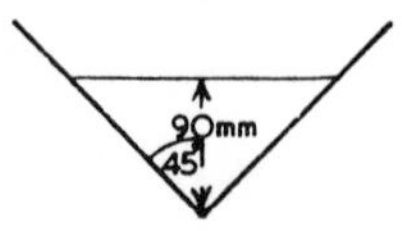

Fig. 234

Half-breadth of water level (see Fig. 232)

$$= 90 \times \tan 45° = 90 \text{ mm}$$

$$\text{Full breadth} = 180 \text{ mm}$$

Area of triangular water section in m²

$$= \tfrac{1}{2}(0{\cdot}18 \times 0{\cdot}09) = 0{\cdot}0081 \text{ m}^2$$

$$\text{Vol. flow } [\text{m}^3/\text{h}] = \text{area } [\text{m}^2] \times \text{velocity } [\text{m/h}]$$

$$= 0{\cdot}0081 \times 0{\cdot}2 \times 3600$$

$$= 5{\cdot}832 \text{ m}^3/\text{h} \quad \text{Ans. (i)}$$

Vol. flow in litres per minute

$$= \frac{5{\cdot}832 \times 10^3}{60} = 97{\cdot}2 \text{ l/m} \quad \text{Ans. (ii)}$$

17. Representing the straight line graph by the equation,

$$y = a + bx$$

where a and b are constants,

the graph is drawn as shown in chapter 7 from which we measure:

constant $a = -10$

" $b = 2$

The equation for this graph is therefore:

$$y = -10 + 2x$$

$$\text{or } y = 2x - 10 \quad \text{Ans.}$$

18. Working in centimetres:

$$\text{Area of sector} = \frac{150}{360} \times \pi \times 10{\cdot}2^2$$

$$= 136{\cdot}2 \text{ cm}^2 \quad \text{Ans. (i)}$$

$$\text{Length of arc} = \frac{150}{360} \times 2\pi \times 10{\cdot}2$$

this is the circumference of the cone base, let its diameter $= d$,

$$\pi d = \frac{150}{360} \times 2\pi \times 10{\cdot}2$$

$$d = \frac{150 \times 2 \times \pi \times 10{\cdot}2}{360 \times \pi}$$

$$= 8{\cdot}5 \text{ cm} = 85 \text{ mm} \quad \text{Ans. (ii)}$$

slant height of cone $=$ radius of sector $= 10{\cdot}2$ cm

radius of cone base $= \frac{1}{2} \times 8{\cdot}5 = 4{\cdot}25$ cm

perpendicular height $= \sqrt{10{\cdot}2^2 - 4{\cdot}25^2}$

$= 9{\cdot}27$ cm $= 92{\cdot}7$ mm Ans. (iii)

19. $19a + 27b = 43$ ... ... (i)

$7a + 4b = -2$ ... ... (ii)

Multiplying (i) by 7 and (ii) by 19,

$133a + 189b = 301$ ... ... (iii)

$133a + 76b = -38$ ... ... (iv)

Subtracting (iv) from (iii),

$$113b = 339$$

$$b = 3$$

Substituting value of b into (ii),

$$7a + 12 = -2$$

$$7a = -14$$

$$a = -2$$

$$10y + 4b = -2x$$

$$3a + x = y$$

Substituting values of a and b,

$$10y + 12 = -2x$$

$$-6 + x = y$$

Re-arranging,

$2x + 10y = -12$ ... ... (v)

$x - y = 6$ ... ... (vi)

Multiplying (vi) by 2 and then subtracting from (v),

$$\begin{aligned} 2x + 10y &= -12 \\ 2x - 2y &= 12 \\ \hline 12y &= -24 \\ y &= -2 \end{aligned}$$

Substituting value of y into (vi),

$$\begin{aligned} x + 2 &= 6 \\ x &= 4 \end{aligned}$$

$$x = 4, \text{ and } y = -2 \text{ Ans.}$$

20. Working in metres:

$$\begin{aligned} \text{Area of elliptical section} &= 0{\cdot}7854Dd \\ &= 0{\cdot}7854 \times 0{\cdot}3 \times 0{\cdot}2 \text{ m}^2 \end{aligned}$$

Centre of ellipse from centre of wheel

$$= 1{\cdot}8 - 0{\cdot}15 = 1{\cdot}65 \text{ m}$$

By ring theorum (Pappus), in one revolution of the elliptical sectional area about centre of wheel:

$$\begin{aligned} \text{Volume swept out} &= \text{area} \times \text{distance its c.g. moves} \\ \text{Volume of rim} &= 0{\cdot}7854 \times 0{\cdot}3 \times 0{\cdot}2 \times 2\pi \times 1{\cdot}65 \\ &= 0{\cdot}4887 \text{ m}^3 \\ \text{Mass [kg]} &= \text{volume [m}^3\text{]} \times \text{density [kg/m}^3\text{]} \\ &= 0{\cdot}4887 \times 7{\cdot}21 \times 10^3 \\ &= 3523 \text{ kg Ans.} \end{aligned}$$

21.

$$y = \frac{x^3}{4} - \frac{x^2}{2}$$

$$\text{Gradient} = \frac{dy}{dx} = \frac{3x^2}{4} - x$$

$$\frac{d^2y}{dx^2} = \frac{6x}{4} - 1$$

For zero gradient,

$$\frac{3x^2}{4} - x = 0$$

$$\therefore \frac{3x^2}{4} = x$$

$$\therefore x^2 = \frac{4}{3}x$$

$\therefore x = 0$ or $x = \frac{4}{3}$

i.e. Gradient is zero at $x = 0$ and $x = \frac{4}{3}$

When $x = 0$,

$$\frac{d^2y}{dx^2} = -1 \text{ (i.e. negative)}$$

When $x = \frac{4}{3}$,

$$\frac{d^2y}{dx^2} = 2 - 1 \text{ (positive)}$$

$\therefore y$ is a minimum at the point $x = 1\frac{1}{3}$ Answer.

The solution is confirmed by the sketch of the graph. (Fig. 233).

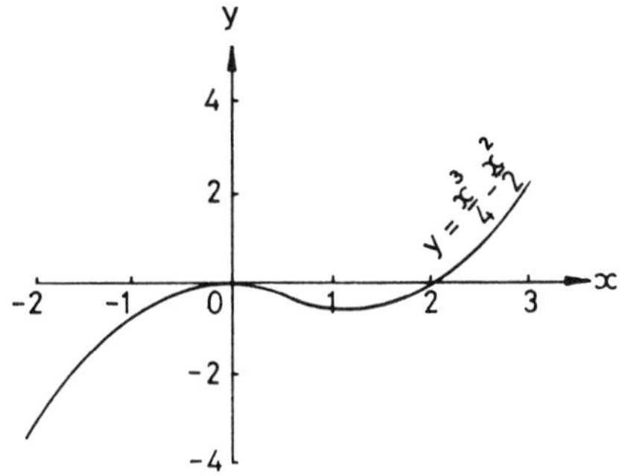

Fig. 235

22. (i) $2x^2 - 2 = 2(x^2 - 1) = 2(x + 1)(x - 1)$ Ans. (i)

(ii) $7x^2 + 9xy - 10y^2 = (7x - 5y)(x + 2y)$ Ans. (ii)

(iii) $14x^3y^2 + 22x^2y^3 - 12xy^4$

$$= 2xy^2(7x^2 + 11xy - 6y^2)$$
$$= 2xy^2(7x - 3y)(x + 2y) \text{ Ans. (iii)}$$

(iv) Dividing first by the given factor:

$$\begin{array}{l} x - y)\,3x^3 + 8x^2y - xy^2 - 10y^3\,(3x^2 + 11xy + 10y^2 \\ \quad\;\; \underline{3x^3 - 3x^2y} \\ \qquad\qquad 11x^2y - xy^2 \\ \qquad\qquad \underline{11x^2y - 11xy^2} \\ \qquad\qquad\qquad 10xy^2 - 10y^3 \\ \qquad\qquad\qquad \underline{10xy^2 - 10y^3} \\ \qquad\qquad\qquad\quad \cdot \qquad \cdot \end{array}$$

$$(x - y)(3x^2 + 11xy + 10y^2)$$
$$= (x - y)(3x + 5y)(x + 2y) \text{ Ans. (iv)}$$

23.

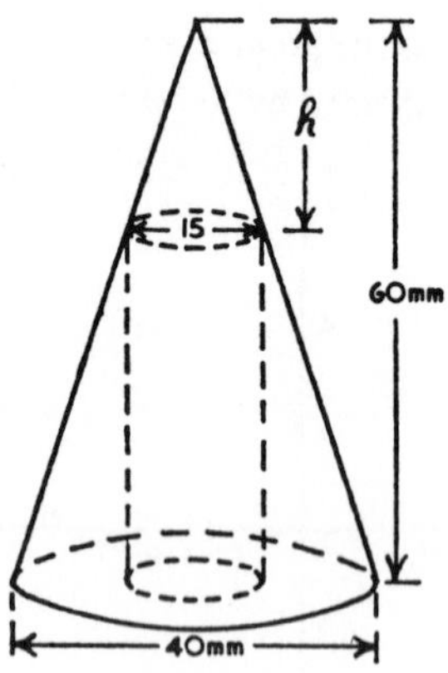

Fig. 236

By similar triangles, working in centimetres:

$$\frac{h}{1{\cdot}5} = \frac{6}{4} \qquad \therefore h = 2{\cdot}25 \text{ cm}$$

$$\text{Height of hole} = 6 - 2{\cdot}25 = 3{\cdot}75 \text{ cm}$$
$$\text{Volume of hole} = \pi \times 0{\cdot}75^2 \times 3{\cdot}75 \text{ cm}^3 \qquad \text{(i)}$$

Volume of conical cap (which drops off)

$$= \tfrac{1}{3} \times \text{ area of base } \times \text{ height}$$
$$= \tfrac{1}{3} \times \pi \times 0{\cdot}75^2 \times 2{\cdot}25 \text{ cm}^3 \quad \text{(ii)}$$

Total volume removed = (i) + (ii)

$$= \pi \times 0{\cdot}75^2 \times 3{\cdot}75 + \tfrac{1}{3} \times \pi \times 0{\cdot}75^2 \times 2{\cdot}25$$
$$= \pi \times 0{\cdot}75^2(3{\cdot}75 + \tfrac{1}{3} \times 2{\cdot}25)$$
$$= \pi \times 0{\cdot}75^2 \times 4{\cdot}5$$
$$= 7{\cdot}954 \text{ cm}^3 \text{ Ans.}$$

24.

$$A_1 = \int_1^5 (10x^2 + x + 4)\,dx$$

$$= \left[\frac{10x^3}{3} + \frac{x^2}{2} + 4x\right]_1^5$$

$$= \left(\frac{1250}{3} + \frac{25}{2} + 20\right) - \left(\frac{10}{3} + \frac{1}{2} + 4\right)$$

$$= 449{\cdot}2 - 7{\cdot}833$$

$$= 441{\cdot}37$$

$$A_2 = \int_1^5 \left(\frac{1}{2}x^2 - \frac{1}{x} - \frac{1}{2}\right)dx$$

$$= \int_1^5 \left(\frac{x^2}{2} - \frac{1}{x} - \frac{1}{2}\right)dx$$

$$= \left[\frac{x^3}{6} - \ln x - \frac{x}{2}\right]_1^5$$

$$= \left(\frac{125}{6} - \ln 5 - \frac{5}{2}\right) - \left(\frac{1}{6} - 0 - \frac{1}{2}\right)$$

$$= (20 \cdot 83 - 1 \cdot 608 - 2 \cdot 5) - (0 \cdot 167 - 0 \cdot 5)$$

$$= 16 \cdot 73 + 0 \cdot 33$$

$$= 17 \cdot 06$$

$$\text{Area} = A_1 - A_2 = 441 \cdot 37$$

$$= 424 \cdot 31 \text{ square units.} \qquad \text{Answer.}$$

25. (a)

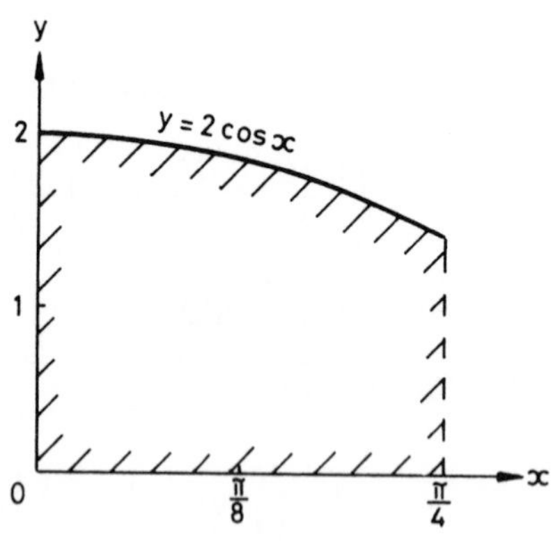

Fig. 237

$$A = \int_a^b y\,dx$$

$$= \int_0^{\frac{\pi}{4}} (2 \cos x)dx$$

$$= [2 \sin x]_0^{\frac{\pi}{4}}$$

$$= [2 \sin 45°] - [2 \sin 0°]$$

$$= 1{\cdot}414 \text{ square units.} \quad \text{Answer.}$$

(b)
$$\frac{dy}{dx} = 3x^2 - 4$$
$$\therefore y = \int(3x^2 - 4)dx$$
$$\therefore y = x^3 - 4x + C$$

when $x = 1$, $y = -2$,

$$\therefore -2 = (1)^3 - 4 \times (1) + C$$
$$\therefore C = 1$$

$\therefore$ The equation is $y = x^3 - 4x + 1$ Answer.

26. (i) See Chapter 6 for proof of formula.

$$x^2 + 5x = 84$$
$$x^2 + 5x + 2{\cdot}5^2 = 84 + 2{\cdot}5^2$$
$$x + 2{\cdot}5 = \pm\sqrt{84 + 6{\cdot}25}$$
$$x + 2{\cdot}5 = \pm\sqrt{90{\cdot}25}$$
$$x = \pm 9{\cdot}5 - 2{\cdot}5$$
$$x = +7 \text{ or } -12 \quad \text{Ans. (ii)}$$

$$2x^2 - 3ax - bx = 0$$
$$x(2x - 3a - b) = 0$$
$$\text{Either } x = 0$$
$$\text{or, } 2x - 3a - b = 0$$
$$2x = 3a + b$$
$$x = \tfrac{1}{2}(3a + b)$$
$$\therefore x = 0 \text{ or } \tfrac{1}{2}(3a + b) \quad \text{Ans. (iii)}$$

27. (a)
$$y = x^2 - 5x + 7 - \frac{1}{x}$$
$$\text{i.e. } y = x^2 - 5x + 7 - x^{-1}$$
$$\therefore \frac{dy}{dx} = 2x - 5 + x^{-2}$$
$$\frac{d^2y}{dx^2} = 2 - 2x^{-3}$$
$$\text{i.e. } \frac{d^2y}{dx^2} = 2 - \frac{2}{x^3} \quad \text{Answer.}$$

(b)

$$y = 2\cos x - 3\sin x$$

$$\therefore \frac{dy}{dx} = -2\sin x - 3\cos x$$

$$\therefore \frac{d^2y}{dx^2} = -2\cos x + 3\sin x \quad \text{Answer.}$$

28.

SECTIONAL AREAS	SIMPSON'S MULTIPLIERS	PRODUCTS
0	1	0
2·9	4	11·6
3·8	2	7·6
4·1	4	16·4
3·7	2	7·4
2·7	4	10·8
1·1	1	1·1
	sum =	54·9

There are 7 cross-sectional areas, therefore 6 spaces.

$$\text{Common interval} = \text{length} \div \text{no. of spaces}$$
$$= 8 \text{ cm} \div 6$$

$$\text{Volume} = \text{sum of products} \times \tfrac{1}{3} \text{ common interval}$$

$$= \frac{54{\cdot}9 \times 8}{3 \times 6} = 24{\cdot}4 \text{ cm}^3 \quad \text{Ans. (i)}$$

$$\text{Mass [g]} = \text{volume [cm}^3] \times \text{density [g/cm}^3]$$
$$= 24{\cdot}4 \times 8{\cdot}4$$
$$= 205 \text{ grammes} \quad \text{Ans. (ii)}$$

29.

$$C = a + bN$$

Inserting given values,

$$8{\cdot}5 = a + b \times 300 \quad \ldots \quad \ldots \quad \text{(i)}$$
$$7{\cdot}5 = a + b \times 200 \quad \ldots \quad \ldots \quad \text{(ii)}$$

$$1{\cdot}0 = 100b \text{ by subtraction}$$
$$b = 0{\cdot}01$$

Substituting value of b into (i),

$$8{\cdot}5 = a + 0{\cdot}01 \times 300$$
$$a = 8{\cdot}5 - 3$$
$$a = 5{\cdot}5$$

$\therefore$ cost equation is:

$$C = 5{\cdot}5 + 0{\cdot}01\,N \text{ pounds} \quad \text{Ans. (a)}$$

$$\text{When } N = 150,$$

$$C = 5{\cdot}5 + 0{\cdot}01 \times 150$$

$$= £7{\cdot}00 \quad \text{Ans. (b)}$$

Minimum cost is when machine is running idly and no articles are being manufactured,

$$\text{When } N = 0, \qquad C = 5{\cdot}5 + 0{\cdot}01 \times 0$$

$$C = 5{\cdot}5$$

$$\therefore \text{ minimum cost} = £5{\cdot}50 \quad \text{Ans. (c)}$$

30. $$y = \sin x + \frac{1}{2}\cos x$$

$$\text{Area} = \int_{\pi/6}^{\pi/3} y\,dx$$

$$= \int_{\pi/6}^{\pi/3} \left(\sin x + \frac{1}{2}\cos x\right) dx$$

$$= \left[-\cos x + \frac{1}{2}\sin x\right]_{\pi/6}^{\pi/3}$$

$$= \left(-\cos\frac{\pi}{3} + \frac{1}{2}\sin\frac{\pi}{3}\right) - \left(-\cos\frac{\pi}{6} + \frac{1}{2}\sin\frac{\pi}{6}\right)$$

$$= (-0{\cdot}5 + 0{\cdot}433) - (-0{\cdot}866 + 0{\cdot}25)$$

$$= -0{\cdot}0067 + 0{\cdot}616$$

$$= 0{\cdot}549 \text{ square units.} \qquad \text{Answer.}$$

31. Let pieces be x and $(1 - x)$ long

$$\text{First area} = \left(\frac{x}{4}\right)^2 = \frac{x^2}{16}$$

$$\text{Second area} = \left(\frac{1-x}{4}\right)^2 = \frac{1}{16}(1 - 2x + x^2)$$

$$\text{Sum of areas A} = \frac{1}{16}(1 - 2x + 2x^2)$$

$$\frac{dA}{dx} = -\frac{1}{8} + \frac{x}{4}$$

$$0 = -1 + 2x \text{ for maximum or minimum}$$

$$x = \frac{1}{2}$$

$$\frac{d^2A}{dx^2} = \frac{1}{4} \text{ i.e. positive, as a minimum}$$

$$A = \frac{1}{16}(1 - 1 + ½) \text{ when } x = \frac{1}{2}$$

$$A = \frac{1}{32}\, m^2 \text{, least possible sum of the two areas}$$

Answer.

32. Working in centimetres.

Solid plate without groove:

$$\text{Area} = 0{\cdot}7854Dd$$

$$= 0{\cdot}7854 \times 42 \times 37 = 1220 \text{ cm}^2$$

$$\text{Volume} = 1220 \times 4{\cdot}5 = 5490 \text{ cm}^3$$

At bottom of groove:

Major axis $= 42 - (2 \times 1{\cdot}3) = 39{\cdot}4$ cm

Minor axis $= 37 - (2 \times 1{\cdot}3) = 34{\cdot}4$ cm

Cross-sect. area of plate at bottom of groove

$= 0{\cdot}7854 \times 39{\cdot}4 \times 34{\cdot}4 = 1064\ \text{cm}^2$

Volume of groove

$= (1220 - 1064) \times 1{\cdot}3 = 202{\cdot}8\ \text{cm}^3$

Net volume of grooved plate

$= 5490 - 202{\cdot}8 = 5287{\cdot}2\ \text{cm}^3$

Mass [kg] $=$ volume [cm^3] $\times$ density [g/cm^3] $\times 10^{-3}$

$= 5287{\cdot}2 \times 7{\cdot}86 \times 10^{-3}$

$= 41{\cdot}56$ kg Ans.

33.

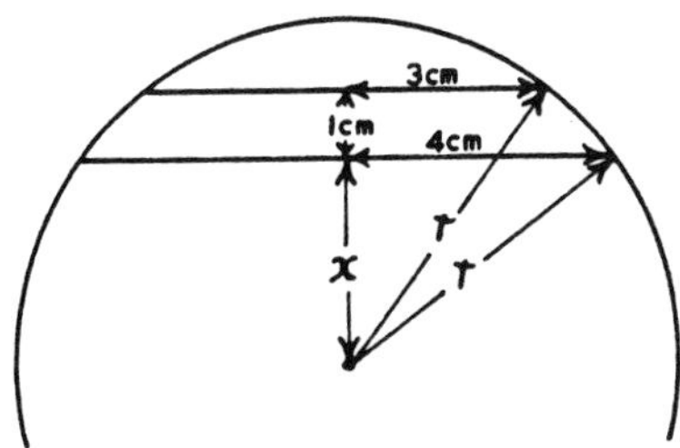

Fig. 238

Referring to Fig. 237, working in centimetres:

$$r^2 = 4^2 + x^2$$
$$= 16 + x^2 \quad \ldots \quad \ldots \quad \ldots \quad \text{(i)}$$

$$r^2 = 3^2 + (x + 1)^2$$
$$= 9 + x^2 + 2x + 1$$
$$= 10 + x^2 + 2x \quad \ldots \quad \ldots \quad \text{(ii)}$$

$$\therefore\ 10 + x^2 + 2x = 16 + x^2$$
$$2x = 6$$
$$x = 3$$

From (i), $r^2 = 4^2 + x^2$

$$r = \sqrt{(16 + 9)}$$
$$r = 5\ \text{cm} = 50\ \text{mm Ans.}$$

34.

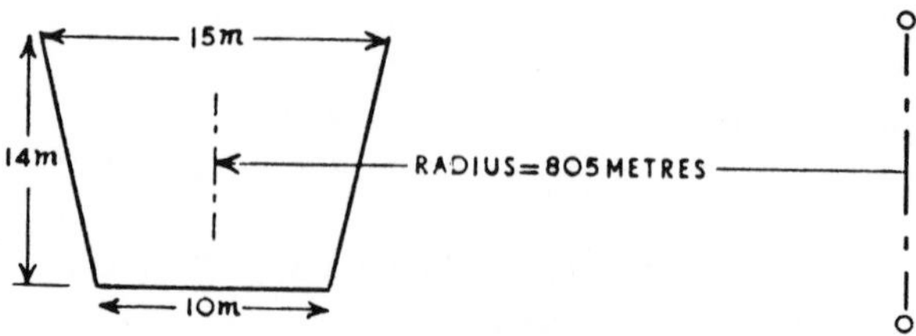

Fig. 239

$$\text{Area of cross section} = \text{average width} \times \text{depth}$$
$$= \tfrac{1}{2}(15 + 10) \times 14$$
$$= 175 \text{ m}^2$$
$$\text{Length of arc} = \frac{6}{360} \times \text{full circumference}$$
$$= \frac{1}{60} \times 2\pi \times 805 \text{ metres}$$

By theorem of Pappus, referring to Fig. 238,

$$\text{Volume swept out} = \text{area of section} \times \text{distance c.g. moves}$$
$$= 175 \times \frac{1}{60} \times 2\pi \times 805$$
$$= 14750 \text{ m}^3 \quad \text{Ans.}$$

35.

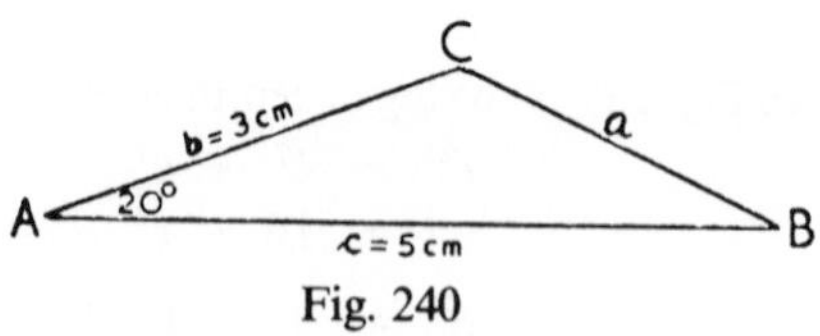

Fig. 240

Referring to Fig. 239 and working in centimetres:

By cosine rule,

$$a^2 = b^2 + c^2 - 2bc \cos A$$
$$= 3^2 + 5^2 - 2 \times 3 \times 5 \times \cos 20°$$
$$= 9 + 25 - 28{\cdot}191$$
$$a = \sqrt{5{\cdot}809}$$
$$= 2{\cdot}41 \text{ cm} = 24{\cdot}1 \text{ mm}$$

By sine rule,

$$\frac{a}{\sin A} = \frac{b}{\sin B}$$

$$\frac{2{\cdot}41}{\sin 20^\circ} = \frac{3}{\sin B}$$

$$\sin B = \frac{3 \times 0{\cdot}342}{2{\cdot}41} = 0{\cdot}4257$$

$$\therefore \text{ Angle } B = 25^\circ\ 12'$$

$$\text{Angle } C = 180 - (20^\circ + 25^\circ\ 12') = 134^\circ\ 48'$$

$$\text{Area} = \frac{bc \sin A}{2}$$

$$= \frac{3 \times 5 \times \sin 20}{2} = 2{\cdot}565 \text{ cm}^2$$

$$\left.\begin{aligned} \text{Remaining side} &= 24{\cdot}1 \text{ mm} \\ \text{Remaining angles} &= 25^\circ\ 12' \text{ and } 134^\circ\ 48' \\ \text{Area} &= 2{\cdot}565 \text{ cm}^2 \end{aligned}\right\} \text{Ans.}$$

36.

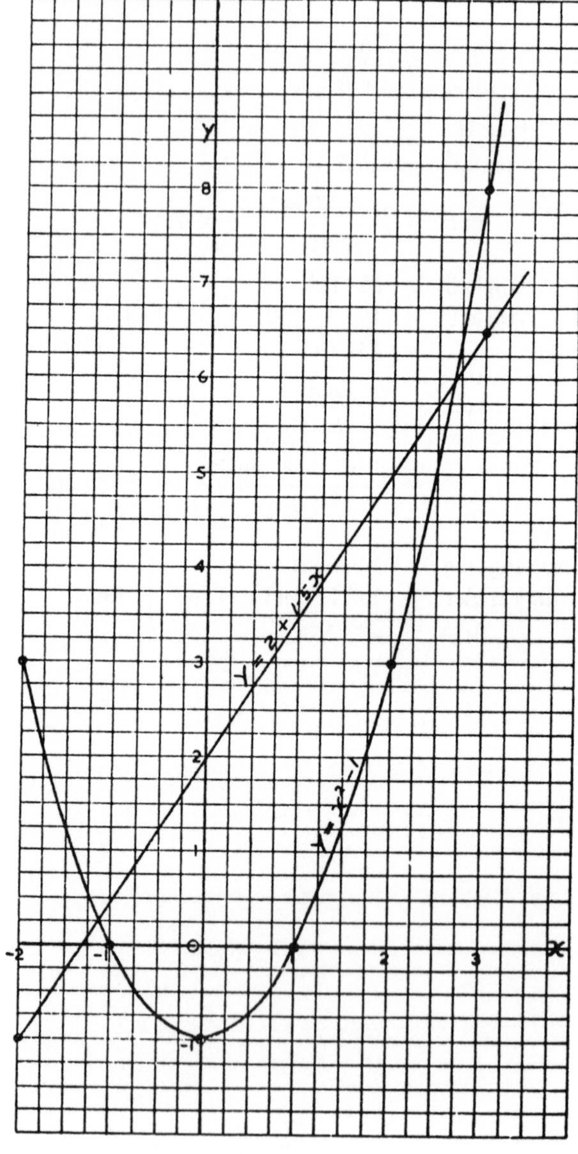

Fig. 241

$$x^2 - y = 1 \qquad \therefore y = x^2 - 1$$

Calculating plotting points for this curve:

x =	−2	−1	0	+1	+2	+3
y =	+3	0	−1	0	+3	+8

$$2y - 3x = 4 \qquad \therefore y = 2 + 1{\cdot}5x$$

Being a straight line, two values only are required to plot this graph:

$$x = -2 \qquad x = +3$$
$$y = -1 \qquad y = +6{\cdot}5$$

The graphs are now plotted as in Fig. 240 from which values of x and y are read thus:

$$\left.\begin{array}{l} x = 2{\cdot}64 \quad \text{and } y = 5{\cdot}96 \\ \text{or } x = -1{\cdot}14 \quad \text{and } y = 0{\cdot}3 \end{array}\right\} \text{Ans.}$$

37. $$\text{Area } A_1 = \int_2^5 \frac{78\,dV}{V^{1\cdot 2}}$$

$$= 78 \int_2^5 V^{-1\cdot 2}\,dV$$

$$= 78\left[-\frac{V^{-0\cdot 2}}{0\cdot 2}\right]_2^5$$

$$= -390\left[\left(\frac{1}{5^{0\cdot 2}}\right) - \left(\frac{1}{2^{0\cdot 2}}\right)\right]$$

$$= -390\left(\frac{1}{0\cdot 138} - \frac{1}{0\cdot 1149}\right)$$

$$= -390\,(7{\cdot}276 - 8{\cdot}703)$$

$$= -390 \times -1{\cdot}427$$

$$= 556{\cdot}5$$

$$A_2 = \int_2^5 \frac{7\,dV}{V^{1 \cdot 15}}$$

$$= \frac{-7}{0 \cdot 15}\left[\left(\frac{1}{5^{0 \cdot 15}}\right) - \left(\frac{1}{2^{0 \cdot 15}}\right)\right]$$

$$= -46 \cdot 67\left(\frac{1}{0 \cdot 1279} - \frac{1}{0 \cdot 1110}\right)$$

$$= -46 \cdot 67\,(7 \cdot 843 - 9 \cdot 009)$$

$$= -46 \cdot 67 \times -1 \cdot 166$$

$$= 54 \cdot 42$$

$A = A_1 - A_2$ is area between the two curves

$$= 556 \cdot 5 - 54 \cdot 42$$

$= 502 \cdot 08$ square units. Answer.

38.

$$s = t^3 - 6t^2 + 9t + 3$$

$$v = \frac{ds}{dt} = 3t^2 - 12t + 9$$

$$a = \frac{dv}{dt} = 6t - 12$$

(i) When the body is stationary, velocity = zero,

$$\therefore 3t^2 - 12t + 9 = 0$$

$$\therefore t^2 - 4t + 3 = 0$$

$$\therefore (t - 1)(t - 3) = 0$$

$$\therefore t = 1 \text{ or } t = 3$$

∴ The body is stationary when $t = 1$ and $t = 3$. Answer.

(ii) when $t = 1$,

$$\text{acceleration} = 6 - 12$$
$$= -6 \text{ m/s}^2. \quad \text{Answer.}$$

when $t = 3$,

$$\text{acceleration} = 18 - 12$$
$$= 6 \text{ m/s}^2. \quad \text{Answer.}$$

39. $$\frac{x+2}{x-1} - \frac{x-1}{3x-4} = \frac{7}{2}$$

Common denominator is $2(x - 1)(3x - 4)$,

Multiplying throughout by this:

$$2(3x-4)(x+2) - 2(x-1)(x-1) = 7(x-1)(3x-4)$$
$$2(3x^2 + 2x - 8) - 2(x^2 - 2x + 1) = 7(3x^2 - 7x + 4)$$
$$6x^2 + 4x - 16 - 2x^2 + 4x - 2 = 21x^2 - 49x + 28$$
$$6x^2 - 2x^2 - 21x^2 + 4x + 4x + 49x - 16 - 2 - 28$$
$$= 0$$
$$-17x^2 + 57x - 46 = 0$$
$$17x^2 - 57x + 46 = 0$$

Solving by formula,

$$x = \frac{-b \pm \sqrt{b^2 - 4ac}}{2a}$$
$$= \frac{57 \pm \sqrt{57^2 - 4 \times 17 \times 46}}{2 \times 17}$$
$$= \frac{57 \pm \sqrt{121}}{34} = \frac{57 \pm 11}{34}$$
$$= \frac{68}{34} \text{ or } \frac{46}{34}$$
$$= 2 \text{ or } 1{\cdot}353 \quad \text{Ans.}$$

40.

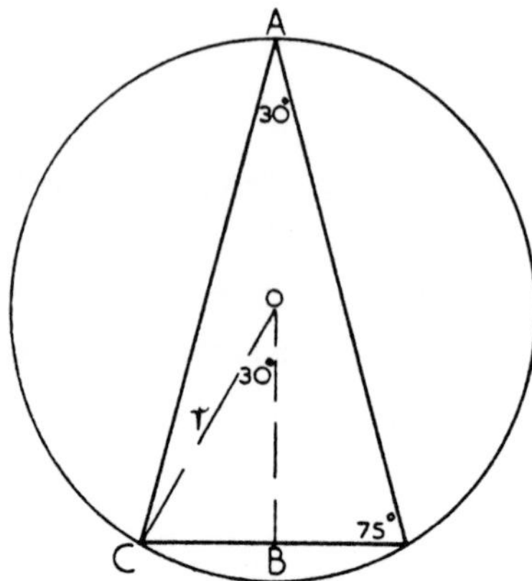

Fig. 242

$$\text{Area of circle} = \pi r^2$$

$$\therefore r = \sqrt{\frac{804 \cdot 2}{\pi}} = 16 \text{ cm}$$

$$\text{Angle at apex of triangle} = 180 - (2 \times 75) = 30^\circ$$

Angle at centre of circle for a triangle on the same base is twice angle at apex,

$$= 2 \times 30 = 60^\circ$$

Referring to Fig. 242,

$$\text{Angle COB} = 30^\circ$$

$$\text{CB} = r \sin 30^\circ = 16 \times 0 \cdot 5 = 8 \text{ cm}$$

$$\text{Base of triangle} = 2 \times 8 = 16 \text{ cm}$$

$$\text{OB} = r \cos 30^\circ = 16 \times 0 \cdot 866 = 13 \cdot 856 \text{ cm}$$

$$\text{AB} = \text{AO} + \text{OB} = 16 + 13 \cdot 856 = 29 \cdot 856 \text{ cm}$$

$$\text{Area of triangle} = \tfrac{1}{2}(\text{base} \times \text{perp. height})$$

$$= \tfrac{1}{2} \times 16 \times 29 \cdot 856$$

$$= 238 \cdot 8 \text{ cm}^2 \quad \text{Ans.}$$

41. (a). (i) $y = Ei - Ri^2$

$$\frac{dy}{di} = E - 2Ri$$ Answer.

(a). (ii) $$y = \left[\frac{C}{1-n}\right]\left[v^{1-n} - p^{1-n}\right]$$

$$\frac{dy}{dv} = \frac{C(1-n)}{1-n}\, v^{-n}$$

$$= \frac{C}{v^n}$$ Answer.

(b). (iii) $$\int_0^1 12\,x^7\,dx$$

$$= \left[1.5\,x^8\right]_0^1$$

$$= 1.5$$ Answer.

(b). (iv) $$\frac{3}{2\pi r^3}\int_0^r \pi(r^2 - x^2)\,dx$$

$$= \frac{3}{2r^3}\left[r^2x - \frac{x^3}{3}\right]_0^r$$

$$= \frac{3}{2r^3}\left[\left(r^3 - \frac{r^3}{3}\right) - 0\right]$$

$$= 1\cdot 0$$ Answer.

42. General equation for a straight line can be represented by:

$$y = a + bx$$

Plotting the pair of points and drawing a straight line through them as in Fig. 243 we read:

$$a = 2$$
$$b = -3$$

∴ Equation for this line is:

$$y = 2 - 3x$$ Ans.

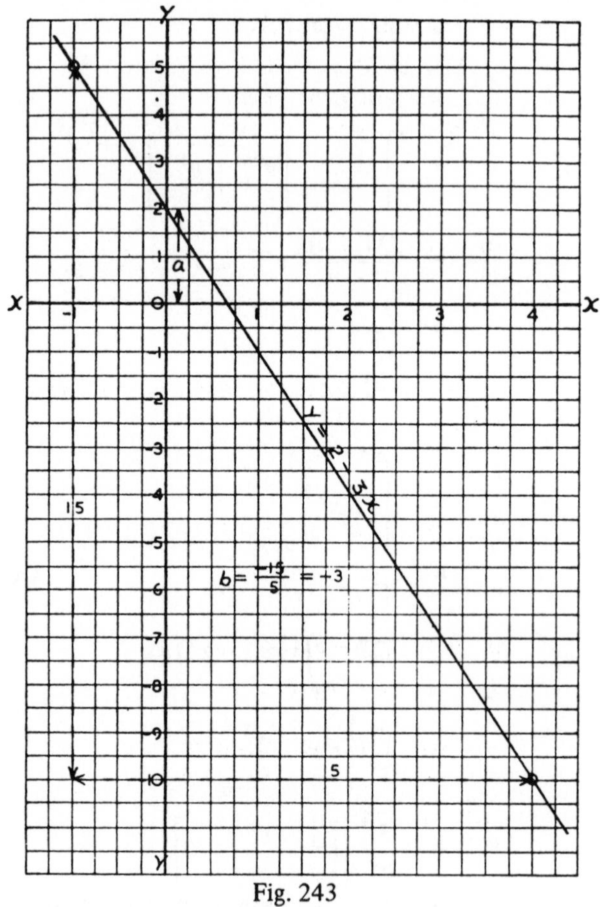

Fig. 243

43. See Chapter 8 for proof of $\sin^2\theta + \cos^2\theta = 1$

$$2\sin^2\theta - 2\cos^2\theta - 2 = 0$$

$$\sin^2\theta - \cos^2\theta - 1 = 0$$

substituting, $\cos^2\theta = 1 - \sin^2\theta$,

$$\sin^2\theta - (1 - \sin^2\theta) - 1 = 0$$

$$\sin^2\theta - 1 + \sin^2\theta - 1 = 0$$

$$2\sin^2\theta - 2 = 0$$

$$2\sin^2\theta = 2$$

$$\sin^2\theta = 1$$

$$\sin\theta = \sqrt{1} = \pm 1$$

$$\therefore \theta = 90^\circ \text{ and } 270^\circ \text{ Ans.}$$

44. See Chapter 10 for proof of formula.

$$\text{Sum of sides} = 5 + 6 + 7 = 18 \text{ cm}$$
$$\text{Semisum} = 9 \text{ cm}$$
$$\text{Area of triangle} = \sqrt{s(s-a)(s-b)(s-c)}$$
$$= \sqrt{9 \times 4 \times 3 \times 2}$$
$$= 14{\cdot}7 \text{ cm}^2$$

$$\text{Radius of circle} = \frac{2 \times \text{area of triangle}}{\text{perimeter of triangle}}$$
$$= \frac{2 \times 14{\cdot}7}{18}$$
$$= 1{\cdot}633 \text{ cm} = 16{\cdot}33 \text{ mm Ans.}$$

45.

$$y = 4x^3 - 3x^2 - 18x$$
$$\frac{dy}{dx} = 12x^2 - 6x - 18$$
$$0 = 12x^2 - 6x - 18$$
$$0 = (4x - 6)(3x + 3)$$
$$4x = 6 \quad \text{i.e. } x = 1{\cdot}5$$
or
$$3x = -3 \quad \text{i.e. } x = -1{\cdot}0$$
$$\frac{d^2y}{dx^2} = 24x - 6$$
$$x = 1{\cdot}5, \ \frac{d^2y}{dx^2} \text{ is positive i.e. minimum}$$
$$x = -1{\cdot}0, \ \frac{d^2y}{dx^2} \text{ is negative i.e. maximum}$$
$$x = 1{\cdot}5, y = 13{\cdot}50 - 6{\cdot}75 - 27 = -20{\cdot}25$$
$$x = -1{\cdot}0, y = -4 - 3 + 18 = 11{\cdot}0$$
$$y = -20{\cdot}25, (x = 1{\cdot}5), \text{ minimum} \quad \text{Answer.}$$
$$y = 11{\cdot}00, (x = -1{\cdot}0), \text{ maximum} \quad \text{Answer.}$$

46. Working in centimetres:

$$\text{Volume of cone} = \tfrac{1}{3}(\text{area of base} \times \text{perp. height})$$

$$= \frac{1}{3} \times \frac{\pi}{4} \times 10^2 \times 10 = 261{\cdot}8 \text{ cm}^3$$

$$\text{Volume of hemisphere} = \frac{1}{2} \text{ of } \frac{\pi}{6} d^3$$

$$= \frac{1}{2} \times \frac{\pi}{6} \times 10^3 = 261{\cdot}8 \text{ cm}^3$$

$$\text{Total volume} = 261{\cdot}8 + 261{\cdot}8 = 523{\cdot}6 \text{ cm}^3 \quad \text{Ans. (i)}$$

Curved surface area of cone

$$= \pi r \times \text{slant height}$$
$$= \pi \times 5 \times \sqrt{10^2 + 5^2} = 175{\cdot}6 \text{ cm}^2$$

Curved surface area of hemisphere

$$= \text{surface area of circumscribing cylinder}$$
$$= \pi \times \text{diameter} \times \text{height}$$
$$= \pi \times 10 \times 5 = 157{\cdot}1 \text{ cm}^2$$

$$\text{Total surface area} = 175{\cdot}6 + 157{\cdot}1 = 332{\cdot}7 \text{ cm}^2 \quad \text{Ans. (ii)}$$

Volumes of similar objects vary as the cube of their corresponding dimensions. As the large object has linear dimensions twice that of the smaller, then:

$$\text{Volume of larger solid} = 523{\cdot}6 \times 2^3$$
$$= 4188{\cdot}8 \text{ cm}^3 \quad \text{Ans. (iii)}$$

Areas of similar figures vary as the square of their corresponding dimensions, therefore:

$$\text{Surface area of larger solid} = 332{\cdot}7 \times 2^2$$
$$= 1330{\cdot}8 \text{ cm}^2 \quad \text{Ans. (iv)}$$

47. $$y = x^2 - 8x + 7 \quad \ldots \quad \ldots \quad \text{(i)}$$

Calculate plotting points for this equation:

x:	0	1	2	3	4	5	6	7	8	9
y:	7	0	−5	−8	−9	−8	−5	0	7	16

The graph is shown in Fig. 244.

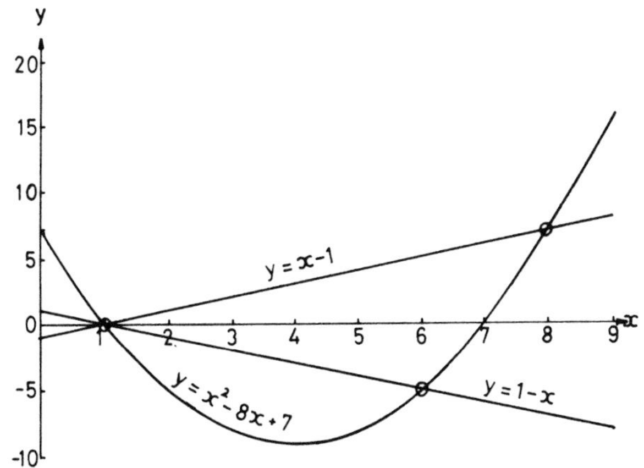

Fig. 244

(a) To solve the equation $x^2 - 7x + 6 = 0$:

From equation (i), by adding $x - 1$, we have:

$$x^2 - 8x + 7 + x - 1 = x^2 - 7x + 6$$

$$\therefore x^2 - 8x + 7 + x - 1 = 0$$

$$\therefore x^2 - 8x + 7 = 1 - x$$

$$\therefore y = 1 - x$$

Calculate plotting points for $y = 1 - x$

x:	0	9
y:	1	−8

The straight-line graph for this equation is also shown in Fig. 244.

At the points of intersection, $x = 1$, and $x = 6$. Answer.

(b) To solve the equation $x^2 - 9x + 8 = 0$

Adding $-x + 1$ to equation (i):

$$\text{i.e. } x^2 - 8x + 7 - x + 1 = x^2 - 9x + 8$$

$$\therefore x^2 - 8x + 7 - x + 1 = 0$$

$$\therefore x^2 - 8x + 7 = x - 1$$

$$\therefore y = x - 1$$

Calculate plotting points,

x:	0	9
y:	−1	8

The graph is shown on Fig. 244. At the points of intersection between this graph and the curve $y = x^2 - 8x + 7$.

$x = 1$ and $x = 8$ Answer.

48.

$$P = V(a + bV^2)$$

$$\text{or, } P = aV + bV^3$$

Inserting given values to form simultaneous equations:

$$13408 = a \times 16 + b \times 16^3 \quad \ldots \quad \text{(i)}$$

$$6024 = a \times 12 + b \times 12^3 \quad \ldots \quad \text{(ii)}$$

Multiplying (i) by 3, and (ii) by 4,

$$40224 = 48a + 12288b \quad \ldots \quad \ldots \quad \text{(iii)}$$

$$24096 = 48a + 6912b \quad \ldots \quad \ldots \quad \text{(iv)}$$

Subtracting (iv) from (iii),

$$16128 = 5376b$$

$$b = 3$$

Substituting, $b = 3$ into (ii),

$$6024 = 12a + 5184$$

$$12a = 840$$

$$a = 70$$

$\therefore$ Law is, $P = 70V + 3V^3$

inserting V = 14 knots and evaluating,

$$P = 70 \times 14 + 3 \times 14^3$$

$$= 980 + 8232$$

$= 9212$ kW Ans.

49.

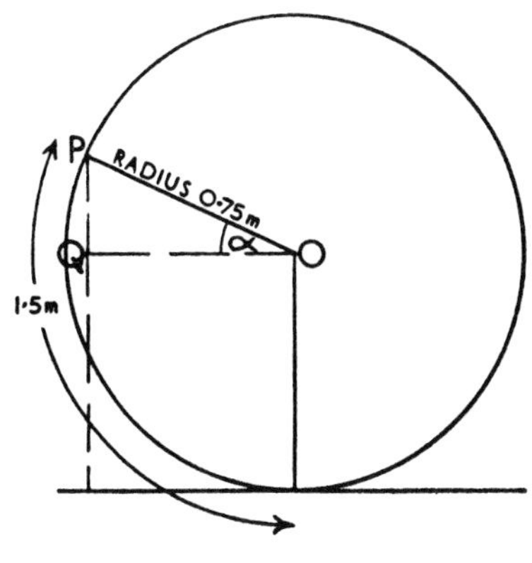

Fig. 245

$$\text{Angle turned through in radians} = \frac{\text{length of arc}}{\text{radius}}$$

$$= \frac{1 \cdot 5}{0 \cdot 75} = 2 \text{ radians (see Fig. 245)}$$

$$2 \times 57 \cdot 3 = 114 \cdot 6 \text{ degrees.}$$

$$\alpha = 114 \cdot 6 - 90 = 24 \cdot 6^\circ$$

$$PQ = OP \times \sin \alpha$$

$$= 0 \cdot 75 \times \sin 24 \cdot 6^\circ = 0 \cdot 3122 \text{ m}$$

$$\text{Height above ground} = 0 \cdot 3122 + 0 \cdot 75$$

$$= 1 \cdot 0622 \text{ m Ans.}$$

50. (i)

$$y = \frac{x^3}{3} - \frac{3}{x^3} + 3x^2 - 2x$$

$$\text{rearranging: } y = \frac{x^3}{3} - 3x^{-3} + 3x^2 - 2x$$

$$\frac{dy}{dx} = x^2 + 9x^{-4} + 6x - 2$$

$$\frac{d^2y}{dx^2} = 2x - 36x^{-5} + 6$$

$$= 2x - \frac{36}{x^5} + 6 \quad \text{Answer.}$$

(ii)

$$y = 3 \sin x + 4 \cos x$$

$$\frac{dy}{dx} = 3 \cos x - 4 \sin x$$

$$\frac{d^2y}{dx^2} = -3 \sin x - 4 \cos x \quad \text{Answer.}$$

51. Referring to the cone (see Fig. 92):

Radius of base $= 1{\cdot}22$ m

Perp. height $=$ total height $-$ height of hemisphere
$= 4{\cdot}27 - 1{\cdot}22$
$= 3{\cdot}05$ m

Slant height $= \sqrt{3{\cdot}05^2 + 1{\cdot}22^2} = 3{\cdot}285$ m

Curved surface area $= \pi r l$
$= \pi \times 1{\cdot}22 \times 3{\cdot}285$
$= 12{\cdot}6 \text{ m}^2$ Ans. (i)

The slant height is the radius of the developed sector, angle subtended at centre

$$= \frac{\text{arc of sector}}{\text{circum. of whole circle}} \times 360°$$

$$= \frac{\pi \times 2{\cdot}44}{2\pi \times 3{\cdot}285} \times 360$$

$= 133{\cdot}7°$ Ans. (ii)

Curved surface area of hemisphere

$=$ curved surface of circumscribing half cylinder
$= \frac{1}{2}\pi d^2$
$= \frac{1}{2} \times \pi \times 2{\cdot}44^2$
$= 9{\cdot}354 \text{ m}^2$ Ans. (iii)

Total volume $=$ vol. of cone $+$ vol. of hemisphere

$$= \tfrac{1}{3}(\text{area of base} \times \text{perp. ht.}) + \frac{1}{2} \times \frac{\pi}{6} d^3$$

$$= \frac{1}{3} \times \pi \times 1{\cdot}22^2 \times 3{\cdot}05 + \frac{\pi}{12} \times 2{\cdot}44^3$$

$$= 4{\cdot}755 + 3{\cdot}804$$

$$= 8{\cdot}559 \text{ m}^3 \quad \text{Ans. (iv)}$$

52.

$$3a + 7b + 2c = 25 \quad \ldots \quad \text{(i)}$$

$$2a + 6b + c = 18 \quad \ldots \quad \text{(ii)}$$

$$a + b + c + d = 14 \quad \ldots \quad \text{(iii)}$$

$$2a + d = 9 \quad \ldots \quad \text{(iv)}$$

From equation (iv),

$$d = 9 - 2a$$

Substitute $d = 9 - 2a$ in equation (iii),

$$a + b + c + 9 - 2a = 14$$

$$\therefore -a + b + c = 5 \quad \ldots \quad \text{(v)}$$

To eliminate c, subtract (v) from (ii),

$$\begin{array}{r} 2a + 6b + c = 18 \\ -a + b + c = 5 \\ \hline 3a + 5b = 13 \end{array} \quad \ldots \quad \text{(vi)}$$

Multiply (v) by 2 and subtract from (i),

$$\begin{array}{r} 3a + 7b + 2c = 25 \\ -2a + 2b + 2c = 10 \\ \hline 5a + 5b = 15 \end{array} \quad \ldots \quad \text{(vii)}$$

To solve: subtract (vii) from (vi),

$$\begin{array}{r} 3a + 5b = 13 \\ 5a + 5b = 15 \\ \hline -2a = -2 \end{array}$$

$$\therefore a = 1$$

Substitute $a = 1$ in (vii),

$$5 + 5b = 15$$

$$\therefore b = 2$$

Substitute $a = 1, \quad b = 2$ in (v),

$$-1 + 2 + c = 5$$

$$\therefore c = 4$$

Substitute $a = 1$ in (iv),

$$2 + d = 9$$

$$\therefore d = 7$$

$\therefore a = 1, \quad b = 2, \quad c = 4, \quad d = 7$ Answer.

53.

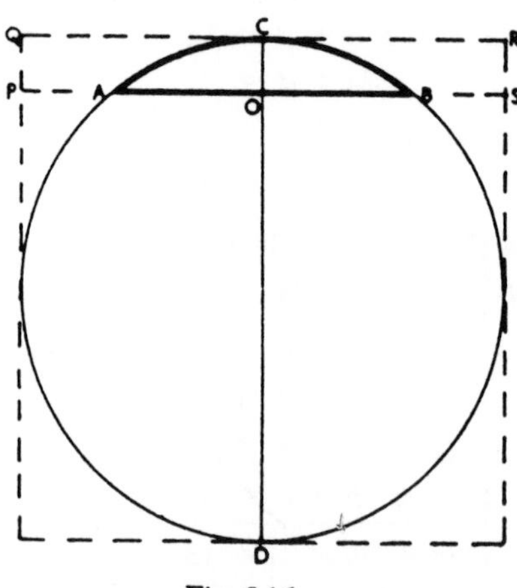

Fig. 246

$$AO \times BO = CO \times DO$$

$$30 \times 30 = 4 \times DO$$

$$DO = 225 \text{ mm}$$

$$\text{Dia. of circle} = CO + DO$$

$$= 4 + 225 = 229 \text{ mm}$$

Curved surface area of lens = curved surface area of slice of sphere of 22·9 cm dia. and 0·4 cm thick

= curved surface area of slice PQRS of circumscribing cylinder

$$= \pi \times 22{\cdot}9 \times 0{\cdot}4$$

$$= 28{\cdot}78 \text{ cm}^2 \text{ Ans.}$$

54. (a)

$$\frac{dy}{dx} = -\cos x - 2\sin x$$

$$\therefore y = \int(-\cos x - 2\sin x)dx$$

$$\therefore y = -\sin x + 2\cos x + C$$

when $y = 3, \quad x = 0$

(Note: $\sin 0 = 0, \quad \cos 0 = 1$)

$$\therefore 3 = 0 + 2 + C$$

$$\therefore C = 1$$

$$\therefore y = -\sin x + 2\cos x + 1 \quad \text{Answer (a).}$$

(b)

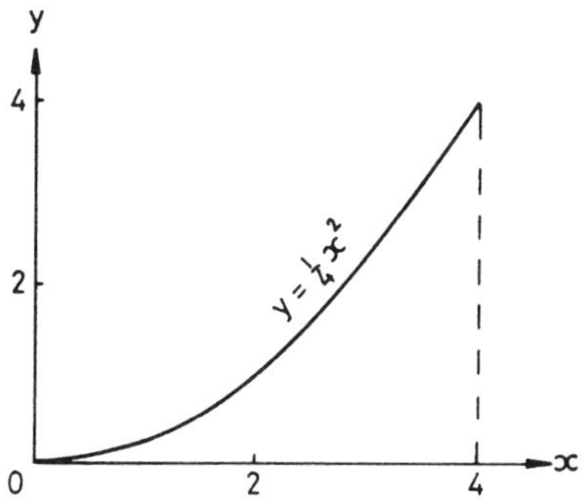

Fig. 247

$$\text{Vol.} = \pi\int_0^4 y^2\,dx$$

$$= \pi\int_0^4 \left(\frac{x^2}{4}\right)^2 dx$$

$$= \pi\int_0^4 \frac{x^4}{16}\,dx$$

$$= \pi\left[\frac{x^5}{80}\right]_0^4$$

$$= \pi\left[\frac{1024}{80}\right] - 0$$

$$\therefore \text{Volume} = 40{\cdot}212 \text{ cubic units} \quad \text{Answer (b)}$$

55.
$$p = \left[\frac{wu^2}{g\left\{\dfrac{K+a}{K(2a+K)+a^2} + \dfrac{D}{Et}\right\}}\right]^{\frac{1}{2}}$$

$$\therefore p^2 g = \frac{wu^2}{\dfrac{K+a}{K(2a+K)+a^2} + \dfrac{D}{Et}}$$

$$\therefore \frac{K+a}{2aK + K^2 + a^2} + \frac{D}{Et} = \frac{wu^2}{p^2 g}$$

$$\therefore \frac{K+a}{(K+a)^2} = \frac{wu^2}{p^2 g} - \frac{D}{Et}$$

$$\therefore \frac{1}{K+a} = \frac{wu^2 Et - Dp^2 g}{p^2 gEt}$$

$$\therefore K + a = \frac{p^2 gEt}{wu^2 Et - Dp^2 g}$$

$$\therefore K = \frac{p^2 gEt}{wu^2 Et - Dp^2 g} - a \quad \text{Answer.}$$

56. $x \propto \dfrac{z}{y^2}$ $\qquad \therefore \dfrac{xy^2}{z} = \text{constant}$

$$\therefore \frac{x_1 {y_1}^2}{z_1} = \frac{x_2 {y_2}^2}{z_2}$$

Let each of the original values of x, y and z be represented by unity, then:

$$\text{new value of } y = 1 - 0{\cdot}2 = 0{\cdot}8$$
$$\text{,, ,, ,, } z = 1 + 0{\cdot}12 = 1{\cdot}12$$
$$x_2 = \frac{1 \times 1{\cdot}12}{0{\cdot}8^2} = 1{\cdot}75$$

$$\text{Fractional increase in } x = 1{\cdot}75 - 1 = 0{\cdot}75$$
$$\text{Percentage increase} = 0{\cdot}75 \times 100 = 75\% \quad \text{Ans. (i)}$$

$$\text{time} \propto \sqrt{\text{length}} \;\therefore \frac{\text{time}}{\sqrt{\text{length}}} = \text{constant}$$

$$\therefore \frac{t_1}{\sqrt{l_1}} = \frac{t_2}{\sqrt{l_2}}$$

$$\text{time for one beat of 1st pendulum} = \frac{44}{105}\text{ second}$$

$$\text{time for one beat of 2nd pendulum} = 1\text{ second}$$

$$\frac{44}{105 \times \sqrt{174 \cdot 4}} = \frac{1}{\sqrt{l_2}}$$

$$\sqrt{l_2} = \frac{105 \times \sqrt{174 \cdot 4}}{44}$$

$$l_2 = \frac{105^2 \times 174 \cdot 4}{44^2}$$

$$= 993\text{ mm Ans.}$$

57.

$$3\cos x + 4\sin y = 2\sqrt{2} - 1 \cdot 5 \quad \ldots \quad \ldots \quad \text{(i)}$$

$$5\cos x - \sqrt{2}\sin y = -3 \cdot 5 \quad \ldots \quad \ldots \quad \ldots \quad \text{(ii)}$$

Multiplying (i) by 5,

$$15\cos x + 20\sin y = 6 \cdot 64 \ldots \quad \ldots \quad \ldots \quad \ldots \quad \text{(iii)}$$

Multiplying (ii) by 3,

$$15\cos x - 4 \cdot 242\sin y = -10 \cdot 5 \quad \ldots \quad \ldots \quad \ldots \quad \text{(iv)}$$

Subtracting (iv) from (iii),

$$24 \cdot 242\sin y = 17 \cdot 14$$

$$\sin y = 0 \cdot 7071$$

$$y = 45^\circ$$

Substituting $\sin y = 0 \cdot 7071$ into (i)

$$3\cos x + 2 \cdot 828 = 1 \cdot 328$$

$$3\cos x = -1 \cdot 5$$

$$\cos x = -0 \cdot 5$$

$$x = 180^\circ - 60^\circ = 120^\circ$$

$$x = 120^\circ,\ y = 45^\circ\text{ Ans.}$$

58.

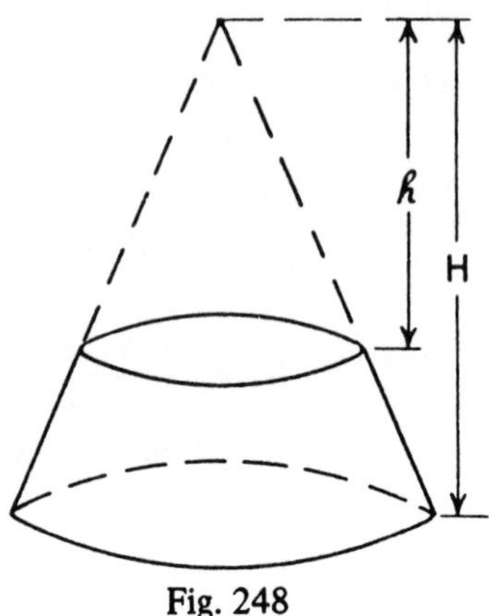

Fig. 248

$$\text{Vol. of frustum} = \frac{19}{27} \times \text{vol. of complete cone}$$

Let volume of complete cone = 27
,, ,, frustum = 19
then ,, ,, top cut off = 27 − 19 = 8

Volumes of similar objects vary as the cube of their corresponding dimensions,

$$\therefore \frac{\text{Vol. of complete cone}}{\text{Vol. of top cut off}} = \frac{\text{Height}^3 \text{ of complete cone}}{\text{height}^3 \text{ of top cut off}}$$

$$\frac{27}{8} = \frac{H^3}{h^3}$$

$$h = H \times \frac{\sqrt[3]{8}}{\sqrt[3]{27}}$$

$$h = \tfrac{2}{3}H$$

$$\therefore \text{Height of frustum} = \tfrac{1}{3}H \quad \text{Ans.}$$

59. (a)

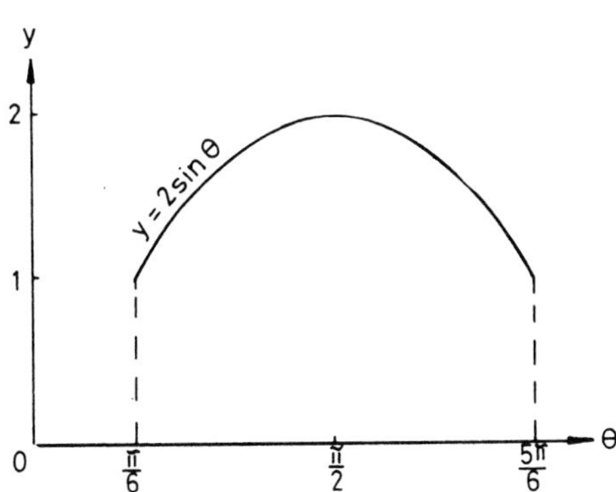

Fig. 249

$$\text{Area} = \int_{\frac{\pi}{6}}^{\frac{5\pi}{6}} (2\sin\theta)d\theta$$

$$= [-2\cos\theta]_{\frac{\pi}{6}}^{\frac{5\pi}{6}}$$

$$= [-2 \times \cos 150°] - [-2\cos 30°]$$

$$= 1{\cdot}732 + 1{\cdot}732$$

$= 3{\cdot}464$ square units Answer.

(b)

$$\int_{\frac{\pi}{2}}^{\pi} (3\cos\theta)d\theta = [3\sin\theta]_{\frac{\pi}{3}}^{\pi}$$

$$= [3\sin 180°] - [3\sin 90°]$$

$= -3$ Answer.

$$\int_1^2 (x^2 - 6x)dx = \left[\frac{x^3}{3} - 3x^2\right]_1^2$$

$$= \left[\frac{8}{3} - 12\right] - \left[\frac{1}{3} - 3\right]$$

$$= [-9\tfrac{1}{3}] - [-2\tfrac{2}{3}]$$

$= -6\frac{2}{3}$ Answer.

$$\int_1^3 (x^4 - 7x + 8)dx = \left[\frac{x^5}{5} - \frac{7x^2}{2} + 8x\right]_1^3$$

$$= \left[\frac{243}{5} - \frac{63}{2} + 24\right] - \left[\frac{1}{5} - \frac{7}{2} + 8\right]$$

$$= 41{\cdot}1 - 4{\cdot}7$$

$$= 36{\cdot}4 \quad \text{Answer.}$$

60.

$$K = \left\{\frac{V + v}{V - v}\right\}^2$$

Firstly, take square root of both sides to get rid of the power. Secondly, multiply both sides by the least common denominator to eliminate fractions. Then simplify and proceed to get terms containing V one one side with all other terms on other side of equation.

$$\sqrt{K} = \frac{V + v}{V - v}$$

$$\sqrt{K}(V - v) = V + v$$

$$\sqrt{K}V - \sqrt{K}v = V + v$$

$$\sqrt{K}V - V = \sqrt{K}v + v$$

$$V(\sqrt{K} - 1) = v(\sqrt{K} + 1)$$

$$V = v\frac{\sqrt{K} + 1}{\sqrt{K} - 1}$$

$$= v\left\{\frac{\sqrt{\frac{f_1}{f_2}} + 1}{\sqrt{\frac{f_1}{f_2}} - 1}\right\} \quad \text{Ans. (a)}$$

Inserting values,

$$\sqrt{\frac{f_1}{f_2}} = \sqrt{\frac{12}{18}} = \pm 0{\cdot}8165$$

$$V = 20 \times \frac{\pm 0{\cdot}8165 + 1}{\pm 0{\cdot}8165 - 1}$$

$$= \frac{20 \times 1{\cdot}8165}{-0{\cdot}1835} = -198$$

$$\text{or} \quad \frac{20 \times 0{\cdot}1835}{-1{\cdot}8165} = -2{\cdot}02$$

Values of V are,

$$-198 \text{ and } -2{\cdot}02 \text{ Ans.}$$

61. Plotting points for $y = x^2 + 2x - 8$:

x	-4	-3	-2	-1	0	$+1$	$+2$
y	0	-5	-8	-9	-8	-5	0

Fig. 257 shows the plotted graph

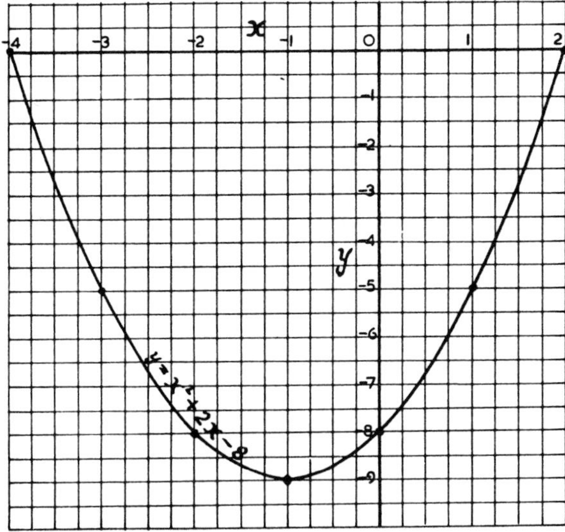

Fig. 250

When the area bounded by the graph is swept one complete revolution about the x-axis, the y ordinates of the graph become the radii at regular intervals along the length of the solid swept out. Putting the cross-sectional areas of these radii through Simpson's rule:

RADII r	CROSS-SECT. AREAS πr^2	SIMPSON'S MULTIPLIERS	PRODUCTS
0	$\pi \times 0$	1	0
5	$\pi \times 25$	4	$\pi \times 100$
8	$\pi \times 64$	2	$\pi \times 128$
9	$\pi \times 81$	4	$\pi \times 324$
8	$\pi \times 64$	2	$\pi \times 128$
5	$\pi \times 25$	4	$\pi \times 100$
0	$\pi \times 0$	1	0
		sum =	$\pi \times 780$

Common interval between ordinates

$$= 1$$

$$\text{Volume generated} = \pi \times 780 \times \tfrac{1}{3} \times 1$$

$$= 816{\cdot}9 \text{ units}^3 \quad \text{Ans.}$$

62. (i)

$$A = x^2 + \frac{16000}{x}$$

$$\therefore A = x^2 + 16000x^{-1}$$

$$\frac{dA}{dx} = 2x - \frac{16000}{x^2}$$

for max.^{m} or min.^{m} values,

$$\frac{dA}{dx} = 0$$

$$\text{i.e. } 2x - \frac{16000}{x^2} = 0$$

$$\therefore 2x = \frac{16000}{x^2}$$

$$\therefore x^3 = 8000$$

$\therefore$ For a max.^{m} or min.^{m} value of A,

$$x = 20$$

Differentiating again, $\dfrac{d^2A}{dx^2} = 2 + \dfrac{16000}{x^3}$

$$= 2 + \frac{16000}{20^3} \text{ (i.e. positive)}$$

hence, when $x = 20$, A has a minimum value. Answer (i).

(ii) For minimum value,

$$A = 20^2 + \frac{16000}{20}$$

$\therefore$ Minimum area $=$ 1200 square units. Answer (ii).

63.

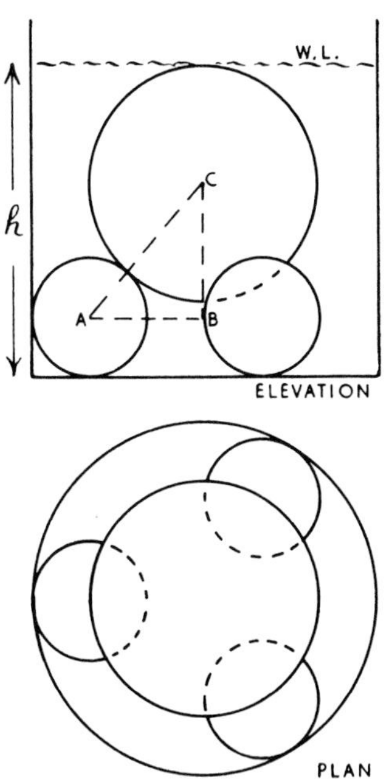

Fig. 251

Referring to triangle ABC in elevation of Fig. 251,

AC = rad. of 2 cm ball + rad. of 4 cm ball
= 1 + 2 = 3 cm
AB = rad. of cyl. − rad. of 2 cm ball
= 3 − 1 = 2 cm

$$BC = \sqrt{(AC)^2 - (AB)^2} = \sqrt{3^2 - 2^2}$$
$$= \sqrt{5} = 2{\cdot}236 \text{ cm}$$

h = rad. of 2 cm ball + BC + rad. of 4 cm ball
= 1 + 2·236 + 2 = 5·236 cm

Internal vol. of cylinder up to height h

$$= 0{\cdot}7854 \times d^2 \times h$$
$$= 0{\cdot}7854 \times 6^2 \times 5{\cdot}236 = 148 \text{ cm}^3$$

$$\text{Volume of sphere} = \frac{\pi}{6} d^3$$

$$\text{Volume of 3 small balls} = 3 \times \frac{\pi}{6} \times 2^3$$

$$\text{Volume of the large ball} = \frac{\pi}{6} \times 4^3$$

$$\text{Total volume of balls} = \frac{\pi}{6}(3 \times 2^3 + 4^3)$$
$$= \frac{\pi}{6} \times 88 = 46{\cdot}08 \text{ cm}^3$$

Volume of space for water to fill

$$= \text{vol. of cyl.} - \text{vol. of balls}$$
$$= 148 - 46{\cdot}08$$
$$= 101{\cdot}92 \text{ cm}^3$$

Volume of water $= 101{\cdot}92$ millilitres Ans. (i)

If all linear dimensions were half of those above,

Volume required $= (\tfrac{1}{2})^3 = \tfrac{1}{8}$ of above

because similar volumes vary as the cube of their corresponding linear dimensions.

Volume $= \tfrac{1}{8} \times 101{\cdot}92 = 12{\cdot}74$ ml Ans. (ii)

64.

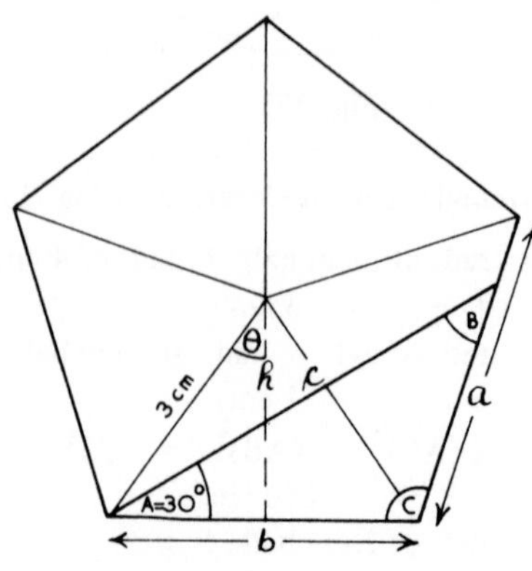

Fig. 252

Pentagon is composed of 5 isosceles triangles,

$$\text{Angles at centre} = 360 \div 5 = 72^\circ$$
$$\theta = \tfrac{1}{2} \text{ of } 72 = 36^\circ$$
$$h = 3 \times \cos 36^\circ = 2{\cdot}427$$
$$b = 2 \times 3 \sin 36^\circ = 3{\cdot}527$$

Area of isosceles triangle

$$= \tfrac{1}{2} bh$$
$$= \tfrac{1}{2} \times 3{\cdot}527 \times 2{\cdot}427$$
$$= 4{\cdot}28 \text{ cm}^2$$
$$\text{Area of pentagon} = 5 \times 4{\cdot}28$$
$$= 21{\cdot}4 \text{ cm}^2 \quad \text{Ans. (a)}$$

$$\text{Angle C} = 2 \times (90 - \theta) = 108^\circ$$
$$\text{Angle B} = 180 - \text{A} - \text{C} = 42^\circ$$

By sine rule:

$$\frac{a}{\sin A} = \frac{b}{\sin B}$$

$$a = \frac{3{\cdot}527 \times \sin 30^\circ}{\sin 42^\circ} = 2{\cdot}636 \text{ cm}$$

$$\text{Area of triangle } abc = \tfrac{1}{2} ab \sin C$$
$$= \tfrac{1}{2} \times 2{\cdot}636 \times 3{\cdot}527 \times \sin 108^\circ$$
$$= 4{\cdot}421 \text{ cm} \quad \ldots \quad \ldots \quad \ldots \quad \ldots \quad \text{(i)}$$

Remaining piece of pentagon

$$= 21{\cdot}4 - 4{\cdot}421 = 16{\cdot}979 \text{ cm} \quad \text{(ii)}$$

Ratio of (i) and (ii)

$$= 4{\cdot}421 : 16{\cdot}979$$

Dividing both by 4·421

$$\text{Ratio} = 1 : 3{\cdot}84 \quad \text{Ans. (b)}$$

65. (a) $2 \sin^2 \theta - \cos \theta = 1$

From the identity,

$$\sin^2 \theta + \cos^2 \theta = 1$$
$$\sin^2 \theta = 1 - \cos^2 \theta$$

Substitute for $\sin^2 \theta$ into original equation and simplify,

$$2 \sin^2 \theta - \cos \theta = 1$$
$$2(1 - \cos^2 \theta) - \cos \theta = 1$$

$$2 - 2\cos^2\theta - \cos\theta = 1$$
$$-2\cos^2\theta - \cos\theta + 1 = 0$$
$$2\cos^2\theta + \cos\theta - 1 = 0$$

Factorising,

$$(2\cos\theta - 1)(\cos\theta + 1) = 0$$

Either $2\cos\theta - 1 = 0$, then $\cos\theta = 0{\cdot}5$
or $\cos\theta + 1 = 0$, then $\cos\theta = -1$

When $\cos\theta = 0{\cdot}5$, $\theta = 60°$ and $(360 - 60) = 300°$
,, $\cos\theta = -1$, $\theta = 180°$

∴ Angles are 60°, 180° and 300° Ans. (a)

(b) $$3\cos 2\theta + 2\cos\theta = 0$$
$$3(2\cos^2\theta - 1) + 2\cos\theta = 0$$
$$6\cos^2\theta - 3 + 2\cos\theta = 0$$
$$6\cos^2\theta + 2\cos\theta - 3 = 0$$

Solving by quadratic formula,

$$\cos\theta = \frac{-b \pm \sqrt{b^2 - 4ac}}{2a}$$
$$= \frac{-2 \pm \sqrt{4 + 72}}{12}$$
$$= \frac{-2 \pm 8{\cdot}718}{12}$$
$$= 0{\cdot}5598 \text{ or } -0{\cdot}8932$$

When $\cos\theta = 0{\cdot}5598$, $\theta = 55°\ 57'$ and $360° - 55°\ 57' = 304°\ 3'$
,, $\cos\theta = -0{\cdot}8932$, $\theta = 180° - 26°\ 44' = 153°\ 16'$
and $180° + 26°\ 44' = 206°\ 44'$

Angles are,

55° 57′, 153° 16′, 206° 44′ and 304° 3′ Ans. (b)

66.

V	5	10	15	20	25
V^2	25	100	225	400	625
R	10	25	50	85	130

By plotting the points of R and V^2 we see that they lie on a straight line, therefore the equation is of the form $y = a + bx$ where x represents V^2, see Fig. 253.

Taking two points on the line such as:

when $V^2 = 400, \quad R = 85$

,, $\quad V^2 = 100, \quad R = 25$

Inserting these into $R = k + cV^2$,

$$85 = k + c \times 400 \quad \ldots \quad \ldots \quad \text{(i)}$$
$$25 = k + c \times 100 \quad \ldots \quad \ldots \quad \text{(ii)}$$

$$60 = c \times 300 \text{ by subtraction}$$
$$c = \frac{60}{300} = 0{\cdot}2$$

From (i),
$$85 = k + 0{\cdot}2 \times 400$$
$$k = 85 - 80 = 5$$

Constants are,
$$k = 5 \text{ and } c = 0{\cdot}2 \quad \text{Ans. (i)}$$

$$\text{when } V = 40,$$
$$R = k + cV^2$$
$$= 5 + 0{\cdot}2 \times 40^2$$
$$= 325 \text{ newtons} \quad \text{Ans. (ii)}$$

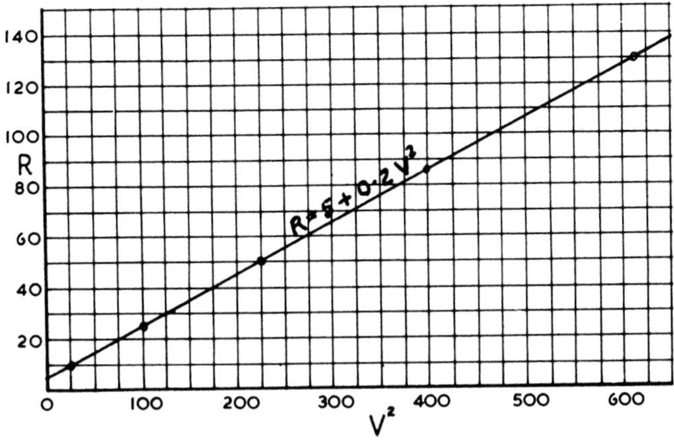

Fig. 253

67.
$$p = V + \frac{k(p + V)}{k + p + V}$$

$$p - V = \frac{k(p + V)}{k + p + V}$$

$$(p - V)(k + p + V) = k(p + v)$$
$$kp + p^2 - kV - V^2 = kp + kV$$
$$p^2 - V^2 = kp - kp + kV + kV$$
$$p^2 - V^2 = 2kV$$
$$\frac{p^2 - V^2}{2V} = k$$
$$\text{or, } k = \frac{p^2 - V^2}{2V} \quad \text{Ans. (i)}$$

When $p = 9{\cdot}2$ and $V = 0{\cdot}42$,

$$k = \frac{p^2 - V^2}{2V} = \frac{(p + V)(p - V)}{2V}$$
$$= \frac{(9{\cdot}2 + 0{\cdot}42)(9{\cdot}2 - 0{\cdot}42)}{2 \times 0{\cdot}42}$$
$$= \frac{9{\cdot}62 \times 8{\cdot}78}{0{\cdot}84}$$
$$= 100{\cdot}55 \quad \text{Ans. (ii)}$$

68.

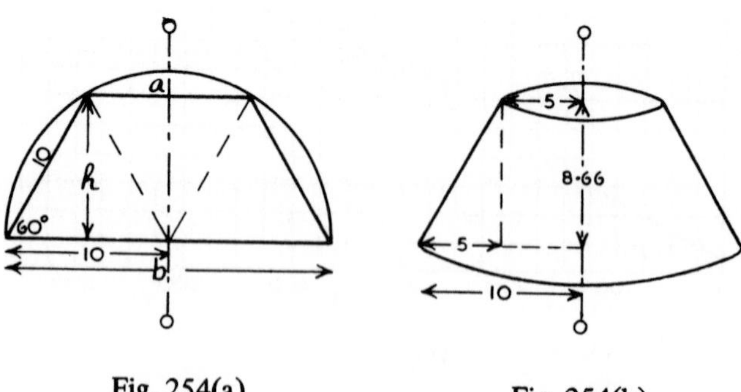

Fig. 254(a) Fig. 254(b)

The length of each non-parallel side being equal to the radius, the trapezium is composed of three equilateral triangles, therefore angle at the base is 60°, length of short side of trapezium = 10 cm, length of long side = 2 × 10 = 20 cm, and $h = 10 \sin 60° = 8{\cdot}66$ cm.

$$\text{Area of trapezium} = \tfrac{1}{2}(a + b) \times \text{perp. ht.}$$
$$= \tfrac{1}{2}(10 + 20) \times 8{\cdot}66$$
$$= 129{\cdot}9 \text{ cm}^2 \quad \text{Ans. (a)}$$

Alternatively, area of this trapezium

$$= \text{area of 3 equilateral triangles}$$
$$= 3 \times 0{\cdot}433 \times \text{side}^2$$
$$= 3 \times 0{\cdot}433 \times 10^2$$
$$= 129{\cdot}9 \text{ cm}^2 \text{ as above}$$

$$\text{Volume of frustum} = \tfrac{1}{3}\pi h(R^2 + Rr + r^2)$$
$$= \tfrac{1}{3} \times \pi \times 8{\cdot}66(10^2 + 10 \times 5 + 5^2)$$
$$= \tfrac{1}{3} \times \pi \times 8{\cdot}66 \times 175$$
$$= 1587 \text{ cm}^3 \quad \text{Ans. (b)}$$

As an alternative to the latter part, the volume can be found by the theorem of Pappus, referring to **Fig. 254(b)**:

$$\text{Area of triangle} = \tfrac{1}{2} \times 5 \times 8{\cdot}66 = 21{\cdot}65 \text{ cm}^2$$
$$\text{c.g. from axis} = 5 + \tfrac{1}{3} \times 5 = \tfrac{4}{3} \times 5 \text{ cm}$$
$$\text{Area of rectangle} = 5 \times 8{\cdot}66 = 43{\cdot}3 \text{ cm}^2$$
$$\text{c.g. from axis} = \tfrac{1}{2} \times 5 = 2{\cdot}5 \text{ cm}$$
$$\text{Volume swept out} = \text{area} \times \text{distance c.g. moves}$$

Volume swept out by triangle

$$= 21{\cdot}65 \times 2\pi \times \tfrac{4}{3} \times 5 = 907 \text{ cm}^3$$

Volume swept out by rectangle

$$= 43{\cdot}3 \times 2\pi \times 2{\cdot}5 = 680 \text{ cm}^3$$

$$\text{Total volume} = 907 + 680 = 1587 \text{ cm}^3$$
(as above)

As a further alternative, volume of frustum may be calculated from volume of whole cone minus volume of top cone removed. All these methods are explained in the text.

69.

$$4a + 3b + 2w = 2 \quad \ldots \quad \text{(i)}$$
$$2a + 8b + 3w = 11 \quad \ldots \quad \text{(ii)}$$
$$3a + 5b - 6w = -36 \quad \ldots \quad \text{(iii)}$$

Proceeding to eliminate w:

Multiply (i) by 3 and (ii) by 2 and subtract,

$$12a + 9b + 6w = 6$$
$$4a + 16b + 6w = 22$$
$$8a - 7b = -16 \quad \ldots \quad \text{(iv)}$$

Multiply (i) by 3 and add (iii),

$$\begin{aligned} 12a + 9b + 6w &= 6 \\ 3a + 5b - 6w &= -36 \\ \hline 15a + 14b &= -30 \quad \ldots \quad \ldots \quad \ldots \quad \ldots \quad \text{(v)} \end{aligned}$$

Proceeding to eliminate b and find the value of a:

Multiply (iv) by 2 and add (v),

$$\begin{aligned} 16a - 14b &= -32 \\ 15a + 14b &= -30 \\ \hline 31a &= -62 \\ a &= -2 \end{aligned}$$

Substitute value of a into (iv) to find b,

$$\begin{aligned} 8 \times (-2) - 7b &= -16 \\ -16 - 7b &= -16 \\ -7b &= 0 \\ b &= 0 \end{aligned}$$

Substitute values of a and b into (i) to find w,

$$\begin{aligned} 4 \times (-2) + 3 \times 0 + 2w &= 2 \\ -8 + 0 + 2w &= 2 \\ 2w &= 10 \\ w &= 5 \end{aligned}$$

Values of a, b and w are, respectively,

-2, 0 and 5 Ans. (a)

$$x^2 + y^2 = 18{\cdot}5 \quad \ldots \quad \ldots \quad \ldots \quad \ldots \quad \text{(i)}$$

$$x - y = 1 \quad \ldots \quad \ldots \quad \ldots \quad \ldots \quad \text{(ii)}$$

From (ii), $y = x - 1$, substitute this into (i),

$$\begin{aligned} x^2 + (x - 1)^2 &= 18{\cdot}5 \\ x^2 + x^2 - 2x + 1 &= 18{\cdot}5 \\ 2x^2 - 2x - 17{\cdot}5 &= 0 \\ x^2 - x - 8{\cdot}75 &= 0 \end{aligned}$$

Solving this by quadratic formula,

$$\begin{aligned} x &= \frac{-b \pm \sqrt{b^2 - 4ac}}{2a} \\ &= \frac{1 \pm \sqrt{1 + 35}}{2} = \frac{1 \pm 6}{2} \\ &= 3{\cdot}5 \text{ or } -2{\cdot}5 \end{aligned}$$

From (i), $y = x - 1$

If $x = 3{\cdot}5$, $y = 3{\cdot}5 - 1 = 2{\cdot}5$

„ $x = -2{\cdot}5$, $y = -2{\cdot}5 - 1 = -3{\cdot}5$

Therefore, $x = 3{\cdot}5$ and $y = 2{\cdot}5$
or, $x = -2{\cdot}5$ and $y = -3{\cdot}5$ } **Ans. (b)**

$$\frac{100}{x} + \frac{100}{x + 5} = \frac{50}{3}$$

Divide every term by 50,

$$\frac{2}{x} + \frac{2}{x + 5} = \frac{1}{3}$$

Multiply every term by the least common denominator which, in this case, is the product of the denominators, $3x(x + 5)$,

$$\begin{aligned} 2 \times 3(x + 5) + 2 \times 3x &= x(x + 5) \\ 6x + 30 + 6x &= x^2 + 5x \\ -x^2 + 7x + 30 &= 0 \\ x^2 - 7x - 30 &= 0 \end{aligned}$$

By factors,

$$\begin{aligned} (x - 10)(x + 3) &= 0 \\ \therefore x &= 10 \text{ or } -3 \quad \text{Ans. (c)} \end{aligned}$$

70. (a) Simplifying left hand side:

$$\frac{1 - \tan^2\theta}{1 + \tan^2\theta} = \frac{1 - \dfrac{\sin^2\theta}{\cos^2\theta}}{1 + \dfrac{\sin^2\theta}{\cos^2\theta}}$$

$$= \frac{\dfrac{\cos^2\theta - \sin^2\theta}{\cos^2\theta}}{\dfrac{\cos^2\theta + \sin^2\theta}{\cos^2\theta}} = \frac{\cos^2\theta - \sin^2\theta}{\cos^2\theta + \sin^2\theta}$$

$$= \frac{\cos^2\theta - \sin^2\theta}{1} = \cos^2\theta - \sin^2\theta = \text{r.h.s. Ans (a)}$$

(b) Simplifying left hand side:

$$\sin\theta + \sin\theta\cot^2\theta + \frac{\tan\theta}{\cos\theta}$$

$$= \sin\theta + \sin\theta \times \frac{1}{\tan^2\theta} + \frac{\sin\theta}{\cos\theta \times \cos\theta}$$

$$= \sin\theta + \frac{\sin\theta \times \cos^2\theta}{\sin^2\theta} + \frac{\sin\theta}{\cos^2\theta}$$

$$= \sin\theta + \frac{\cos^2\theta}{\sin\theta} + \frac{\sin\theta}{\cos^2\theta}$$

$$= \frac{\sin^2\theta\cos^2\theta + \cos^4\theta + \sin^2\theta}{\sin\theta\cos^2\theta}$$

$$= \frac{\cos^2\theta(\sin^2\theta + \cos^2\theta) + \sin^2\theta}{\sin\theta\cos^2\theta}$$

$$= \frac{\cos^2\theta \times 1 + \sin^2\theta}{\sin\theta\cos^2\theta}$$

$$= \frac{1}{\sin\theta\cos^2\theta} = \text{r.h.s. Ans. (b)}$$

71. $4y - x - 20 = 0, \therefore y = 5 + 0{\cdot}25x$

When $x = -2,\ y = 5 - 0{\cdot}5 = 4{\cdot}5$

,, $x = +8,\ y = 5 + 2 = 7$

$y - x + 1 = 0, \therefore y = -1 + x$

When $x = -2,\ y = -1 - 2 = -3$

,, $x = +8,\ y = -1 + 8 = 7$

$2y + x - 7 = 0, \therefore y = 3{\cdot}5 - 0{\cdot}5x$

When $x = -2,\ y = 3{\cdot}5 + 1 = 4{\cdot}5$

,, $x = +8,\ y = 3{\cdot}5 - 4 = -0{\cdot}5$

The graphs are shown plotted in Fig. 255 and enclose a triangle. Let the lengths of the sides of the triangle be a, b and c.

Graph forming side a rises 2·5 over a horizontal length of 10, therefore

$$a = \sqrt{2{\cdot}5^2 + 10^2} = 10{\cdot}31$$

Graph forming side b rises 5 over a horizontal length of 5, therefore,

$$b = \sqrt{5^2 + 5^2} = 7{\cdot}071$$

Graph forming side c drops 2·5 over a horizontal length of 5, therefore,

$$c = \sqrt{2{\cdot}5^2 + 5^2} = 5{\cdot}59$$

$$\text{Area of triangle} = \sqrt{s(s-a)(s-b)(s-c)}$$

where a, b and c are the lengths of the three sides and s is their semisum.

$$\begin{aligned} s &= \tfrac{1}{2}(10{\cdot}31 + 7{\cdot}071 + 5{\cdot}59) \\ &= 11{\cdot}486 \\ \text{Area} &= \sqrt{11{\cdot}486 \times 1{\cdot}176 \times 4{\cdot}415 \times 5{\cdot}896} \\ &= \sqrt{351{\cdot}6} \\ &= 18{\cdot}75 \text{ units}^2 \quad \text{Ans.} \end{aligned}$$

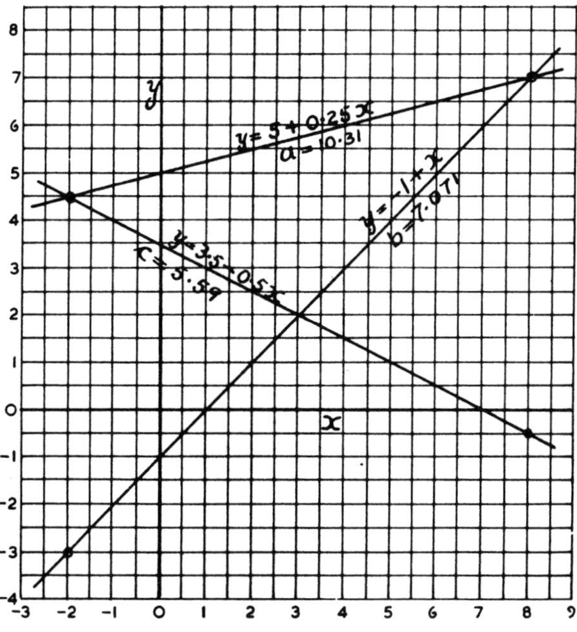

Fig. 255

72.

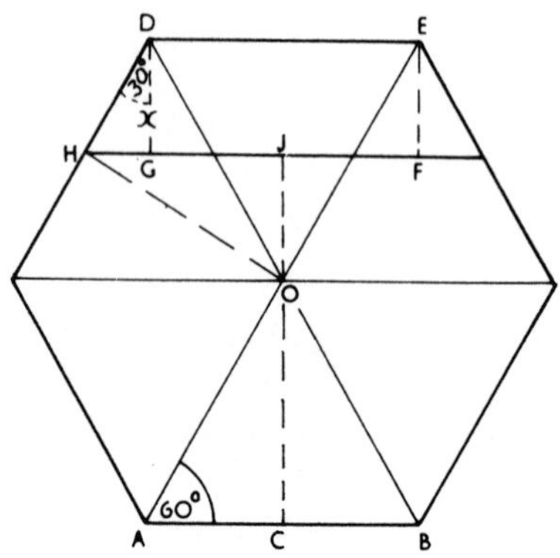

Fig. 256

Hexagon is composed of 6 equilateral triangles with all angles of 60°, each side equal to unity.

$$OC = OA \sin 60° = 1 \times 0{\cdot}866 = 0{\cdot}866$$

$$\text{Area of triangle OAB} = \tfrac{1}{2} \times 1 \times 0{\cdot}866 = 0{\cdot}433$$

(or obtain this from the formula, area of equilateral triangle

$$= 0{\cdot}433 \text{ side}^2)$$

$$\text{Area of hexagon} = 6 \times \text{area of triangle}$$

$$= 6 \times 0{\cdot}433 = 2{\cdot}598$$

Area to be cut off is the shape of a trapezium consisting of a central rectangle and a triangle at each end.

Let thickness DG be represented by x

$$\text{Area cut off} = \text{area of rect.} + \text{two triangles}$$

$$= \text{DE} \times x + 2 \times \tfrac{1}{2} \times \text{HG} \times x$$

$$= 1 \times x + 1 \times (x \tan 30°) \times x$$

$$= x + 0{\cdot}5774x^2$$

and this is to be equal to 20% of the area of the hexagon,

$$x + 0{\cdot}5774x^2 = 0{\cdot}2 \times 2{\cdot}598$$

$$0{\cdot}5774x^2 + x - 0{\cdot}5196 = 0$$

a quadratic equation, divide by coefficient of x^2,

$$x^2 + 1{\cdot}732x - 0{\cdot}9 = 0$$

Apply the formula

$$x = \frac{-b \pm \sqrt{b^2 - 4ac}}{2a}$$

$$= \frac{-1{\cdot}732 \pm \sqrt{3 + 3{\cdot}6}}{2}$$

$$= \frac{-1{\cdot}732 \pm 2{\cdot}569}{2}$$

$$= 0{\cdot}4185 \quad \text{Ans. (i)}$$

$$\begin{aligned}
OJ &= OC - x \\
&= 0{\cdot}866 - 0{\cdot}4185 = 0{\cdot}4475 \\
HJ &= HG + GJ \\
&= x \tan 30^\circ + \tfrac{1}{2} \text{ side} \\
&= 0{\cdot}4185 \times 0{\cdot}5774 + 0{\cdot}5 = 0{\cdot}7416 \\
OH &= \sqrt{(HJ)^2 + (OJ)^2} \\
&= \sqrt{0{\cdot}7416^2 + 0{\cdot}4475^2} \\
&= 0{\cdot}8662 \quad \text{Ans. (ii)}
\end{aligned}$$

73.

$$y = e^{-2x}$$

$$\text{i.e. } y = \frac{1}{e^{2x}} \quad \text{(N.B. } e = 2{\cdot}718 \text{ approx.)}$$

when	$x = 0$	$y = 1{\cdot}0$
	$x = 0{\cdot}2$	$y = 0{\cdot}67$
	$x = 0{\cdot}4$	$y = 0{\cdot}449$
	$x = 0{\cdot}6$	$y = 0{\cdot}301$
	$x = 0{\cdot}8$	$y = 0{\cdot}202$
	$x = 1{\cdot}0$	$y = 0{\cdot}135$
	$x = 1{\cdot}2$	$y = 0{\cdot}091$
	$x = 1{\cdot}4$	$y = 0{\cdot}061$
	$x = 1{\cdot}6$	$y = 0{\cdot}041$
	$x = 1{\cdot}8$	$y = 0{\cdot}027$
	$x = 2{\cdot}0$	$y = 0{\cdot}018$

ORDINATES	SIMPSONS MULTIPLIERS	PRODUCTS
1·0	1	1·0
0·67	4	2·68
0·449	2	0·898
0·301	4	1·204
0·202	2	0·404
0·135	4	0·54
0·091	2	0·182
0·061	4	0·244
0·041	2	0·082
0·027	4	0·108
0·018	1	0·018
	Sum =	7·39

$$\text{Area} = \frac{1}{3} \times \text{common interval} \times \text{sum of products}$$

$$= \frac{1}{3} \times 0{\cdot}2 \times 7{\cdot}36$$

$$= 0{\cdot}4907 \text{ square units}$$

$$\text{Mean ordinate} = \frac{\text{area}}{\text{length of base}} \text{ (see Fig. 257)}$$

$$= \frac{0{\cdot}4907}{2}$$

$$= 0{\cdot}245 \text{ units.} \quad \text{Answer.}$$

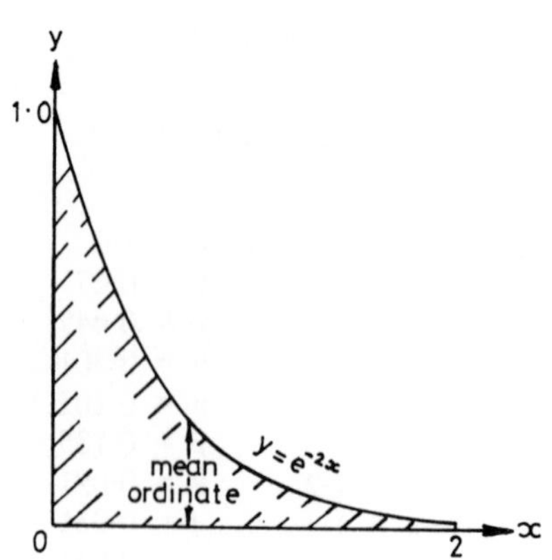

Fig. 257

74. By Pythagoras, referring to section of cone,

$$\text{Perp. height} = \sqrt{(\text{slant ht.})^2 - (\text{radius})^2}$$
$$= \sqrt{50^2 - 30^2}$$
$$= 40 \text{ cm}$$
$$\text{Volume of cone} = \tfrac{1}{3}\,\text{area of base} \times \text{perp. height}$$
$$= \tfrac{1}{3} \times \pi \times 30^2 \times 40 \text{ cm}^3 \qquad \text{(i)}$$

$$\text{Volume of hollow sphere} = \frac{\pi}{6}(D^3 - d^3)$$

$$\text{Substituting } D = d + (2 \times \text{thickness}) = d + 3,$$

$$\text{Volume} = \frac{\pi}{6}\left\{(d+3)^3 - d^3\right\}$$

$$= \frac{\pi}{6}\left\{d^3 + 9d^2 + 27d + 27 - d^3\right\}$$

$$= \frac{\pi}{6}\left\{9d^2 + 27d + 27\right\}$$

$$= \frac{\pi \times 9}{6}(d^2 + 3d + 3) \ \ldots \qquad \text{(ii)}$$

Equating (i) and (ii),

$$\frac{\pi \times 9}{6}(d^2 + 3d + 3) = \tfrac{1}{3} \times \pi \times 30^2 \times 40$$

$$d^2 + 3d + 3 = \frac{\pi \times 30^2 \times 40 \times 6}{3 \times \pi \times 9}$$

$$d^2 + 3d = 8000 - 3$$
$$d^2 + 3d - 7997 = 0$$

Solving this quadratic by formula,

$$d = \frac{-b \pm \sqrt{b^2 - 4ac}}{2a}$$

$$= \frac{-3 \pm \sqrt{9 + 31988}}{2}$$

$$= \frac{-3 \pm 178{\cdot}9}{2} = 87{\cdot}95 \text{ cm}$$
$$D = 87{\cdot}95 + 3 = 90{\cdot}95 \text{ cm}$$

Ans.

75.

$$a = \frac{c \times h^2}{2s \times (d - h)}$$

$$2as(d - h) = ch^2$$

$$2asd - 2ash = ch^2$$

This is a quadratic equation. Set down in the usual way and solve by quadratic formula:

$$ch^2 + 2ash - 2asd = 0$$

$$h = \frac{-b \pm \sqrt{b^2 - 4ac}}{2a}$$

where, $a = c,\ b = 2as,\ c = -2asd$

$$h = \frac{-2as \pm \sqrt{4a^2s^2 - 4 \times c \times (-2asd)}}{2c}$$

$$= \frac{-2as \pm \sqrt{4a^2s^2 + 8casd}}{2c}$$

$$= \frac{-2as \pm \sqrt{4as(as + 2cd}}{2c}$$

$$= \frac{-as \pm \sqrt{as(as + 2cd)}}{c} \quad \text{Ans. (i)}$$

Substituting numerical values:

$$h = \frac{-1{\cdot}5 \times 16 \pm \sqrt{1{\cdot}5 \times 16(1{\cdot}5 \times 16 + 2 \times 15 \times 17)}}{15}$$

$$= \frac{-24 \pm \sqrt{12\,816}}{15}$$

$$= \frac{89{\cdot}2}{15} \text{ or } \frac{-137{\cdot}2}{15}$$

$$= 5{\cdot}947 \text{ or } -9{\cdot}147 \quad \text{Ans. (ii)}$$

76.

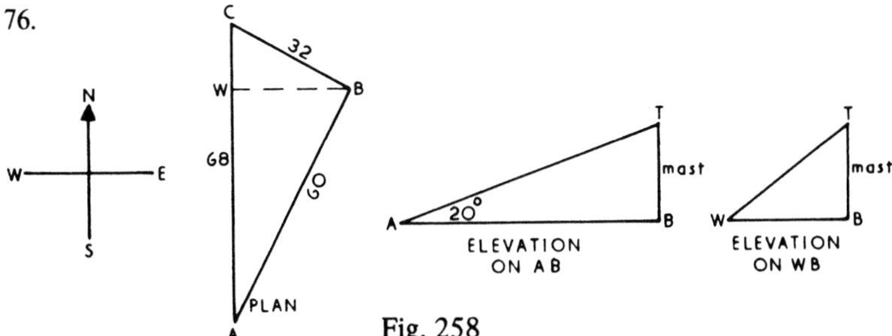

Fig. 258

Referring to plan view of Fig. 258, if the angle opposite the longest side is a right angle, then by Pythagoras, the longest side is the hypotenuse and the square of this side will be equal to the sum of the squares of the other two sides.

$$(AC)^2 = 68^2 = 4624 \quad \ldots \quad \ldots \quad \text{(i)}$$

$$(AB)^2 + (BC)^2 = 60^2 + 32^2$$

$$= 3600 + 1024 = 4624 \quad \text{(ii)}$$

(i) and (ii) are equal, therefore angle B is a right angle. Ans. (a)

$$\tan A = \frac{32}{60} = 0{\cdot}5333$$

$$\text{Angle A} = 28^\circ\ 4' \quad \text{Ans. (bi)}$$

$$\text{Angle C} = 90^\circ - 28^\circ\ 4' = 61^\circ\ 56' \quad \text{Ans. (bii)}$$

Referring to elevation on AB,

$$\text{Height of mast TB} = \text{AB} \tan A$$

$$= 60 \tan 20^\circ = 21{\cdot}84 \text{ m} \quad \text{Ans. (c)}$$

Referring to plan view,

$$\text{WB} = \text{AB} \sin A$$

$$= 60 \sin 28^\circ\ 4' = 28{\cdot}23 \text{ m}$$

Referring to elevation on WB,

$$\tan \text{W} = \frac{\text{TB}}{\text{WB}}$$

$$= \frac{21{\cdot}84}{28{\cdot}23} = 0{\cdot}7736$$

$$\text{Angle W} = 37^\circ\ 44' \quad \text{Ans. (d)}$$

77. Plotting points for $y = x^2 + 1$:

x	-2	-1	0	1	2	3
y	5	2	1	2	5	10

The graph is shown in Fig. 259

Difference between equations

$$\begin{aligned} y &= x^2 + 1 \\ \text{and } 0 &= x^2 - x - 2 \\ \hline \text{is } y &= x + 3 \end{aligned}$$

Drawing this straight line graph on the same axes using the plotting points:

When $x = -2$, $y = -2 + 3 = 1$

,, $x = +3$, $y = +3 + 3 = 6$

The straight line cuts the curve at $x = -1$ and $x = +2$, therefore $x = -1$ and $+2$ Ans.

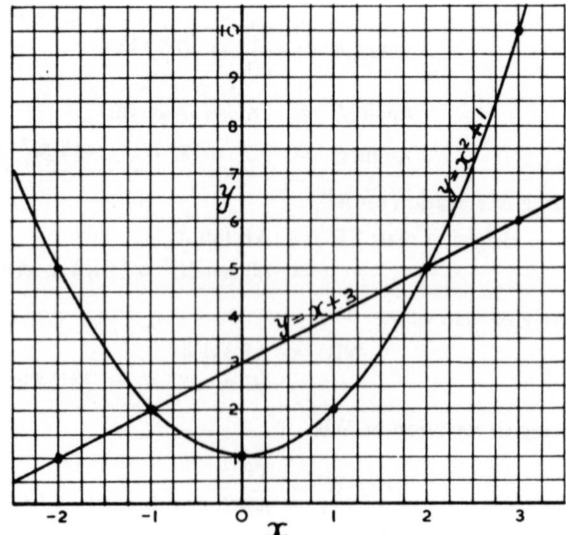

Fig. 259

78. (a) See Chapter 11 for Area of Segment.

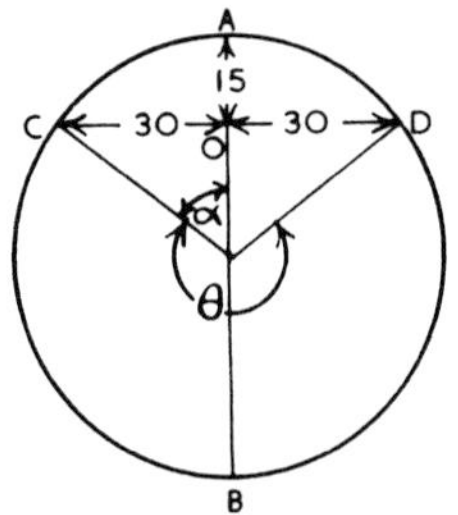

Fig. 260

Referring to Fig. 260, by crossed chords,

$$AO \times BO = CO \times DO$$
$$15 \times BO = 30 \times 30$$
$$BO = 60 \text{ cm}$$
$$\text{Diameter of pipe} = AO + BO$$
$$= 15 + 60 = 75 \text{ cm} \quad \text{Ans. (i)}$$

$$\sin\alpha = \frac{CO}{\text{radius}} = \frac{30}{37{\cdot}5} = 0{\cdot}8$$

$$\alpha = 53^\circ\,8'$$
$$\theta = 360^\circ - 2 \times 53^\circ\,8' = 253^\circ\,44'$$
$$\sin\theta = \sin 253^\circ\,44'$$
$$= -\sin(253^\circ\,44' - 180^\circ)$$
$$= -\sin 73^\circ\,44' = -0{\cdot}96$$

$$\theta \text{ radians} = 253^\circ\,44' \times \frac{2\pi}{360} \text{ or } \frac{253{\cdot}73^\circ}{57{\cdot}3}$$

$$= 4{\cdot}428 \text{ radians}$$

$$\text{Area of segment} = \tfrac{1}{2}r^2(\theta - \sin\theta)$$
$$= \tfrac{1}{2} \times 37{\cdot}5^2\{4{\cdot}428 - (-0{\cdot}96)\}$$
$$= \tfrac{1}{2} \times 37{\cdot}5^2 \times 5{\cdot}388$$
$$= 3788 \text{ cm}^2 \quad \text{Ans. (ii)}$$

1 dm = 10 cm, 1 dm^2 = 10^2 cm^2, 1 m = 10 dm:

$$\text{Volume } [dm^3] = \text{area } [dm^2] \times \text{length } [dm]$$

$$= \frac{3788}{10^2} \times 2 \times 10$$

$$= 757{\cdot}6 \text{ litres} \quad \text{Ans. (iii)}$$

79.
$$WZ = x^{a+b} \times x^{a-b}$$

Adding the indices,

$$a + b + a - b = 2a$$
$$\therefore WZ = x^{2a} \text{ Ans.}$$

$$W^2Z = x^{2(a+b)} \times x^{a-b}$$

Adding the indices,

$$2a + 2b + a - b = 3a + b$$
$$\therefore W^2Z = x^{3a+b} \text{ Ans.}$$

$$\frac{W^2}{Z} = \frac{x^{2(a+b)}}{x^{a-b}}$$

Subtracting the indices,

$$2a + 2b - (a - b) = a + 3b$$
$$\therefore \frac{W^2}{Z} = x^{a+3b} \text{ Ans.}$$

$$\frac{Z^3}{W} = \frac{x^{3(a-b)}}{x^{a+b}}$$

Subtracting the indices,

$$3a - 3b - (a + b) = 2a - 4b = 2(a - 2b)$$
$$\therefore \frac{Z^3}{W} = x^{2(a-2b)} \text{ Ans.}$$

$$4x + y = 10 \quad \dots \quad \dots \quad \dots \quad \dots \quad \text{(i)}$$
$$\frac{2}{x} + \frac{7}{y} = 3 \quad \dots \quad \dots \quad \dots \quad \dots \quad \text{(ii)}$$

From (i), $y = 10 - 4x$, substitute into (ii),

$$\frac{2}{x} + \frac{7}{10 - 4x} = 3$$

Multiply every term by least common denominator, which is $x(10 - 4x)$,

$$2(10 - 4x) + 7x = 3x(10 - 4x)$$
$$20 - 8x + 7x = 30x - 12x^2$$
$$12x^2 - 31x + 20 = 0$$

By quadratic formula,

$$x = \frac{-b \pm \sqrt{b^2 - 4ac}}{2a}$$

$$= \frac{31 \pm \sqrt{961 - 960}}{24}$$

$$= 1\tfrac{1}{3} \text{ or } 1\tfrac{1}{4}$$

From (i),

When $x = 1\frac{1}{3}$, $y = 10 - 4 \times 1\frac{1}{3} = 4\frac{2}{3}$

When $x = 1\frac{1}{4}$, $y = 10 - 4 \times 1\frac{1}{4} = 5$

$$\left.\begin{aligned} x &= 1\tfrac{1}{3} \text{ and } y = 4\tfrac{2}{3} \\ \text{or, } x &= 1\tfrac{1}{4} \text{ and } y = 5 \end{aligned}\right\} \text{ Ans. (b)}$$

80. (a)

$$Q = \sqrt{\frac{2ghDk^2}{d(s^2 - k^2)}}$$

$$\therefore Q^2 = \frac{2ghDk^2}{d(s^2 - k^2)}$$

$$\therefore Q^2 d(s^2 - k^2) = 2ghDk^2$$

$$\therefore Q^2 ds^2 - Q^2 dk^2 = 2ghDk^2$$

$$\therefore 2ghDk^2 + Q^2 dk^2 = Q^2 ds^2$$

$$\therefore k^2(2ghD + Q^2 d) = Q^2 ds^2$$

$$\therefore k^2 = \frac{Q^2 ds^2}{2ghD + Q^2 d}$$

$$\therefore k = \sqrt{\frac{Q^2 ds^2}{2ghD + Q^2 d}}$$

$$\therefore k = Qs\sqrt{\frac{d}{2ghD - Q^2 d}} \quad \text{Answer.}$$

(b)

$$F = \frac{1}{2\pi}\sqrt{\frac{1}{Lk}}$$

$$\therefore 2\pi F = \sqrt{\frac{1}{Lk}}$$

$$\therefore 4\pi^2 F^2 = \frac{1}{Lk}$$

$$\therefore 4\pi^2 F^2 Lk = 1$$

$$\therefore k = \frac{1}{4\pi^2 F^2 L} \qquad \text{Answer.}$$

(c)

$$L = \frac{\pi}{2}(k + b) + \frac{(k - b)}{4c} + 2c$$

$$\therefore \frac{\pi}{2}(k + b) + \frac{(k - b)}{4c} = L - 2c$$

$$\therefore \frac{4c\pi}{2}(k + b) + (k - b) = 4c(L - 2c)$$

$$\therefore 2c\pi(k + b) + (k - b) = 4cL - 8c^2$$

$$\therefore 2c\pi k + 2c\pi b + k - b = 4cL - 8c^2$$

$$\therefore 2c\pi k + k = 4cL - 8c^2 + b - 2c\pi b$$

$$\therefore k(2c\pi + 1) = 4cL - 8c^2 + b - 2c\pi b$$

$$\therefore k = \frac{4cL - 8c^2 + b - 2c\pi b}{2c\pi + 1} \qquad \text{Answer.}$$

81.

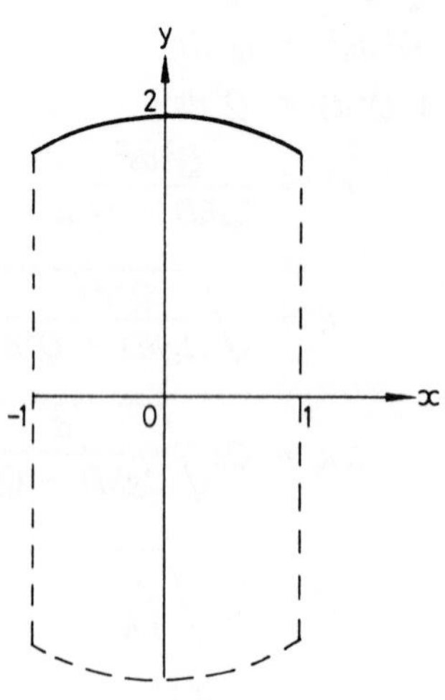

Fig. 261

$$7x^2 - 28 + 7y^2 = 0$$

$$\therefore x^2 - 4 + y^2 = 0$$

$$\therefore y = \sqrt{4 - x^2} \text{ (see fig. 261)}$$

$$\text{Volume of revolution} = \pi\int_a^b y^2\,dx$$

$$= \pi\int_{-1}^{+1} \left(\sqrt{4 - x^2}\right)^2 dx$$

$$= \pi\int_{-1}^{+1} (4 - x^2)dx$$

$$= \pi\left[4x - \frac{x^3}{3}\right]_{-1}^{+1}$$

$$= \pi[4 - \tfrac{1}{3}] - \pi[-4 + \tfrac{1}{3}]$$

$$= 4\pi - \frac{\pi}{3} + 4\pi - \frac{\pi}{3}$$

$$= 7\tfrac{1}{3}\pi$$

$$= 23{\cdot}04 \text{ cm}^3$$ Answer.

INDEX